MW01156827

A photograph of Korbinian Brodmann that served as frontispiece in the original 1909 edition (and is also featured on the cover of this edition. The cover also shows two of the best known of Brodmann's maps of the human brain (Figs. 85 and 86), found on page 110 of this edition).

Vergleichende Lokalisationslehre der Großhirnrinde

in ihren Prinzipien dargestellt auf Grund des Zellenbaues

Von

Dr. K. Brodmann
Assistenten am neurobiologischen Laboratorium der Universität zu Berlin.

Mit 150 Abbildungen im Text.

Leipzig
Verlag von Johann Ambrosius Barth
1909

Brodmann's Localisation in the Cerebral Cortex

The Principles of Comparative Localisation in the Cerebral Cortex
Based on Cytoarchitectonics

By

Dr. K. Brodmann
Assistant in the Neurobiological Laboratory
of the University of Berlin

Translated with editorial notes and an introduction

By

Laurence J. Garey
Centre for Psychiatric Neuroscience
Lausanne, Switzerland

With 150 text figures

The third edition of this work is published by permission of World Scientific Publishing Co Pte Ltd, Singapore. *http://www.wordsscibooks.com/meds/sci/p151.html.* The original English-language edition was published in the United Kingdom by Smith-Gordon Company Limited (London: 1994).

Library of Congress Control Number: 2001012345

ISBN-10: 0-387-26917-7
ISBN-13: 978-0387-26917-7

Printed on acid-free paper.

Printed in the United States of America. (EB)

9 8 7 6 5 4 3 2 1

springeronline.com

Contents

Translator's note: Brodmann's Contents list is a curious mixture of accurate references to titles of actual chapters and sections in the text, altered section titles, and short descriptions of section contents. I have tried to retain some of the inconsistency, while attempting to make the Contents useful to the modern reader. For this reason, I have incorporated in it the new sections that I have added: Translator's Introduction, References and Notes, and Glossary of Species Names.

Translator's Introduction

In 1985 the Johann Ambrosius Barth Verlag in Leipzig reprinted the first (1909) edition of Brodmann's famous book "Vergleichende Lokalisationslehre der Grosshirnrinde in ihren Prinzipien dargestellt auf Grund des Zellenbaues".

This book is one of the major "classics" of the neurological world. To this day it forms the basis for "localisation" of function in the cerebral cortex. Brodmann's "areas" are still used to designate cortical functional regions, such as area 4 for motor cortex, area 17 for visual cortex, and so on. This nomenclature is used by clinical neurologists and neurosurgeons in man, as well as by experimentalists in various animals. Indeed, Brodmann's famous "maps" of the cerebral cortex of man, monkeys and other mammals must be among the most commonly reproduced figures in neurobiological publishing (see, for example, Zilles, 1990). There can be few textbooks of neurology, neurophysiology or neuroanatomy in which Brodmann is not cited, and his concepts pervade most research publications on systematic neurobiology.

In spite of this, few people have ever seen a copy of the 1909 monograph, and even fewer have actually read it! There had never been a full English translation available, and the original book has been almost unavailable for years, the few antiquarian copies still around commanding high prices. Interestingly, von Bonin produced a translation of Chapter 9 in 1960, in spite of his criticism of Brodmann (see below). As I, too, use Brodmann's findings and maps in my neurobiological work, and have the good fortune to have access to a copy of the book, I decided to read the complete text and soon discovered that this was much more than just a report of laboratory findings of a turn-of-the-twentieth-century neurologist. It was an account of neurobiological thinking at that time, covering aspects of comparative neuroanatomy, neurophysiology and neuropathology, as well as giving a fascinating insight into the complex relationships between European neurologists during the momentous times when the neuron theory was still new.

I think it important to tell something of Brodmann's own somewhat unusual, and rather sad, professional career, as well as to define his place in neurological history.

Korbinian Brodmann was born on 17 November 1868 in Liggersdorf, Hohenzollern, the son of a farmer, Joseph Brodmann. He studied medicine at the Universities of Munich, Würzburg, Berlin and Freiburg-im-Breisgau, where he received his medical degree (the "Approbation") on 21 February 1895.

After this Brodmann worked in the University Paediatric Clinic and Policlinic in Munich, with the intention of perhaps establishing himself as a practitioner in the Black Forest.

But he contracted diphtheria and, as Oskar Vogt wrote in 1959, "convalesced" in 1896 by working as an Assistant in the Neurological Clinic in Alexanderbad in the Fichtelgebirge region, then directed by Vogt himself.

Under his influence, Brodmann turned to neurology and psychiatry, and Vogt described him as having "broad scientific interests, a good gift of observation and great diligence in widening his knowledge". Vogt was preoccupied with the idea of founding an Institute for Brain Research, that finally materialised in Berlin in 1898 as the Neurobiological Laboratory. In order to prepare for a scientific career Brodmann first went to Berlin, and then studied pathology in Leipzig where, in 1898, he took his Doctorate with a thesis entitled "A Contribution to the Understanding of Chronic Ependymal Sclerosis". After this he worked in the University Psychiatric Clinic in Jena, directed by Otto Binswanger, before transferring to the Municipal Mental Asylum in Frankfurt-am-Main from 1900 to 1901, where meeting Alzheimer inspired an interest in the neuroanatomical problems that determined the whole of his further scientific career.

In Autumn 1901 Brodmann joined Vogt and until 1910 worked with him in the Neurobiological Laboratory in Berlin where he undertook his famous studies on comparative cytoarchitectonics of mammalian cortex. Vogt suggested to Brodmann that he "undertake a systematic study of the cells of the cerebral cortex", using sections stained with the new method of Nissl. Cécile and Oskar Vogt were engaged on a parallel study of myeloarchitectonics, and experiments using physiological cortical stimulation. In April 1903, Brodmann and the Vogts gave a beautifully coordinated presentation, each of their own architectonic results, to the annual meeting of the German Psychiatric Society in Jena. Brodmann described the totally different cytoarchitectonic structure of the pre- and postcentral gyri in man and the sharp border between them.

Brodmann's major results were published between 1903 and 1908 as a series of communications in the "Journal für Psychologie und Neurologie". The best known is his sixth communication, of 1908, on histological localisation in the human cerebral cortex. He edited this prestigious journal until his early death in 1918, a journal which lived on as the "Journal für Hirnforschung", and later became the "Journal of Brain Research". His communications served as a basis for his 1909 monograph on comparative cortical localisation, which is the subject of the present translation, but he did not live to see its second edition in 1925.

Brodmann's career in Berlin was marred by the surprise rejection by the Medical Faculty of his "Habilitation" thesis on the prosimian cortex. So when, as Oskar Vogt admitted, the Neurobiological Laboratory did not seem to be developing as well as he had expected, Brodmann went to work with Robert Gaupp in Tübingen where, on Gaupp's recommendation, he was appointed Profesor by the Faculty of Medicine. The behaviour of the Berlin Faculty remains incomprehensible. Brodmann, indeed, comments on their negative attitude to his research in his Foreword to his monograph (page XIV of the present translation). Vogt (1919) himself complained about the harm done by faculty members, not only to Brodmann but to the development of the whole Laboratory. In contrast, the anatomist Froriep welcomed Brodmann warmly to membership of the Faculty in Tübingen in a speech of greetings, and the Academy of Heidelberg honoured his work with the award of a prize.

On 1 May 1916 Brodmann took over the Prosectorship at the Nietleben Mental Asylum in Halle an der Saale, directed by Berthold Pfeiffer. For the first time he was assured of reasonable material security and here he met Margarete Francke, who became his wife on 3 April 1917. In 1918 their daughter Ilse was born.

During his time in Berlin Brodmann had lectured in postgraduate courses in Munich organised by Kraepelin who anticipated an important contribution to neuroanatomical research from architectonics and neurohistology. Nissl joined the Psychiatric Research Institute in Munich, and in 1918 Brodmann also received a prestigious appointment to the newly formed Munich Institute and took charge of the Department of Topographical Anatomy. Thus began a harmonious collaboration with Nissl, although Brodmann was only to live for less than a year.

Oskar Vogt published Brodmann's obituary in their beloved Journal für Psychologie und Neurologie in 1919 and wrote: "Just at the moment when he had begun to live a very happy family life and when, after years of interruption because of war work, he was able to take up his research activities again in independent and distinguished circumstances, just at the moment when his friends were looking forward to a new era of successful research from him, a devastating infection snatched him away after a short illness, on 22 August 1918". In 1959 Vogt wrote a biography of Brodmann, in which his high estimation of the man and scientist is obvious.

Kraepelin declared at Brodmann's graveside that science had lost an inspired researcher, and in 1924 Spielmeyer spoke critically about the contemporary German academic world: "If we look at his career, we are painfully aware that little provision was made in German universities for a researcher of Brodmann's stature ... Until his 48th year Brodmann had to be content with subordinate posts that in no way corresponded to his importance, and he watched with some bitterness as officious mediocrity led to the most distinguished posts while he, the successful and recognised researcher, in spite of all his lack of pretension, could never attain the most modest permanent university position."

We might ask ourselves if things have changed much in the meantime!

Before Brodmann, a certain confusion reigned concerning the laminar structure of the cortex. In 1858, Meynert's pupil, Berlin, gave a first description of the six layers of the human isocortex as distinguished by variations in cell size and type, including pyramidal and granule cells. Meynert himself, starting in 1867, described the subdivision of the human cortex into numerous functional regions. An important early cortical localisational study was that of Betz in 1874, in which he pointed out "nests" of unusually large cells, his so-called "giant pyramids", in the human motor cortex of the precentral gyrus, an area separated by the central sulcus from the sensory cortex of the postcentral gyrus which contained no such giant cells. In 1878 Ferrier devoted his Croonian Lecture to cerebral localisation. Later, human cortical maps based on fibre architecture (myeloarchitectonics) were published, notably those of Kaes (1893), Bechterew (1896) and Flechsig (1898).

Before the end of the nineteenth century, numerous publications followed on the laminar pattern of the cerebral cortex in various mammals, including man, of which the best known are those of Lewis (1878, 1881), Lewis and Clarke (1878) and Hammarberg (1895). Brodmann considers these in some detail, pointing out the many inconsistencies. The year 1900 marked the beginning of the publication of Cajal's studies on human cortex, as well as Bolton's treatise on the human visual cortex. In particular, Brodmann had little respect for Cajal's "erroneous" views on cortical lamination. Elliot Smith published a detailed atlas of human cortical localisation in 1907, referring to the preceding work of Flechsig, Campbell and Brodmann.

In 1905 Campbell's major work entitled "Histological studies on the localisation of cerebral function" appeared. He was an Australian, and had studied in Edinburgh, as well as with Krafft-Ebbing in Vienna. He investigated eight human cerebral hemispheres, as well as brains of the chimpanzee, orang-utan, cat, dog and pig. In 1953 von Bonin commented that Campbell's division of the primate brain was not as "fine as those of the German school", referring particularly to the work of Brodmann.

Myeloarchitectonics also progressed in the first part of the twentieth century. Notable contributions, of special importance to Brodmann because of his professional relations with them, were those of Cécile and Oskar Vogt between 1900 and 1906, and their colleagues from the Berlin Neurobiological Laboratory, such as Mauss (1908) and Zunino (1909).

Brodmann refined and extended these observations, integrating ideas on phylogenetic and ontogenetic influences with his theories of adult cortical structure, function and even pathology. The basis of Brodmann's cortical localisation is its subdivision into "areas" with similar cellular and laminar structure. He compared localisation in the human cortex with that in a number of other mammals, including primates, rodents and marsupials. In man, he distinguished 47 areas, each carrying an individual number, and some being further subdivided. The Vogts described some four times as many areas from their myeloarchitectonic work. An important support for Brodmann's concepts of functional localisation was provided by Foerster's electrical stimulations of human cortex in 1926, work based on Brodmann's structural studies.

Brodmann is sometimes criticised for drawing general conclusions from small numbers of brains. It is even said that he studied only one human brain. Although he does not specifically tell us how many brains he used, he does "thank Professor Benda for kindly providing human brains" at the end of his Foreword (p. 2). Certainly later, for example when he turned to more "anthropological" aspects of human cortex (1913), he used many brains.

Later work was to a great extent elaboration of Brodmann's observations. In the cytoarchitectonic atlas published by von Economo and Koskinas in 1925, Brodmann's numbers were replaced by letters. In 1962 Hassler commented that "von Economo and Koskinas describe almost exclusively Brodmann's cortical areas ... there is therefore no justification for replacing Brodmann's numbers". Bailey and von Bonin (1951) were among the few people to accept von Economo's parcellation; they criticised Brodmann and the Vogts, and only differentiated some 19 areas themselves. Others, including Karl Kleist (1934) and Lashley and Clark (1946), were also against a too vigorous subdivision of the cortex. However, since then a number of atlases have appeared, essentially vindicating Brodmann's view, among which is that of Sarkissov and his colleagues in 1955.

Modern experimental methods have also supported cortical localisation, both anatomical and functional. There was a growing tendency in the 1950s and 1960s to concentrate much experimental effort on the, until then, somewhat neglected cerebral cortex. Physiologists became interested in the results of cortical cytoarchitectonic studies. This approach was largely due to the fact that physiological investigations were showing that the cortex was divisible into functionally distinct and localised areas. One need only consider the exquisite correspondence found in the visual and somatosensory systems between individual cortical areas and subtle variations in physiological function (Hubel and Wiesel, 1962, 1977; Powell and Mountcastle, 1959). In many cases Brodmann's areas have been further subdivided, but no major objections to his pioneering work have been upheld for long. Histochemistry and "chemical neuroanatomy" (the study of specific chemicals, particularly neurotransmitters, in neurons) have confirmed sharply circumscribed cortical areas corresponding to those seen with classic cell stains. The reader looking for an in-depth treatment of the varied facets of cerebral cortical structure and function should refer to the series of volumes on the "Cerebral Cortex" edited by Peters and Jones from 1984.

Furthermore, clinical observation in cases with localised pathology, together with descriptions of deficits due to wounds in two World Wars and even more relatively local conflicts, pointed to the same conclusion. More recently enormous improvements in medical imaging, including functional studies with positron emission tomography and magnetic resonance imaging, have paved the way for direct in vivo visualisation of the human brain, and the results are unequivocally in favour of an exquisite cortical localisation, not always in perfect harmony with Brodmann's views, but in overall agreement with them.

Interest in Brodmann's localisational theories has accelerated together with the modern enthusiasm for a search for the neuroanatomical basis for human consciousness and intelligence (eg Semendeferi et al. 2002; Schoenemann et al. 2005). Such work demands a solid basis for localisation within a brain, a knowledge of variability, and a means to identify the same brain areas in different individuals. In this context, the recent discovery of some later, forgotten works by Brodmann, including his work on anthropological aspects of localisation (1913, translated by Elston and Garey, 2004), has provided a mass of new information on human cortical organisation. By 1913 Brodmann's attention had moved toward systematic study of human brains of different races, not with the aim of elucidating any differences in "quality" between races, but rather with a view to collecting as many data about a given species as possible. In his 1913 paper Brodmann not only presented large amounts of new, quantitative data on the human cortex, but also unpublished results on the cortex from a wide range of primates and non-primates. The data presented are as useful today as when first published. The variation in cortical topography that he reported in human brains is of essential importance for the interpretation of present-day functional imaging studies, particularly those involving visual or prefrontal cortex. The latter has attracted much attention as being a possible "site" for making man's brain different from that of other primates, in terms of intelligence and consciousness. To this day, Brodmann's comparative data remain unsurpassed.

In reading the "Localisation" one is struck by the many forward-looking references to concepts and techniques that emerged only much later, such as multiple representations of functional areas, the chemical anatomy of the brain, and ultrastructure. What might Brodmann have discovered if he had lived beyond the age of 49?

Brodmann's language has the typical rather heavy style of early twentieth century German. I have tried to retain it as far as possible, only turning the phrases and sometimes breaking the sentences when it was necessary to produce readable English. In general, I have anglicised the mixture of Latin and German terms for such things as cortical layers and areas, as well as animal species, except that in the figure legends species names are given in the Latin form used by Brodmann. The Glossary of Species Names provides modern equivalents.

There are two types of footnote in this English Edition. Those that appear at the actual foot of the relevant page are those that Brodmann used, while the Translator's Notes are grouped at the end of the book and are preceded by an asterix in the text.

Brodmann used several, rather inconsistent methods for giving bibliographic references. Some of them appear at the end of his original monograph, others as footnotes, and others as parentheses in the text. I have adhered to his methods and format in each case, retaining his versions literally, even when there are mistakes. I have, however, added a complete new reference list including Brodmann's citations (given in full, and corrected, as was often necessary!), as well as other references pertaining to the Translator's Introduction and my own additional notes.

Since the First Edition (by Smith-Gordon, 1994) and the Second (by Imperial College Press, 1999), Springer Verlag has agreed to publish this Third Edition, with some minor corrections and a few additions, particularly in relation to the prefrontal cortex. It is a pleasure to acknowledge the support given by Kathy Lyons and Claire Wynperle and their collaborators, as well as Steve Gallant and Joe Kuhns at Narragansett Graphics.

Laurence Garey, Lausanne, 2005

Foreword

When I began my work in the Neurobiological Laboratory (*1) of the University of Berlin eight years ago the task befell me to undertake a topographic analysis of the human cerebral cortex based on its cellular structure, in the context of the research programme of this institute. The practical goal of this task was essentially to provide a description of the normal histological structure of the whole human cerebral cortex, which had long been felt necessary by neuroanatomists and pathologists. In contrast to earlier research of a similar nature, the emphasis would be not only on gross divisions of the brain, such as lobes and gyral complexes, but also on the smallest gyri and parts of gyri, in order to obtain a complete picture of cortical structure and all its local modifications, and thus try to describe topographical parcellation and localisation in the cortex that would also be of value for clinicians.

But during the course of these investigations the need soon became apparent to place the whole work on a much broader, developmental, and above all comparative anatomical, basis if it was to help us understand the structural plan of the cerebral cortex and explain the astonishing structural complexity based on common organisational principles that were revealed as one penetrated more deeply into the subject. Therefore to begin with, I was obliged to abandon the extremely complicated and unfathomable human brain and try to obtain an insight into the structure of the cerebral cortex using an ontogenetic approach in simpler forms. This means that the originally envisaged task was modified, and at the same time broadened. The object of the research was no longer merely the human central nervous system, but that of the whole mammalian class, and in the course of the years material from all the main groups of mammals was included in the investigation, each with at least a few of their important representatives.

The following descriptions contain the results and a synthesis of these studies. Respecting the aim of the book to expose the essentials of a theory of comparative localisation in the mammalian cortex, detailed data are only reproduced in so far as they appear indispensable to the establishment of the principles of topical cortical development. Thus anyone looking for practical information on the human will often be disappointed. It will be my next priority to correct this deficiency as soon as possible, if circumstances permit. Many details are already published in a current series of communications on histological localisation in the cerebral cortex in the Journal für Psychologie und Neurologie (1903-1908) (*2). But I am also conscious of only being able to offer incomplete data in other respects. The theory of anatomical localisation is, like physiological and clinical localisation, still in a state of development. The directions to take to reach them are not immediately attainable. So many problems must remain unresolved, some can be only provisionally unravelled, and for yet others the way ahead can only be sketched out roughly. Thus this book should not, and cannot, be more than a first draft or outline of the new theory.

The publication of the results of my research in their present form was made possible by the generosity of the trustees of the Berlin Municipal Benefaction (*3) who awarded me a substantial grant towards my research costs with the help of the Jagor Foundation (*4). My duty to express my gratitude publically to the trustees of this Foundation is even greater because my repeated attempts to obtain support for the same purpose from the Science Research Fund of Berlin University failed due to the opposition of the authorities of the Medical Faculty (*5).

I owe particular thanks to Professor Heck and Dr. Heinroth of the Berlin Zoological Garden for the great amiability with which they constantly supported and encouraged my work with donations of valuable animal specimens.

I thank Professor Benda for kindly providing human brains.

Finally, my duty to express my gratitude to my collaborators in the Neurobiological Laboratory, and especially to its Director, Dr Oskar Vogt, is obvious from the years of work in common that unite us. Without the technical organisation of this institute and without the continued active participation of my collaborators the achievement of my goals would have remained entirely impossible.

Berlin, August 1909

K. Brodmann.

Introduction

The aims and methods of histological cortical localisation.

The subject of the following treatise is histological localisation in the cerebral cortex, that is to say localisation which uses exclusively anatomical features as the basis for investigation, in contrast to physiological or clinical aspects. The first and most important task of such a localisational study is the parcellation of the cerebral cortex according to common anatomical features, that is the systematic grouping of structurally similar neural components and complexes and the separation of structurally dissimilar ones. In particular it is the aim of comparative localisation to identify similar or *homologous* parts of the cerebral cortex in different animals or animal groups on the basis of their structure. Our goal is thus to produce a comparative organic theory of the cerebral cortex based on anatomical features, as first imagined by Theodor Meynert.

In this definition of our task two fields of research are excluded from further consideration from the very beginning, namely fibre architecture on the one hand and myelogenesis on the other. Neither of these, although undoubtedly major factors in anatomical localisation and of the greatest heuristic value, belong to histological methodology in the strictest sense.

Fibre architecture deals with the conduction pathways between different parts of the cerebral cortex on the one hand and between the cortex and lower levels of the central nervous system on the other, that is fibre paths that run mainly outside the cerebral cortex. Further, they do not represent *histological* entities as such, as, from a methodological point of view, one has to strictly distinguish between fibre pathways and fibre-free systems (*6).

The myelogenetic method is based on differences in timing of myelinisation of nerve fibres in the early stages of an individual's development and allows a subdivision of the cerebral cortex into myelogenetically different structural zones. They can be referred to as early and late myelinising areas, on the basis of topographical differences in myelinisation. However, these processes are active for only a very short time, at least as far as the cortex is concerned, in man for only a few weeks before and after term. A myelogenetic study is thus an exquisite *developmental* method, but not a histological one, and it is still very debatable whether localisational data obtained with it can be applied directly to the mature brain and physiological conclusions drawn.

In contrast to these two methods, histological parcellation depends entirely on the nervous components making up the cortex in mature, adult individuals. According to whether, of these components, neurons, myelinated or unmyelinated nerve fibres form the substrate for an investigation, one can distinguish three forms of histological localisation: cytoarchitectonics, myeloarchitectonics and fibrilloarchitectonics (*7).

In the following description we shall deal exclusively with the first. The reasons are as follows.

The study of the fibrillary features of the cerebral cortex, or *fibrilloarchitectonics*, still remains in its infancy. Those studies of the local neurofibrillar structure of the cerebral cortex that are available (Bielschowsky and Brodmann [1]), Doinikow [2])) lead us to suppose that after the preliminary localisational work is complete, a systematic study of these features might reveal topographic differences that remain more or less undetected with the other methods. To begin with, we must be content with gross topographic information.

On the other hand, *myeloarchitectonics* have already produced satisfactory independent localisational results. In general they agree with those from cytoarchitectonics.

One may just recall that Campbell accepts such an absolute agreement between the cell and fibre architecture in man and the anthropoid apes that he only gives a single brain map for both. Following that, Mauss in our Neurobiological Laboratory recently (1908) carried out a myeloarchitectonic localisation in lower monkeys that equally confirmed an extensive agreement with my earlier cytoarchitectonic subdivisions of 1905. Zunino has just shown the same for the rabbit cortex. The corresponding brain maps agree in all essential points with mine. In relation to this, O. Vogt has derived a cortical parcellation in man using myeloarchitectonics that is much more detailed than

---

[1]) Bielschowsky and Brodmann, Zur feineren Histologie und Histopathologie der Grosshirnrinde mit besonderer Berücksichtigung der Dementia paralytica, Dementia senilis und Idiotie. Journal f. Psycholog. u. Neurolog. V. 1905. (*8)

[2]) Doinikow, Beitrag zur vergleichenden Histologie des Ammonshorns. Journal f. Psycholog. u. Neurolog. XIII. 1908. (*9)

my cellular localisation, as far as can be judged from as yet unpublished studies. However, I have no reason to foresee finding any major divergences. Indeed, in man the fibre structure of the cortex is often more finely differentiated than the cell architecture, especially in the outer layers (I-III), so it is possible by using it to subdivide larger cytoarchitectonic zones into smaller fields of specific fibre structure. Thus one obtains a greater number of individual topographical fields without the parcellation into the more extensive major cytological areas losing its value. Thus, basically, it is only a matter of differences in degree in spatial localisation and not a major divergence. It should be noted here that in lower animals - and even in monkeys - the cell architecture is often superior in terms of clarity of regional differentiation, such that myeloarchitectonics do not in general reveal a more extensive parcellation than cytoarchitectonics. This reveals a very important difference in the structural differentiation of the cerebral cortex between man and animals (*10), the importance of which is not to be minimised for the moment. A discussion of this is pointless as long as there does not exist a definitive map of myeloarchitectonic localisation, at least for man.

In view of these considerations, and as here only the general principles of localisation will be described, I will make no attempt to treat these two localisational methods, myeloarchitectonics and cytoarchitectonics, comparatively. We shall therefore concentrate exclusively on the latter.

Cytoarchitectonic localisation can also follow different directions; it can concentrate on the individual cellular elements, or can be based on particular local cell groupings such as those forming layers, or finally it can select as its criterion for parcellation the overall structure of a segment of cortex, as long as it is of homogeneous structure. Thus one must distinguish three types of cytoarchitectonic division of the cerebral cortex:

1. localisation according to individual histological elements - histological elemental localisation;
2. localisation according to cell layers (or also fibre layers) - laminar localisation or *stratigraphic parcellation*;
3. localisation according to tangentially organised fields of homogeneous structure - areal localisation or topographic cortical parcellation.

The basis for **elemental localisation** is the, in itself valid, concept that tissue elements of uniform specific structure, whether they are limited to a large or small cortical field or diffusely distributed over the whole cortex, must also have a uniform physiological function, and thus that such elements are to be regarded as not only morphologically but also functionally equivalent. This consideration enables one to establish, at least in principle, the feasibility and practicability of subdividing the cerebral cortical surface according to individual nervous elements, such as cell types with particularly characteristic features - such as intrinsic shape, size, internal structure, axonal connections, fibrillar organisation and so on. However, what has been achieved to this end so far is not exactly encouraging. The difficulties in achieving such a subdivision by elements are considerably greater than may appear at first sight. First and foremost we still lack clear criteria for the recognition of anatomically equivalent cellular elements.

W. Betz provided an important, and perhaps the only lasting, advance in this direction in the oldest work on cortical localisation that we, at least, possess. Already in 1874, he showed that two different "anatomical centres", one anterior and one posterior, were separated by the sulcus of Rolando (*11) on the brain surface. The anterior domain, that Betz termed "motor centre" [3]) is characterised, as he described, by the presence of unusually large cells, grouped into clusters, the so-called "*giant pyramids*", that were completely absent in the posterior "sensory centre". According to Betz, the anterior centre can be spatially segregated from the posterior merely on the basis of these cells, and we have here an example (at that time perhaps the only valid one) of histological elemental localisation.

Kolmer was guided by the same basic concept in his establishment of the "motor cortical region". He also proceeds from the same supposition that, to determine the extent of an anatomical cortical centre, that is for the establishment of homologous (anatomically equivalent) regions in different animals, "the concordant appearance of distinctly characteristic cells is the most useful reference point". As a sort of test of this proposal, he describes the spatial delimitation of the cortical zone that he considers to be the extent of the "*motor cortical cells*", as defined by Nissl, that is those neurons that are related to motor function in the physiological sense and that can be distinguished anatomically from all other cells by their own peculiar structure. Thus Kolmer and Betz basically pursue the same goal of histological elemental localisation. Let us now compare the results of their localisational studies! Betz places his field on the whole anterior and only the upper one-sixth of the posterior central gyrus, including the paracentral lobule; Kolmer, on the other hand, maintains that such motor cells lie on both sides of the central sulcus (well into the parietal lobe) and take in a wide, clearly delimited strip that becomes narrower inferiorly. Thus, despite the same starting point, quite contradictory results!

The accounts of localisation by means of cortical elements that have been published more recently have not progressed beyond the level of abstracts. There has been occasional talk of "sensory cells" located in particular regions, or of sensitive or sensorial "*special cells*". People have invented acoustic or optic special cells and even a "memory" (*12) cell, and have not shied away from the fantastic "*psychic cell*". Apart from the fact that such so-called "special cells" have only been described in young or foetal brain with the Golgi method and mainly only in animals, and therefore lack confirmation in the adult human brain, and quite apart from the fact that no attempt has been made to determine the precise regional location of the zone within which such cells appear exclusively, it seems to me that to pose this problem is wrong. Not only is it not proven, but it is highly unlikely on general biological considerations, that a

[3]) On the grounds of the fact that he had found the same cells in the dog around the cruciate sulcus, that is to say in the excitomotor zone of Fritsch and Hitzig.

special sensory function is related to a cell type of particular structure. The essential for the elaboration of any cortical function, even the most primitive sensory perception, is not the individual **cell type** but **cell groupings**. Modern anatomy gives no support to the idea that a single specific cell type is necessary, for instance, for specific light sensitivity. So I must refute energetically the concept that is widespread among physiologists that anatomy can only really support and further neurophysiology by determining the distribution of a single cell type (Lewandowsky) (*13). That would mean neuroanatomy giving false leads. This concept undoubtedly arises from an overestimation of the individual cell, but at the same time from an underestimation of what neurophysiology already owes to neuroanatomy. Data have already disproved that the majority of the cell types that can be grossly distinguished with modern methods (pyramidal cells, spindle cells, granule cells, stellate cells etc.) are organised similarly over the whole surface of the cerebral cortex. Rather, their organisation in laminar groupings, in a word their cytoarchitecture, is regionally extraordinarily varied. It is possible that later it will be feasible to further differentiate histologically many grossly morphologically similar cell types according to their fine structure. For this, the main necessity is new histological, and particularly staining, techniques that have a specific affinity for functionally related cells or, what amounts to the same, histochemically related cells, and will reveal them selectively. However, histological technology is still far from this [4]) (*15). Thus, for the present, cortical localisation based on individual histological elements is still too deficient in basic hypotheses for there to be hope of success.

Things are not better as concerns **localisation by layers** or *stratigraphic cortical parcellation*. Certainly, there is at first sight something attractive for the uninitiated in taking the characteristic and striking layers in which the cells (and fibres) are organised, as seen in cross-sections, and that exist throughout the whole cerebral cortex in man as in lower mammals, as representative of particular basic functions and therefore also to use them to divide the cortex anatomically. However, for the moment we know nothing definite about the significance of the individual layers or even about one single layer. Indeed, the little that was once considered as firmly established has proved to be doubtful or untenable upon critical examination and in the light of new facts.

One may recall that the significance of the prominent layer of giant pyramids in the precentral gyrus remains largely obscure in spite of the large number of individual studies related to its function, and although its anatomical localisation has been known for a long time. One can only say it must be closely related to motor function. Support for this comes from pathological observations of various kinds (eg. in amyotrophic lateral sclerosis and tertiary traumatic degeneration - Probst; Campbell; Rossi and Roussy). What form this

---

[4]) In Bielschowsky fibre preparations one sometimes sees an indication of such histochemical affinity and selective staining of particular cell types, eg. of neurons with short axons. Equally one should recall the methylene blue staining of certain similar cells in the abdominal ganglia of invertebrates (Biedermann, Retzius) (*14).

relationship takes, however, is completely unknown. Above all, we have absolutely no proof that this layer represents the only motor component of the cortex, comparable with a specifically sensory one in the granular layers, as is being actively attempted. Many new observations (electrical stimulation) support the idea that cortical motor activity can be produced without the intervention of this giant pyramidal layer (Brodmann). Above all, it is clear that the excitomotor zone stretches anteriorly well beyond the extent of this layer in all animals (C. and O. Vogt; Mott) (*16). Thus we can have cortical motor activity without the direct contribution of the giant pyramidal layer, so it cannot be the only cortical motor centre.

The situation is similar with regard to another specific cortical layer, of which the spatial extent has been known for a long time and is clearly recognisable to the naked eye in the region of the calcarine sulcus as the stria of Vicq d'Azyr (*17). It forms the main architectonic feature of a sharply delimited structural area (the striate area) that is, in addition to this light fibre band, further characterised by two distinct granular layers running parallel to it. Even though it emerges from Henschen's praiseworthy studies that the area occupied by this layer is closely associated with visual activity, and even accepting Henschen's idea that specific zones of this cortical area are related in regular fashion to specific parts of the retina, the principle of stratigraphic localisation is still not proved. One can neither say which of the three particular layers in question in this cortical area - the fibrous stria of Vicq d'Azyr or the two granular layers - represents the specific "visuosensory element" within the area, and above all nor can one say whether they alone do this or whether the whole cortical depth with all its layers participates in the elaboration of cortical visual activity, even at its most primitive.

Finally - to introduce a third example - we know just as little about the significance of the so-called "molecular layer", as it is described in the literature, the outermost cortical layer and our *Lamina zonalis*. It is the most constant layer of the cerebral cortex, is never absent in any animals and has always been the object of painstaking investigation. In spite of this, its study has brought us no further in terms of localisation. Opinions about it are diametrically opposed, in that some take it for a highly developed layer, an "*association organ*", whereas others declare it to be a functionally inferior formation or a "neurologically worthless layer" (Meynert).

In view of this, it seems to me that the time has no more come for a pure stratigraphic consideration that for elemental localisation. This is perhaps only because of deficiencies in our present techniques, but as far as I can see, it has not advanced cortical localisation at all up to now. Those who find it to their taste can dress up the individual layers with terms borrowed from physiology or psychology, such as "*sensitive*" or "*perceptive*" layers, association or projection layers, "*memory*" (*12) or "*psychic*" layers, but they should not claim to be serving scientific progress in so doing. These, and all similar expressions that one encounters repeatedly today, especially in the psychiatric and neurological literature, are utterly devoid of any factual basis; they are purely arbitrary fictions and only destined to cause confusion in uncertain minds [5]).

On account of what has been said, neither parcellation based on cortical layers nor that based on histological elements can be seen at present as promising procedures for cortical localisation. So, for an approach with a firm basis of reality and reflecting the state of our histological technique, there remains for the moment only the third form of cortical parcellation, namely **topographical localisation**. This is local subdivision of the cerebral cortex into structural fields, or in other words division according to tangential, regionally circumscribed zones of the hemispheric surface, each of homogeneous intrinsic structure but heterogeneous compared with the others. We term such different structural zones *anatomical areas* (*19).

The point of departure and the whole basis of such cortical parcellation, whether involving the cellular or the fibre architecture, is the cross-section of the cortex and especially the lamination visible in it. Cytoarchitectonics in particular is linked to the cellular lamination, that is to say the fact that the cellular elements making up the cortex reveal a pattern of layers one above the other in a section taken perpendicular to the cortical surface, with a different composition according to the site, and further that these cell strata often demonstrate regionally highly variable features.

So comparative cortical localisation must first of all investigate the local features of this cellular lamination throughout the mammalian class, and only secondly broach the question of how to achieve a topographic surface parcellation of the cerebral cortex in man and in animals on the basis of cytoarchitectonics.

Thus our treatise will be divided into two main sections:

1. study of the cellular lamination of cortical sections and its modifications throughout the mammalian class - *comparative cortical architectonics*;
2. areal parcellation of the hemispheric surface in various mammals on the basis of cytoarchitectonic differences - *comparative topographical localisation in the cerebral cortex*.

---

[5]) For instance, one has only to read that a dense myelinated fibre plexus in a particular layer of the superior temporal gyrus "enables unconscious sensations to be comprehended as conscious concepts", in order to appreciate the degree of confusion in such minds (Neurol. Zentralbl. 1908, p.546) (*18)

Part I.

The principles of comparative cortical cytoarchitectonics

In Part I we shall deal in turn with:

1. the question of the basic or "primitive" layering pattern of the mammalian cerebral cortex;

2. local differences in the cellular structure of layers within an individual brain;

3. particularities of cortical structure in different animals;

Chapter I

The basic laminar pattern of the cerebral cortex.

Since the first pioneering research of Meynert and Betz, a continuous stream of workers has studied the cellular lamination of the cerebral cortex and its specific modifications in man and in individual animals [1]). It would, therefore, be reasonable to expect that there would now be a solid basis of knowledge and understanding, at least as far as the essential elements are concerned. However, while recognising the numerous valuable individual contributions that have emerged from the competition, one cannot, after critical examination, avoid the conclusion that today we are further from agreement over the basic questions than for decades, especially concerning the common features and the origins of cortical lamination. Wherever we look we see major contradictions, not only in interpretations, but also in observations. There are totally conflicting results concerning the number, organisation and nomenclature of the layers, and a complete terminological confusion reigns, making interpretation quite impossible.

---

[1]) The comprehensive literature is cited in my earlier works; I shall only mention here those authors who have worked independently in this field; they are: Meynert, Betz, Mierzejewsky, Baillarger, Major, Bevan Lewis, Clarke (*20), Arndt, Berliner (*21), Hammarberg, Roncoroni, Nissl, Kolmer, Bolton, Schlapp, Cajal, Farrar, Koppen, Hermanides, Löwenstein (*22), Campbell, 0. Vogt, Mott, Watson, Elliot Smith (*23), Rosenberg, Haller etc.

In Tables 1 and 2 some of the best known layering schemes are summarised; the first column contains my own suggested nomenclature. As can be seen, each author has his own interpretation of the cortical layers, and gives corresponding numbers and names. The number of layers for man varies from five to nine according to different workers, and each uses his own nomenclature. For other animals, between three and ten layers have been suggested. Thus, sometimes completely different layers carry identical names, while on other occasions layers that are anatomically similar, and homologous, are given different names by different authors, although it is a basic prerequisite of scientific logic that similar structures should carry similar names and that homologous patterns should have homonymous designations.

I shall not attempt to explain the origins of the divergent and, at first sight, inexplicable differences in the results of different authors. Undoubtedly the main blame must be attributed to ignorance of developmental data and insufficient attention to comparative anatomy. These are the reasons why either important layers were completely overlooked, allowing heterogeneous structures to be considered as homogeneous, or on the other hand ontogenetically related sublayers, only secondarily differentiated and split off, were taken to be basically independent layers. In this context, it is necessary to discuss in more detail two new reports by Cajal and Haller that deal in depth with the question of lamination in mammals.

S. Ramon y Cajal begins the fifth volume of his "Studien über die Hirnrinde des Menschen" (*30), that deals specifically with "a comparative description of the structure of the cortex", with the words: "In man and the gyrencephalic mammals there is essential conformity in the architecture of the cortical layers. Change, or anatomical simplification, begins essentially with the rodents (rat, guinea pig, rabbit) (*31), is clearly visible in lower mammals, and is most marked in birds, reptiles and anurans ... Structural simplification involves ... the number of different nuclei and regions and the number of layers in the hemispheres" [2]). Concerning the cortex of small mammals, he writes that in rodents, and especially in the mouse, the cortex undergoes a noticeable simplification. "The thickness of the grey matter obviously decreases, the cells become smaller, the number of layers decreases to five as a result of the loss of a granule layer and the large pyramids subsequently forming a single layer" [2]).

According to Cajal there is thus an anatomical simplification of the cortical structure in lissencephalic mammals "from rodents down", expressed as a diminution of the number of layers, and especially as a loss of the granular layer, forming a four- or five-layered cortex, while in rodents (mouse and rabbit) upwards there are six, seven, or up to nine layers as found in man.

An essentially different view is put forward by B. Haller in his work "Die phyletische Entfaltung der Grosshirnrinde" (1908) (*33) that encompasses all chordates.

This is based on his research on the situation in the lowest vertebrates and

[2]) In the original these quotations are separated (*32).

Table 1. (General basic laminar patterns according to various authors.) (*24)

Brodmann (1902) (*25)	**Meynert (1868)**	**B. Lewis (1878) (*27)**	**Betz (1881)**	**Hammarberg (1895)**	**R. y Cajal (1902)**	**Campbell (1905) (*28)**	**Mott (1907)**
I. Lamina zonalis	1. Molecular layer	1. Cell-poor layer	1. Neuroglial layer	1. Plexiform layer	1. Plexiform layer	1. Plexiform layer	1. Zonal layer
II. L. granularis externa	2. Outer granular layer (*26)	2. Small pyramids	2. Small pyramids	2 & 3. Pyramidal layer	2. Small pyramids 3. Medium pyramids 4. Large pyramids	2. Small pyramids 3. Medium pyramids 4. Large pyramids	2. Small, medium and large pyramids
III. L. pyramidalis	3. Pyramidal layer	3. Large pyramids	3. Medium and large pyramids				
IV. L. granularis interna	4. Inner granular layer	4. Inner small pyramids	4. Granular layer (*26)	4. Small irregular cells	5. Granular layer	5. Stellate cells	3. Granular layer
V. L. ganglionaris	5. Spindle cell layer	5. Ganglion layer	Spindle layer	5. Ganglion layer	6. Deep, medium pyramids	6. Inner, large pyramids	4 & 5. Inner line of Baillarger and polymorph layer
VI. L. multiformis		6. Spindle layer		6. Spindle cell layer	7. Spindle cells	7. Spindle cells	

Table 2. (Lamination of the calcarine or "visual" cortex according to various authors.) (*24)

Brodmann (1902) (*25)	Meynert (1868)	Betz (1881)	Hammarberg (1895)	Schlapp (1898)	Bolton (1900) & Mott (1907) (*29)	R. y Cajal (1902)	Campbell (1905) (*28)
I. Lamina zonalis	1. Molecular layer	1. Neuroglial layer	1. Molecular layer	1. Tangential fibre layer	1. Outer layer of nerve fibres	1. Plexiform layer	1. Plexiform layer
II. L. granularis externa	2. Small pyramids	2. Small pyramids	2. Small pyramids (II & III)	2. Outer polymorphic cells	2. Small pyramids	2. Small pyramids (II & III)	2. Small pyramids (II & III)
III. L. pyramidalis				3. Pyramidal cells			
IVa. L. granularis interna superficialis	3. Outer granular layer	3. First granular layer	3. Small cells (IVa) and solitary cells (IVb)	4. Granular layer	3a. Outer layer of granules	3. Medium pyramids (IIIb and IVa)	3. Medium pyramids (IIIb and IVa)
IVb. L. granularis interna intermedia (stria of Gennari)	4. Outer intermediate granular layer and solitary cells	4. Longitudinal fibre layer	4. Dense single cells (IVb)	5. Solitary cells	3b. Middle layer of nerve fibres (Gennari)	4. Large stellate cells (IVb)	4. Stellate cells (IVb)
IVc. L. granularis interna profunda	5. Middle granular layer	5. Second granular layer	5. Small granules (IVc)	6. Granular layer	3c. Inner layer of granules	5. Small stellate cells (IVc)	5. Small stellate cells (IVc)
V. L. ganglionaris	6. Inner intermediate granular layer and large solitary cells	6. Second fibre layer	6. Solitary cells (V)	7. Cell-poor layer	4. Inner layer of nerve fibres	6. Small cells with arciform axons (IVc)	6. Solitary cells
		7. Large pyramids				7. Giant pyramids	
VI. L. multiformis	7. Inner granular layer		7. Polymorphic cells				
VIa. L triangularis		8. Spindle cells		8. Inner polymorphic layer	5. Polymorphic cells	8. Large cells with ascending axons	7. Spindle cells
VIb. L. fusiformis	8. Spindle cells		8. Spindle cells			9. Spindle cells	

strives to construct a consistent phylogenetic progression of cortical
ment for all mammals as far as man. He recognises the "primary phy
mammalian stage" (also marked by the lack of a corpus callosum) as
layered cerebral cortex, as found in the marsupial opossum and musk ka [illegible]oo
(*34). According to Haller, they have a homogeneous cytoarchitecture throughout their dorsal pallium (*35), implying a total lack of local physiological specialisation. "The whole of the dorsal pallium at this stage consists of, apart from the plexiform layer, a dense, narrow upper layer and a far broader lower layer (*36)"... "This stage was common to all mammals of the common mammalian line and from it emerged further developments in the various groups; according to exactly the same cerebrogenetic rules [3])."

According to Haller further differentiation of the dorsal pallium in mammals continued with the microchiropterans (*38), in which, beginning with the pipistrelle, "the primitive homogeneous pattern of the entire dorsal pallium" that is found in monotremes and marsupials is modified by the separation of a new layer from the deepest layer of the primitive cortex at the frontal pole, that he calls the inner zone (*39). Thus Haller proposes that a four-layered cortex is formed containing two separate structural zones, an inner and an outer (except in the piriform cortex) – compare his Figures 17, 18 and 19.

According to Haller the next level is reached in the rodents, in which there is an increase in the number of layers in part of the cortex by development of the middle cortical layers, particularly that containing small stellate cells, our granular layer (compare his Figures 21, 23, 24 and 26). "Only at this stage is the six-layered pattern of the dorsal pallium generally introduced", and this "is of the greatest importance for the further development of the cerebral pallium" (page 441). Haller concludes that "Brodmann's acceptance of the general validity of the six-layered pattern is thus only really valid for a majority of mammals"; the others, on the other hand, are subject to a preliminary stage of three or four layers, and the "primitive three-layered cytoarchitecture" of mammals is supposed to be derived from the situation in amphibia and reptiles.

In contrast to Cajal and Haller I have for years supported the idea that the original, primitive pattern of cortical layering in the whole mammalian class is six-layered, and that this six-layered pattern is visible in all orders, either permanently or at least as a temporary ontogenetic stage in the embryo, even in those cortical zones where it disappears in the mature brain.

I have considered both developmental factors and comparative anatomical findings as my basis for this interpretation. Recent research has confirmed my opinion; in spite of Haller's opposition. I maintain now as before that the original pattern for the whole mammalian order is the **primitive six-layered type** and that all variations in cortical structure are derived from this primitive six-layered cytoarchitecture. The only exceptions are certain "rudimentary" cortical zones. In man these include on the one hand a relatively small part of the rhinencephalon (*40), that undergoes considerable development in

---

[3]) p.440-441 loc. cit. (*37)

macrosmatic animals to form the piriform lobe, and on the other hand the more or less extensive cortical regions of the cingulate gyrus adjacent to the corpus callosum, mainly around its anterior half and at the splenium. As far as I can see, these zones have an atypical laminar pattern, or at least in the material available so far I have not succeeded in finding a six-layered ontogenetic transitional stage. (See also page 20).

In summary, we must distinguish between two different basic cortical cytoarchitectonic patterns:

1. homogenetic formations that in the higher mammals include those cortical types that cover the major part of the hemisphere and which are derived from the basic six-layered type, either possessing this pattern permanently, or temporarily during ontogenesis;

2. heterogenetic formations in which the six-layered embryonic stage cannot be, or has not yet been, demonstrated.

It will now be our task to give a detailed argument for the origin of the homogenetic formations of the neopallium from the basic six-layered pattern. This involves both a developmental and a comparative anatomical approach.

a) The developmental basis for the six-layered cortex.

For an anatomist one of by far the most familiar notions is that to understand the nature of morphological organisation it is essential as a first step to look at its ontogenetic development. According to Haeckel's law of biogenesis, the principles of which are recognised by most experts, ontogeny (embryological development) represents a shortened recapitulation of the succesive developmental stages met in the course of phylogeny (evolution of the species). If we wish to understand the basis for the genesis of cortical lamination, it is useful to study the development of this lamination from its origin in the embryo and follow its subsequent transformations from the early primitive stages.

This task has until now been completely neglected for the critical stages of development, the essential information having been provided only for the earliest, still largely undifferentiated, stages during the first few months of the human brain by W. His [4]). According to his pioneering research, the original cortical Anlage is unlayered (Figure 4). It is the product of the differentiation of the primitive embryonic neural tube and consists of a simple cellular region without further organisation near the superficial surface of the wall of the hemisphere (*42), on the outer edge of which a cell-free border can just be distinguished, the so-called "border veil" (*43). Through unequal growth in thickness, continuous migration of neuroblasts from the matrix (*44) (the inner plate of His) (*45), and through ingrowth of fibre bundles, the architectonic transformations that will ensure the definitive lamination of the mature cortex are prepared. Only from the fifth month onward do the embryonic cortical cells

[4]) W. His, Die Entwicklung des menschlichen Gehirns während der ersten Monate. Leipzig 1904. (*41)

(neuroblasts) become organised into actual layers, that is into dense and sparse zones of cellular elements parallel to the surface. First, layers V and VI are laid down as separate cellular strata in about the sixth month, the former as a cell-poor, light stripe, the latter cell-rich. Layers II, III and IV are not yet distinguishable as entities but form a single dark zone of densely-packed neuroblasts. These processes provide the initial basis for a stratified cortical pattern. It is not the intention of this treatise to enter into the complex intermediate steps that intervene before the final architectonic plan of the cortex is achieved, steps that can be very different for different regions of the hemisphere. We are only interested in the stage that illustrates clearly the general plan of the definitive cortical lamination, and that provides the key for the individual cytoarchitectonic patterns of man and animals.

This developmental stage in man occurs around the sixth to the eighth embryonic months. At this time, that varies slightly with different cortical areas, the cerebral cortex enters a six-layered phase that spreads over its whole surface - with the exception of course of the heterogenetic region of the archipallium and what Meynert called "defective cortex". Everywhere, on the exposed surfaces as well as in the depths of the adjacent sulci, the pattern is extremely clear and systematic. The still mainly undifferentiated neuroblast-like cortical cells are grouped in laminae such that there are three light bands alternating with three dark cellular bands or layers, the "*primary tectogenetic layers*", that are placed one above the other like an onion skin. In Figure 1 the lamination is already clear in a six-month human foetus, and is even more pronounced in Figure 2, at eight foetal months.

The individual layers are, counting from the outside to the inside and incorporating the nomenclature of Meynert, Betz, Clarke, Lewis and Hammarberg (Figure 3):

I. *Lamina zonalis* - molecular layer.

II. *Lamina granularis externa* - outer granular layer.

III. *Lamina pyramidalis* - pyramidal layer.

IV. *Lamina granularis interna* - inner granular layer.

V. *Lamina ganglionaris* - ganglion cell layer.

VI. *Lamina multiformis* - spindle cell layer. (*47)

This six-layered phase does not last for the same length of time over the whole surface of the hemisphere, and does not start at the same time everywhere. Many regions undergo a very precocious development and are significantly ahead of other areas [5]), while some develop more slowly and only progress from the preliminary primitive stage to the typical six-layered pattern later. The process of architectonic remodelling can sometimes be accelerated and shortened, so that the temporary six-layered stage is compressed in time and is so transitory that it is difficult to visualise; this is "*ontogenetic acceleration*"

[5]) This has recently been shown unequivocally for the development of neurofibrils in the cortex. (cf. Brodmann, Bemerkungen über die Fibrillogenie und ihre Beziehungen zur Myelogenie. Neurol. Zentralbl. 1907. No. 8.) *48)

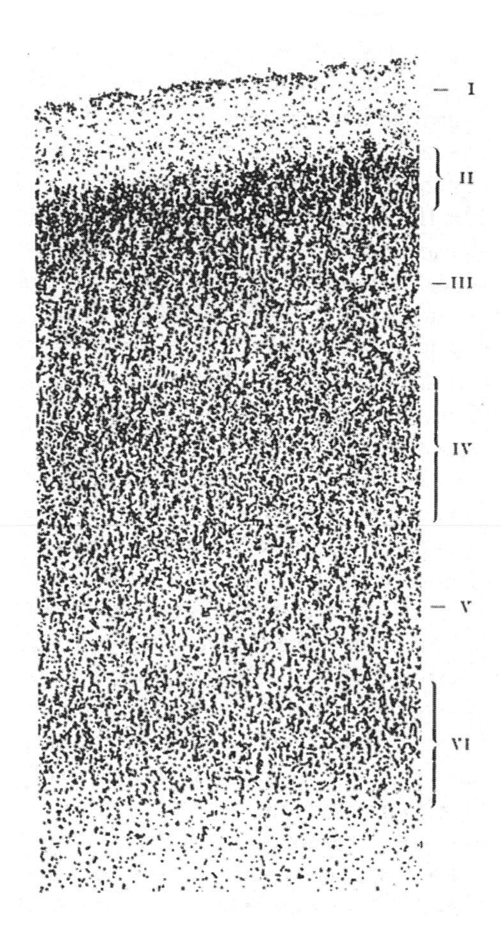

Fig. 1. Human foetus aged 6 months. 66:1, 10μm (*46). The early appearance of the basic tectogenetic layers. Layers II, IV and VI are cell-rich and darker, layers I, III and V lighter and poorer in cells.

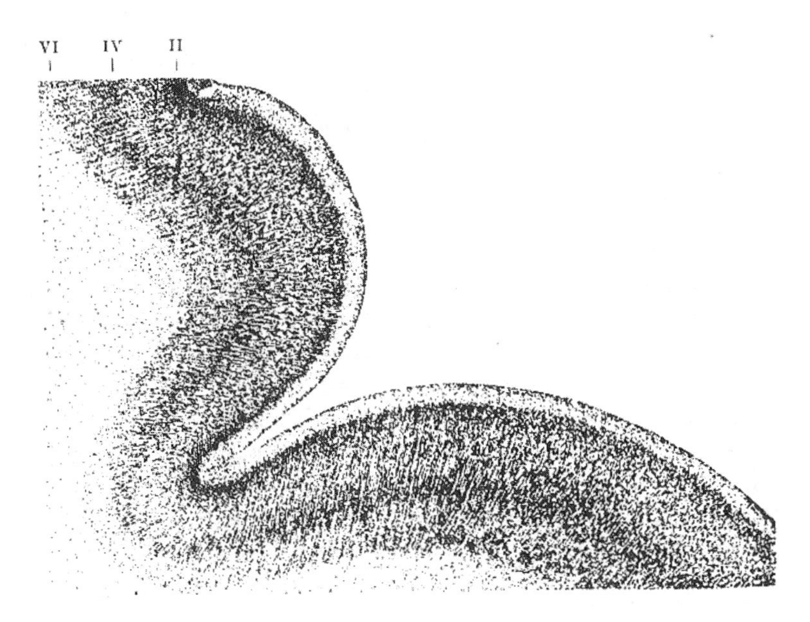

Fig. 2. Basic tectogenetic layers in the eighth foetal month in a survey section through the cortex of the parietal lobe. 20:1, 10μm. The layers stand out clearly from each other and cover the surface of the gyri and the depths of the sulcus equally.

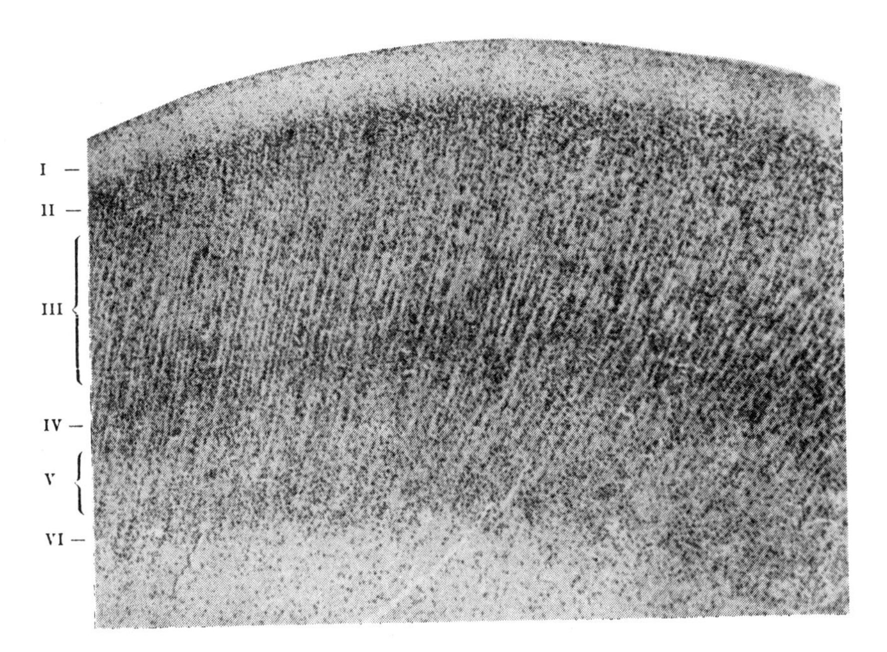

Fig. 3. The same as in Fig. 2 at higher magnification. 46:1, 10μm. The radial arrangement of the cells, reflecting the later fibre bundles, is already well established.

of phylogenetic recapitulation as understood by Haeckel, and the question arises as to whether many of the so-called heterogenetic cortical areas undergo such a shortened development that the transitory stage escapes our detection although it is present as an extremely compressed event. If we accept that, under the influence of ontogenetic acceleration, parts of an organ are, so-to-speak, suppressed during ontogenesis and lost, then we can think of a process that can be interpreted as "*defective homology*" in Gegenbaur's sense.

However this may be, it is certain that all the homogenetic cortex with its modified forms, that is essentially the whole neopallium and therefore, in man, by far the major part of the cerebral cortex, undoubtedly passes through the basic six-layered stage, and is derived from it by secondary transformation. This also applies to structures that do not exhibit a six-layered pattern in the mature brain, including cortical types with both more or less layers.

How is this process of transformation achieved? Even at a time when the cortical plate is otherwise completely undifferentiated, certain local differences can be seen, especially in cortical thickness, as W. His has thoroughly expounded. Growth in thickness of the whole pallium of the hemisphere, and with it the outer layer or cortical plate, progresses at different rates in different areas, being fastest in the basal zones, adjacent to the corpus striatum, and slowest on the medial surface. Thus the wall of the hemisphere appears progressively younger from its basal to its dorsal parts, and from there down to the depths of the sagittal fissure, and the development of the actual cortical layers follows in the same way (Figure 4). At the beginning of the fifth month differences in cell density are added to these regional differences in thickness: frontal regions are on average less dense and occipital regions more so. In this way the first local differences are expressed, but they give little indication of subsequent cortical organisation.

The essential local transformations in the basic six-layered architectonic pattern only start around the beginning of the seventh month, foreshadowing the definitive structure. Basically, two different principle processes are involved in various parts of the cortex, in man as well as in animals, and proceed within sharply defined borders. They are:

first, a diminution or disappearance of lamination, *a loss of layers*,

second, an increase in lamination that results in a division or separation of the primitive layers, leading to an *increase in layers* and the formation of sublayers.

We shall only briefly illustrate these two major forms of ontogenetic differentiation by two examples, for all the subsequent architectonic modifications will be summarised later.

1. The most important and biologically most interesting example of involution and loss of layers is represented by the so-called *agranular cortices*, that is those types that no longer possess an inner granular layer (IV) in the mature brain. Chief among these is the giant pyramidal type, the so-called "motor cortex" of the literature, that lies in the precentral gyrus (area 4 of our brain map) (*54). In its immature form, and even when the central sulcus is

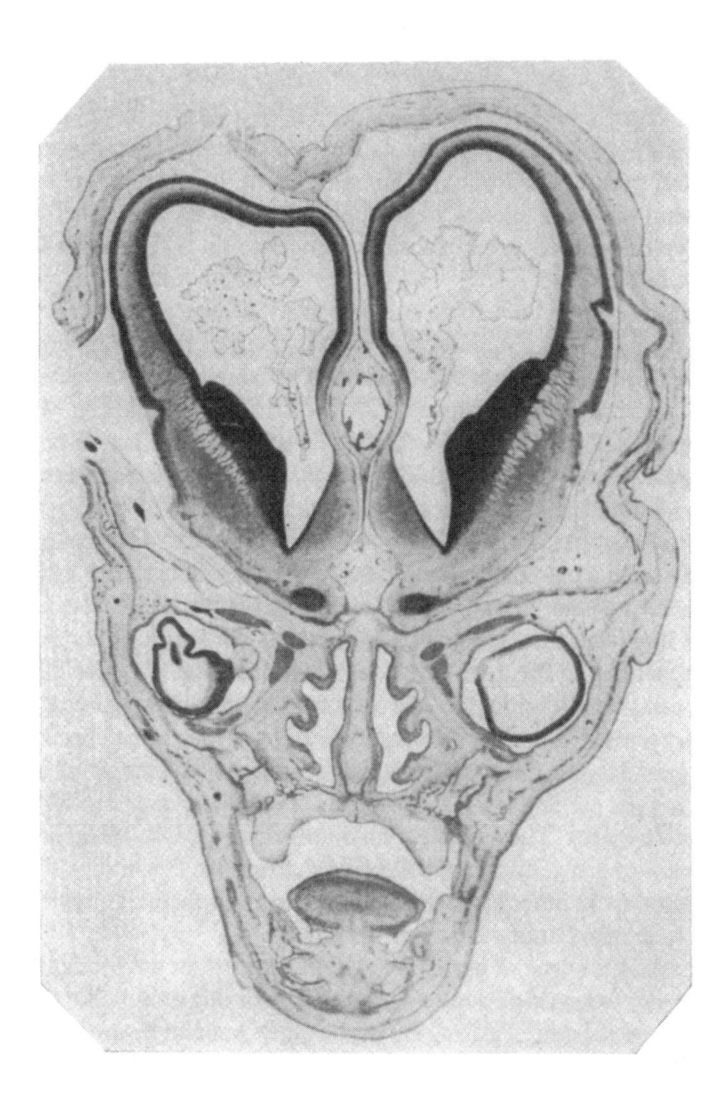

Fig. 4. Human embryo aged 3 months. Coronal section through the forebrain at the anterior end of the third ventricle. The wall of the hemisphere is composed of 3 layers: the ventricular zone (*49), the intermediate zone (*50) and the cortical plate (*51). The last consists of a uniform, thick cellular lamina, narrowest on the medial surface. (From W. His, Die Entwicklung des menschlichen Gehirns, p.85, Fig. 56.) (*52,*53)

already beginning to form, it has a distinct inner granular layer and is typically six-layered like other cortical areas, whereas this layer is totally absent in the adult human.

An illustrative embryonic stage is represented in Figures 5 and 6. It differs from the usual basic type (Figure 3) on the one hand by the fact that the inner granular layer is already less thick and appears less dense and cellular, and on the other hand by the beginning of the formation in layer V of specific large cells that will eventually produce the giant cells of Betz. Also the cell density is rather less overall than in the section in Figure 3 from the parietal lobe. Furthermore the deepest layer, or layer VI, begins to broaden noticeably and become less dense, so that its border with the white matter is more indistinct. Thus already at these early stages features are developing that will typify the mature giant pyramidal type of cortex in the adult brain, namely the disappearance of the inner granular layer and the formation of the Betz cells.

Figure 7 illustrates this. Layer IV is completely absent; the embryonic giant cells that are just appearing in Figure 6 have attained a considerable degree of development, grouped in "nests" and dominating the structure of the whole cortical cross-section (Figure 43). Layers III, V and VI fuse together. Because of this and because of the lack of an inner granular layer the cortex appears almost completely unlaminated, only layer II standing out particularly clearly because of its rich population of small cells, as in the embryonic stages. The extensive involution of the basic ontogenetic layers and the development of a specific giant cell layer characterise this cortical type and allow its differentiation from other areas such as the cortex of the neighbouring postcentral gyrus, not only in man but in most other mammals.

2. The calcarine type of cortex, the "visual cortex" of the literature, qualifies as the most pertinent example of duplication of layers during embryonic development. It is essentially characterised by a division of the inner granular layer into three; and thus has eight layers rather than six. Its derivation from the basic type is illustrated in Figures 8 to 12.

Figure 8 represents a section through the region of the calcarine fissure at the sixth foetal month. The typical embryonic six-layered pattern exists even in the immediate vicinity of the fissure where later the duplication of the layers will take place; the only significant feature here is a slight thickening of layer IV. If this is compared with a section of the same region at the beginning of the eighth month, profound changes in the overall structure are visible.

Figures 10 and 11 show this process of architectonic transformation at the peak of its development. The inner granular layer (IV) splits into two dark, cell-dense strips, a superficial inner granular sublayer (IVa) and a deep inner granular sublayer (IVc), while a light cell-poor stripe, the intermediate inner granular sublayer or stria of Gennari (IVb), extends between them.

The final product of this transformation is seen in Figure 12 from the occipital lobe of an adult brain. In essence this has the same lamination that already differentiated in the embryo, except that individual layers are emphasised differentially. Layer II is almost completely fused with layer III,

Fig. 5. Human foetus in the eighth month. Giant pyramidal cortex. Magnification 40:1. Section thickness 5μm. The embryonic six-layered tectogenetic transitional stage is well developed. The outer and inner granular layers (layers II and IV) are clearly formed. The giant pyramids within the light, cell-poor ganglion cell layer (Vg) are at their first developmental stage. Cortex and white matter are well separated.

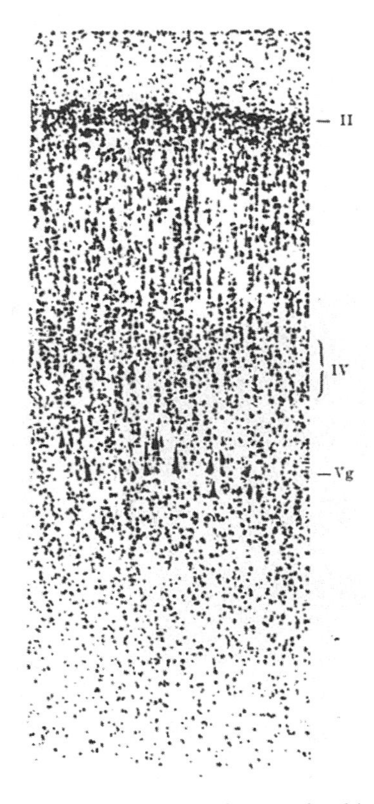

Fig. 6. The same as in Fig. 5 at a magnification of 66:1. The thinning of the inner granular layer (IV) and the development of the giant pyramids in the ganglion cell layer (V) stand out clearly.

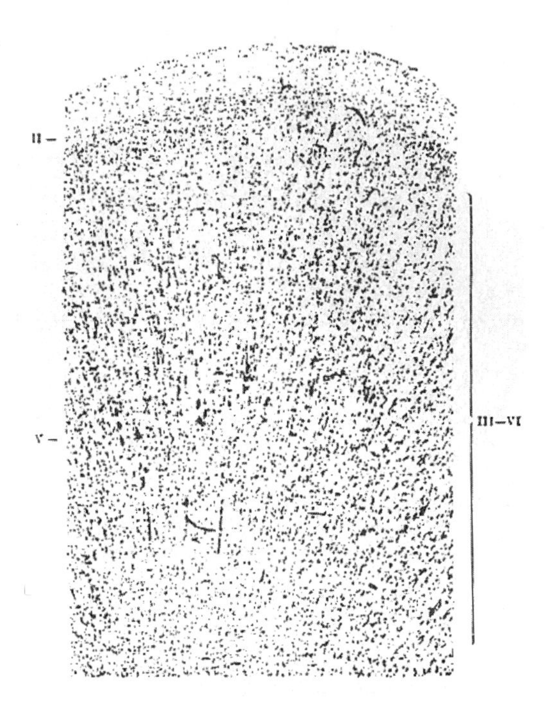

Fig. 7. The same as Figs. 5 and 6 in a human adult (apex of the precentral gyrus). 25:1, 10μm. The inner granular layer has regressed completely; indeed, there is almost no layering in the cortical section. - The Betz giant cells are arranged in "cumulative" nests (see text page 82ff.). Cortex and white matter have an indistinct border. (cf. Fig. 58 from the depths of the central sulcus.)

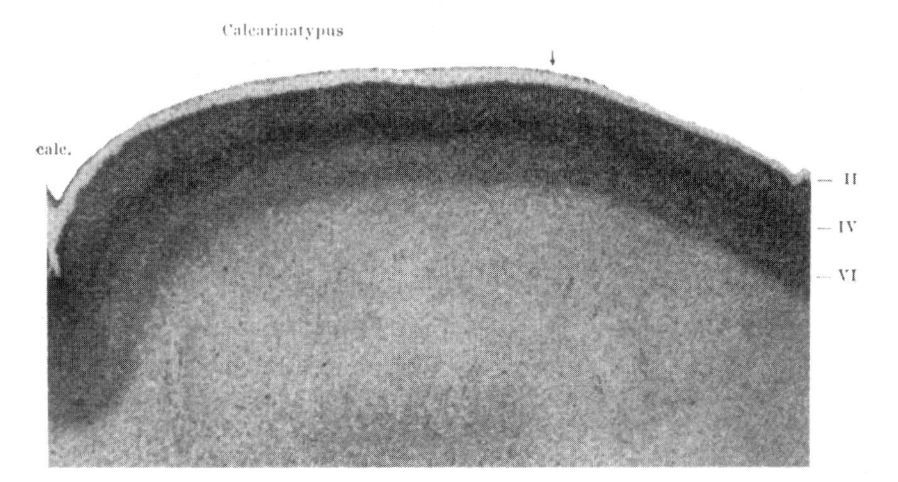

Fig. 8. Human foetus in the sixth month. Survey section through the cuneus at the transition to the calcarine cortex. 15:1, 10μm. The basic six-layered cortex appears over the whole cortical section, even in the vicinity of the calcarine sulcus (calc.). The subsequent site of transition to the calcarine cortex (↓) is marked merely by a denser and wider layer IV. The deeper ganglion cell layer becomes lighter at ↓.

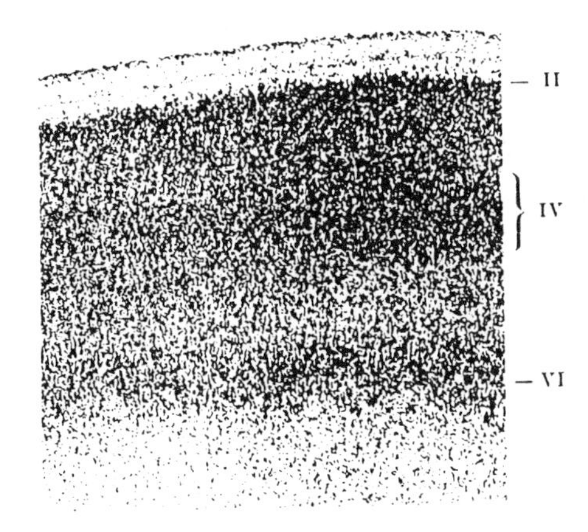

Fig. 9. Human calcarine cortex in the sixth foetal month at higher magnification. 46:1, 10μm. The inner granular layer (IV) is still simple and not divided, but is more solidly built and more densely cellular than in neighbouring cortex. (Part of Fig. 8.)

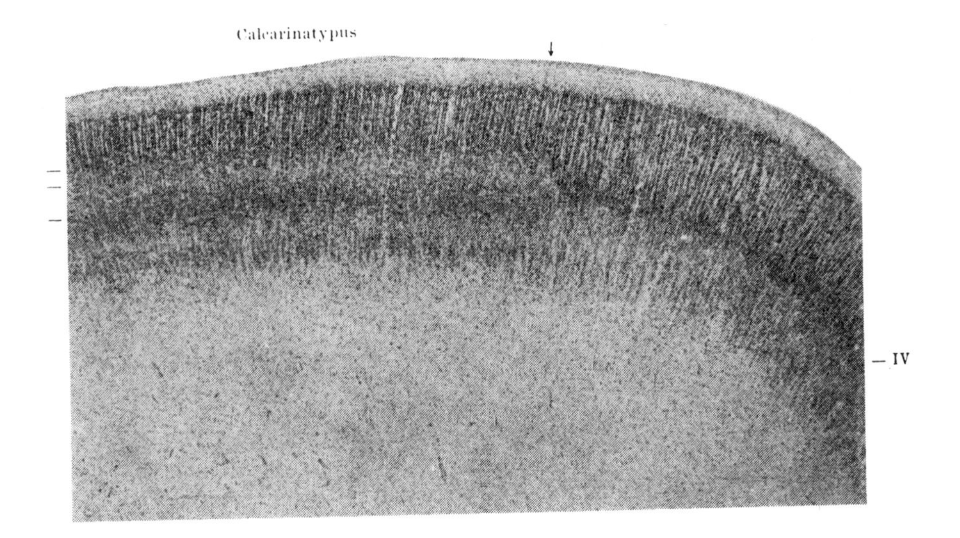

Fig. 10. The same as in Fig. 8 at the beginning of the eighth foetal month. 22:1, 10μm.

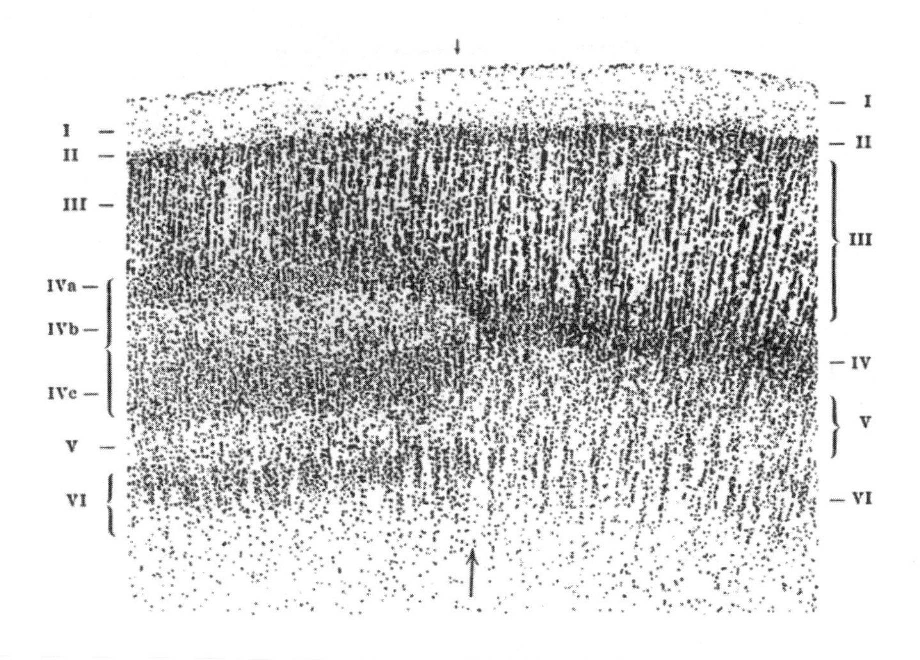

Fig. 11. Human foetus aged 8 months. Section through the cortex of the calcarine sulcus (as in Figs. 8 and 10). The site of transition ↓ at higher magnification. 46:1, 10μm. On the right is the basic six-layered cortex, on the left the calcarine cortex, with a sharp transition ↓ between the two structural types and increase in number of layers. Laminar changes in the calcarine cortex visible from superficial to deep consist of

1. marked narrowing of layer III,
2. splitting of the inner granular layer into 3 sublayers, IVa, IVb and IVc,
3. narrowing of layer V,
4. denser layer VI.

The layers in Figs. 10, 11 and 12 are: (*47)

- I. *Lamina zonalis,*
- II. *Lamina granularis externa,*
- III. *Lamina pyramidalis,*
- IV. *Lamina granularis interna,*
  - IVa. *Sublamina granularis interna superficialis,*
  - IVb. *Sublamina granularis intermedia (stria of Gennari or Vicq d'Azyr),*
  - IVc. *Sublamina granularis interna profunda,*
- V. *Lamina ganglionaris,*
- VI. *Lamina multiformis,*
  - VIa. *Sublamina triangularis,*
  - VIb. *Sublamina fusiformis.*

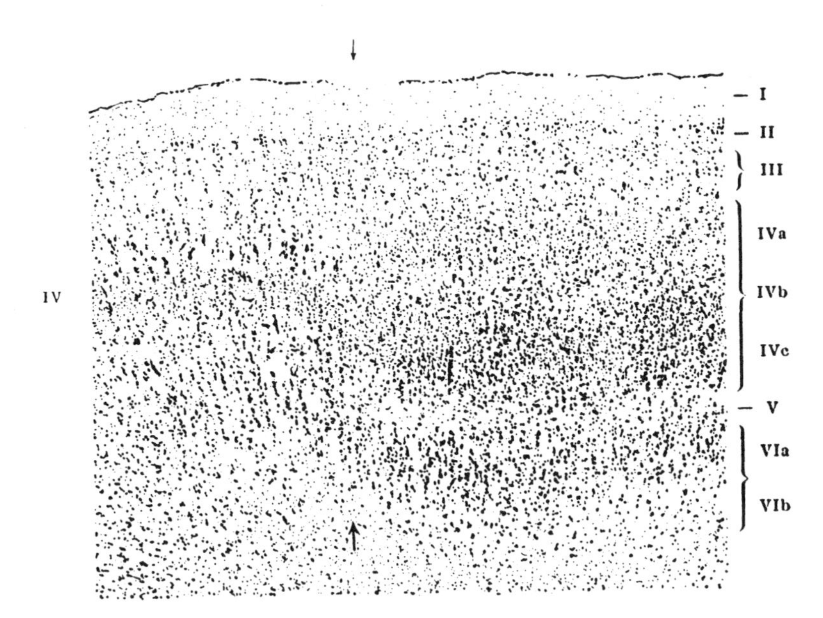

Fig. 12. Adult human. 25:1, 10μm. Transition to the calcarine cortex (↓) as in Fig. 11. The layers are the same: on the right is the eight-layered cortex of the striate area (area 17 of our brain map), on the left the six-layered cortex of the occipital area (area 18).

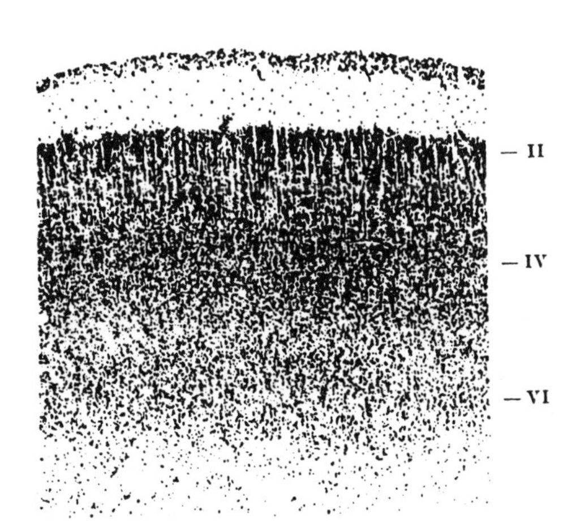

Fig. 13. Cat embryo. 66:1, 10μm. Six-layered tectogenetic transitional stage as in the human foetus in Fig. 9.

which is like a regression to the primitive pattern. On the other hand layer VI has differentiated clearly, in the way described above, into two sublayers, VIa and VIb.

In this case, as seen frequently, there are two parallel and opposite ontogenetic processes: on the one hand an increase in layers by the formation of sublayers from a primitive layer, on the other a loss of layers by fusion of primitive layers.

It is not necessary to cite further examples of architectonic reorganisation of the embryonic lamination as only the principles of the derivation of secondary patterns from a common initial type will be outlined here.

However, it is of prime importance to note that the same principles of differentiation are active during ontogeny in lower animals as in man. A few examples from carnivores and marsupials illustrate this.

Figure 13 is a photomicrograph of a section of cortex from a foetal cat. The similarity of the laminar pattern to that of man represented in Figures 1 and 9 can hardly be contested. The dense stripe of the inner granular layer (IV) stands out clearly, as do layers II and VI. On the other hand obvious lighter areas represent layers III and V. In other words we have the same six layers expressed as in the human foetus.

The next figure shows the transformation of the six-layered pattern at a later stage.

Figure 14 is a frontal section of a hemisphere in the region of the cruciate sulcus of a 14-day-old cat. It illustrates the disappearance of the inner granular layer in agranular cortex, especially the giant pyramidal type. In the rest of the section the basic six-layered pattern is obvious throughout, and is much clearer than in the adult animal. In the giant pyramidal cortex of the cruciate sulcus the compact layer of giant pyramids (V) is conspicuous in place of the inner granular layer.

In Figure 15, a section through the occipital cortex of the foetal wallaby, the same developmental stage of the calcarine type of cortex as in Figures 10 and 11 of man is represented. There is the same process of splitting of the inner granular layer (IV) into three sublayers, IVa, IVb and IVc, except that the last is less conspicuous to the untrained eye, especially as the intermediate layer (IVb) does not have the same marked clarity as in man. The outer granular layer (II) on the other hand is strikingly dense in the embryo. It should be compared with the micrograph of an adult kangaroo (Figure 22).

Thus we have become familiar with the development of those structures that stem directly from the basic six-layered cortex.

b) The comparative anatomical basis for the six-layered cortex.

Comparative anatomy offers rich and varied material and, together with ontogeny, is of foremost importance as a means to visualise the natural structural plan of an organ. As it begins with the lowest forms of organisation in the animal kingdom it leads us to the origins of morphological differentiation, like

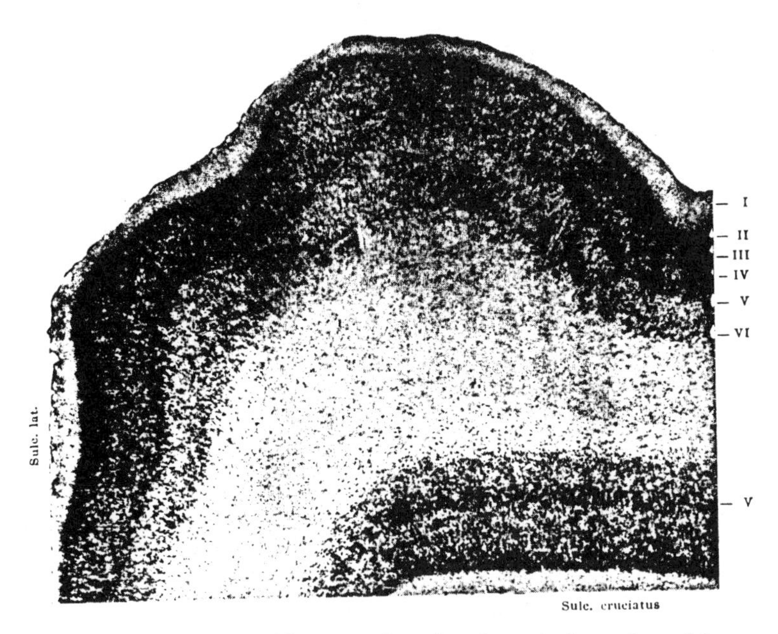

Fig. 14. New-born cat (14 days old). Coronal section through the region of the cruciate sulcus. 46:1, 10μm. Regional modifications of the basic tectogenetic cortex. In the marginal gyrus there are six layers I-VI as in man. Around the cruciate sulcus there is strong development of the giant pyramidal layer (V) and regression of the inner granular layer as in Fig. 7 of man. Layer II is everywhere well demarcated from layer III. (See also Fig. 80 in Part II.)

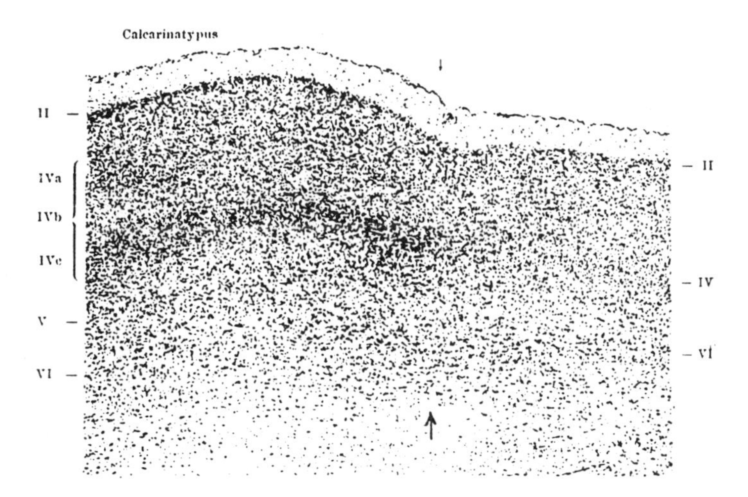

Fig. 15. Foetal wallaby (*Onychogale frenata*). 25:1, 10μm. Transition from the eight-layered calcarine cortex to the basic six-layered type. Left of arrows the inner granular layer of the basic cortex splits. This is the same essential process as in Figs. 10 and 11 of the human foetus. Note also the strong development of the external granular layer (II).

embryology but from another point of view, revealing Nature's designs in a sort of draft form.

As we have seen, there are widely divergent views about the basic or primitive pattern of structural differentiation in the cerebral hemisphere.

Haller derives the higher stages of cortical development, which he interprets like myself as six-layered, from a *primitive three-layered cytoarchitecture* in lower mammals. He finds this primitive developmental stage still clearly expressed in marsupials and to some extent in chiropterans and concludes that we could "also take into account the lack of a corpus callosum, and conclude that this three-layered cerebral cortex, or four with the white matter, which is also found in monotremes, is the most primitive mammalian cerebral cortical cytoarchitectonic pattern".

On the other hand, Cajal draws a fundamental distinction between "lower" lissencephalic animals on the one hand (from rodents downwards), to which he attributes a four- or five-layered pattern because of the lack of an inner granular layer, and "higher" gyrencephalic brains on the other hand, which, as in man, are supposed to possess a six- to nine-layered cortex [6]).

My own interpretation is that all mammals possess a common primitive cortical architectonic form, the six-layered pattern, and that if this is absent it is simply explained by secondary transformation of the original form.

This is supported by new observations on lower animals, particularly those in which Cajal and Haller accept such discrepant simplified cytoarchitectonics. I have already shown in my fifth communication on histological cortical localisation that the marsupial kangaroo has a six-layered cortex, pointing out the presence of a form of calcarine cortex derived from the basic type.

Over the years the material that I use for my research has become very extensive. I now possess a representative of one or several species from all orders of mammals (*55) - except cetaceans. Most are complete series of sections through whole hemispheres, but some from large animals are more or less extensive samples from the hemispheres. Apart from brains of adults I have many examples available of foetal and immature stages. The technical preparation of all was the same as described previously: formalin fixation, paraffin embedding and cell staining with aniline dyes (simplified Nissl method).

Grouped by orders, I have cortical preparations from the following species (*56):

I. Primates: man, orang-utan, chimpanzee, langur, mangabey, mona guenon, rhesus macaque, drill (*57), spider monkey, woolly monkey, capuchin monkey, black saki, squirrel monkey, common marmoset, black-eared marmoset, negro tamarin.

---

[6]) I shall not dwell on the peculiar contrasting of (higher) gyrencephalic mammals with (lower) lissencephalic mammals. It can be assumed that the presence or absence of cerebral sulci is independent of the high or low rank of an animal in the taxonomic series, for there exists in the primates, the highest mammals, a lissencephalic group (the marmosets), and on the other hand the lowest mammalian order has a gyrencephalic representative, the echidna.

II. Prosimians (*58): sifaka, indris, black lemur, mouse lemur, potto, slow loris.
III. Chiropterans: flying fox, pipistrelle.
IV. Insectivores: hedgehog, mole, tenrec.
V. Carnivores: brown bear, kinkajou, stone marten, wolf, dog, fox, mongoose, musang, lion, tiger, cat.
VI. Pinnipeds (*59): common seal.
VII. Rodents: squirrel, mouse, rat, rabbit.
VIII. Ungulates (*60): hyrax, African elephant, pig, peccary, chevrotain, goat.
IX. Edentates: three-toed sloth.
X. Marsupials: phalanger, kangaroo, wallaby, opossum.
XI. Monotremes: echidna (*61).

I have since had the opportunity to study other species of marsupials, particularly the opossum, and I was able to determine that, in full agreement with higher mammals, all possess a typical six-layered cortex. However, in the opossum the extent of the six-layered cortex is relatively limited, in contrast with the kangaroos (*M. rufus*, *M. dorsalis* and *Onychogale*) in which it occupies a wide area, but it is undoubtedly present. Also, I was able to confirm the six-layered pattern, although in substantially different form; in the cerebral hemisphere of a monotreme, the echidna, of which I was recently able to obtain a brain. I unfortunately have no photomicrographs available of the opossum and echidna that could demonstrate this. However, Figures 16, 17, 18 and 19 compare a section of human cortex with the six-layered cortex of a series of other mammals, including a carnivore, a rodent and a kangaroo. (For chiropterans, see Figure 77.) All homologous layers are labelled homonymously and are evident from the illustrations. The basic correspondence between the animals is obvious and so complete that it would be pointless to specify detailed differences at this low magnification. There is little doubt about the presence of an inner granular layer (IV), indeed it seems that this layer is even more clearly developed in the lower animals (kangaroo) than in the human example under consideration. One should also note the insignificant and negligible differences in the thickness of the cortex in the illustrated examples of these widely different orders.

I should like to emphasise particularly, in contrast to Cajal, that the same six-layered cortical structure exists in other lissencephalic animals, such as insectivores and small rodents etc. Thus the series from the highest to the lowest mammals is complete in that various families from all orders (except the cetaceans), including the monotremes, contain illustrations of the concept outlined above.

The very same lower animals to which Cajal and Haller attribute a simple, poorly laminated cortical structure, undoubtedly possess the same basic six-layered pattern as the highest mammals. Where there are modifications to the cortical structure, they can be traced back to it. In rodents the six-layered

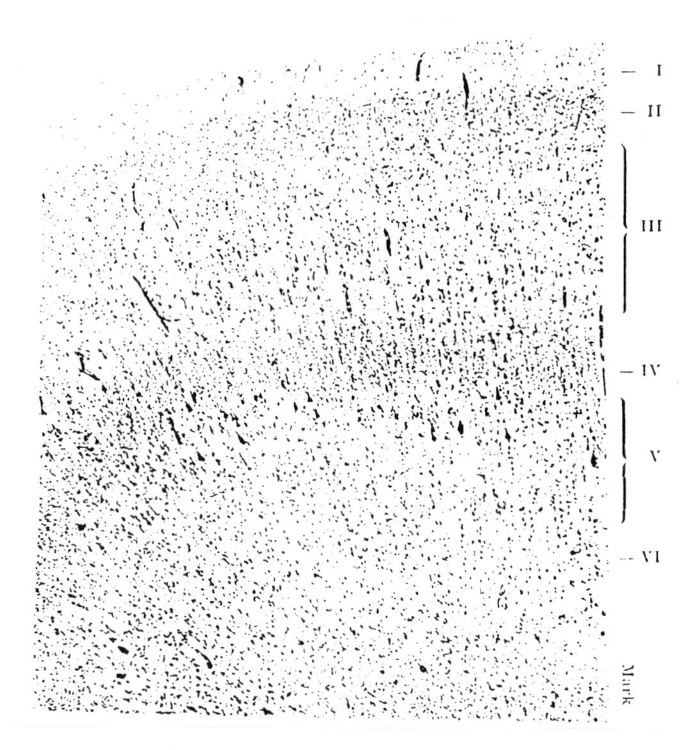

Fig. 16. Adult human. 25:1, 10μm. Preparietal cortex (area 5 of the brain map) as an example of a six-layered homogenetic cortex. Layer IV is strongly developed; in layer V there are large ganglion cells similar to Betz giant pyramids (cf. Fig. 7).

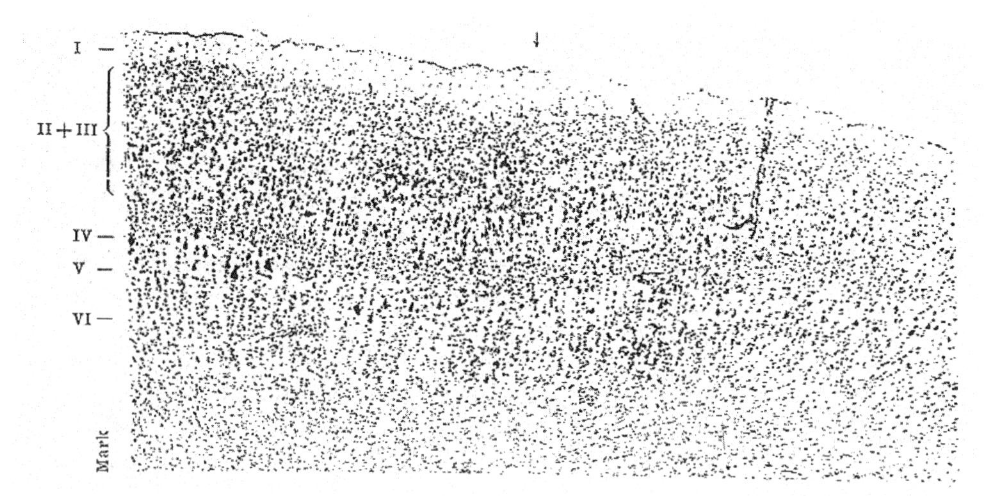

Fig. 17. The same as Fig. 16 in the kinkajou (*Cercoleptus caudivolvulus*). 25:1, 10μm. The homologous preparietal cortex (left of ↓). Gradual transition to parietal cortex (right of ↓).

Fig. 18. Adult rabbit (*Lepus cuniculus*). 25:1. 10μm. Six-layered basic type in the parietal cortex. Layer IV clearly distinguishable. Layer VI very deep; II and III fused.

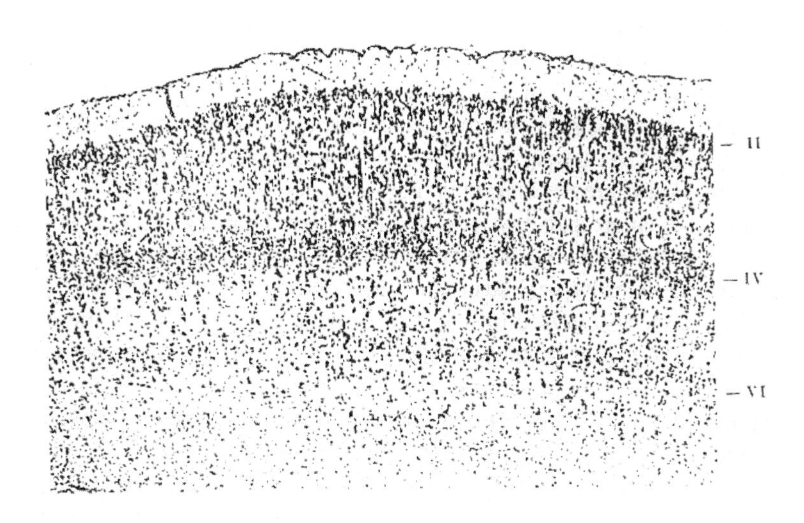

Fig. 19. Adult wallaby (*Macropus dorsalis*). 25:1, 10μm. Typically six-layered as in Figs. 16-18. The inner granular layer is even more distinct than in man. The outer granular layer (II) is also clearly visible. VI is very narrow.

pattern is very typical and occupies a large part of the neopallium, as we shall see in Chapter II when we describe the parcellation of the architectonic areas. Cajal's thesis, that rodents and other lissencephalic mammals possess an anatomically simple cortical structure characterised by a reduction in the number of layers, thus cannot be accepted as correct. Equally, Haller's assumption of a primitive three-layered cytoarchitecture should be rejected as erroneous.

One should not forget - as we learned from developmental studies - that even in man wide areas of regressed cortex with obviously diminished layering are encountered, especially involving loss of the inner granular layer. However, when looked at ontogenetically they belong to the basic homogenetic six-layered type as they are directly derived from it, even if only during a fleeting transitory stage. The definitive reduced lamination does not therefore represent a primitive or phylogentically old condition but rather a higher stage produced by secondary ontogenetic transformation. Naturally somewhere in the ancestral series there is a genuine primitive condition (with three, two or one layer), but this cannot be compared with a transformed six-layered pattern. This is merely "*imitative homology*" according to Fürbringer: the process of regression (to fewer layers) in mammals demonstrates only a purely superficial similarity to a stage that has long been surpassed during phylogeny (the three layers of lower vertebrates) and is therefore "imitatively homologous"; on the other hand regressive or heterotypical structures are "defectively homologous" to six-layered types. The richly specialised forms of cortical structure depend on such secondary transformations, as we shall soon see. There is no doubt that cortical types without the six layers, the so-called agranular varieties, such as arise on a considerable scale in lower mammals. like microchiropterans and the opossum, hedgehog, mouse, and rabbit, correspond to those structures in higher orders (primates, prosimians) that do not have the six-layered structure in the adult due to regression of the inner granular layer, and are thus agranular. The detailed argument for this will be developed in Part II.

Chapter II

Regional variations in cell structure of the cerebral cortex.

1. The general rules of variability.

Developmental studies have shown that the primitive lamination of the cerebral cortex undergoes far-reaching local modifications during ontogeny, by means of which completely new patterns emerge that do not reveal their common origin from a basic six-layered type without detailed examination. They are the source of the numerous and diverse local specialisations that characterise the mature cortex of the adult. The architectonic rearrangements involve either the number and particular structure of individual layers, or the density and size of cellular elements through the whole cortex and within given layers, or the total thickness of the cortex and the relative thickness of the different layers. They sometimes result in sharp linear borders, as shown in Figures 10 to 12, 15, 20 to 22 and 24 to 26, sometimes in subtle transitions (Figures 17, 23 and 28 to 32).

The rules for local variations in cytoarchitecture depend on the above criteria. They will now be discussed in detail and, as far as feasible, explained by illustrations. There are microphotographic records of many of them in my earlier works on histological localisation; especially in my third and seventh communications on the cortical structure of monkeys and prosimians. Firstly, one should distinguish two major categories of architectonic transformation of the primitive cortex that occur throughout the whole mammalian class.

1. Architectonic variations in established six-layered cortex. We call all examples of this type "*homotypical formations*" for they maintain the same basic pattern throughout life.

2. Extreme variants with altered numbers of layers or "*heterotypical formations*", that no longer have the six layers in the mature brain because of the secondary transformations described above.

I. In homotypical formations, where **the number of layers remains the same,** the cellular structure of a cortical section can be modified in the following ways.

a) Through changes of cell packing density or, in other words, cell number per unit volume. It is not rare for the whole depth of the cortex in one part of the hemisphere to become more or less cellular, or the process may be restricted to a single layer. Figures 20, 21 and 22 are typical examples of the former case with a sharply delimited zone of increased cell density appearing over the whole depth. Figure 23 provides an example of the second case; in an area of otherwise rather even cell density the inner granular layer (IV) of the preparietal area (between ↓1 and ↓2) suddenly becomes much more densely cellular. The degree that local differences in cell density can attain can be judged from a comparison of later higher-power micrographs of human, monkey and kinkajou cortex, especially Figures 50 to 52.

b) Through changes of cell size or specific cell type in one or more layers. Two categories of such modifications can be distinguished: either the cells at a particular place take on new forms more or less abruptly, for example their average size decreasing so that granule-like elements predominate, or an entirely new cell type appears in a single layer. Calcarine cortex (Figure 12) is the chief example of the first case, giant pyramidal cortex (Figure 7) and preparietal cortex (Figures 16 and 23) of the second, these last types presenting cells of unusual volume in their ganglion cell layer (V). The surprising differences in cell size in various parts of the cortex are also clearly visible in the high power micrographs of Figures 43 to 57, in which sections from the large-celled giant pyramidal and small-celled calcarine cortex are compared.

c) Through changes in the relative thickness of individual layers. A layer can thicken considerably at the expense of neighbouring layers and vice versa; one sees such an isolated thickening especially frequently in the pyramidal (III) and ganglion cell (V) layers. But the spindle cell layer (VI) also frequently undergoes considerable variations in thickness that are often sudden in onset. A typical example of this is the transition from the occipital type of cortex to the calcarine, as Figures 21 and 22 show clearly. Also the transition from archipallium to neopallium, that is from homogenetic to heterogenetic cortex (*63), illustrates the same process. Three identical transition zones in monkeys and prosimians can be compared in Figures 24 to 26.

d) Through increase or decrease of the whole cortical thickness. The cortex can become thinner or thicker overall while the relative thicknesses of individual layers do not change.

II. Extreme variations in cortical architecture accompanied by an **altered number of layers**, or *heterotypical formations*, may result from an increase or decrease of the basic or primitive layers.

1. An increase in the layers arises:

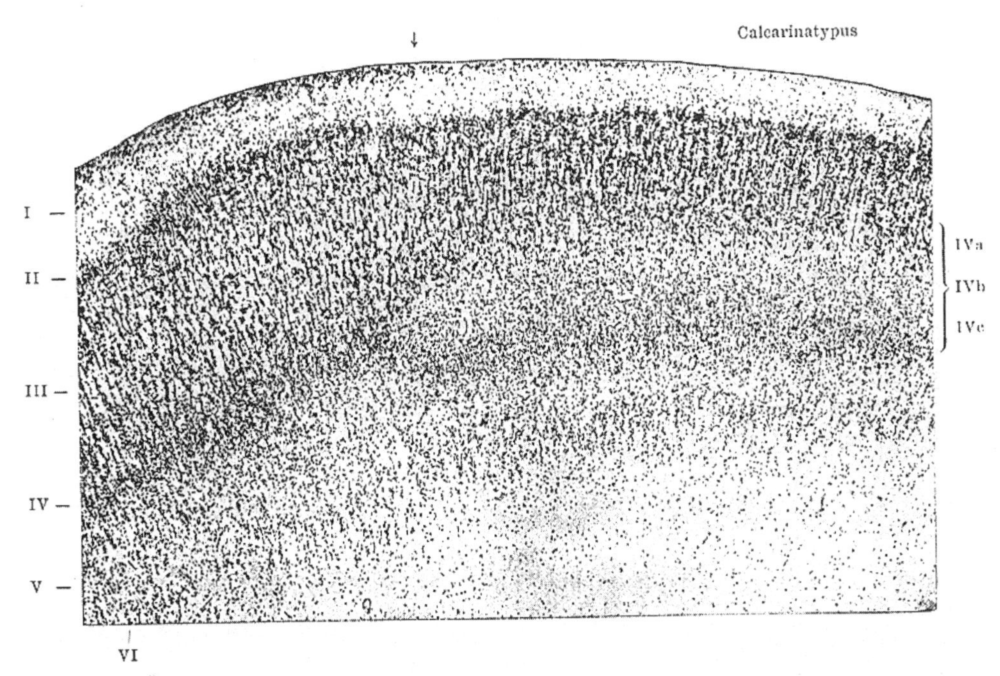

Fig. 20. Human foetus aged 8 months. 46:1, 10_m. Site of transition to calcarine cortex (↓) with sudden increase in cell density and laminar rearrangement. Cell density even changes in layer I at ↓ (See page 46.)

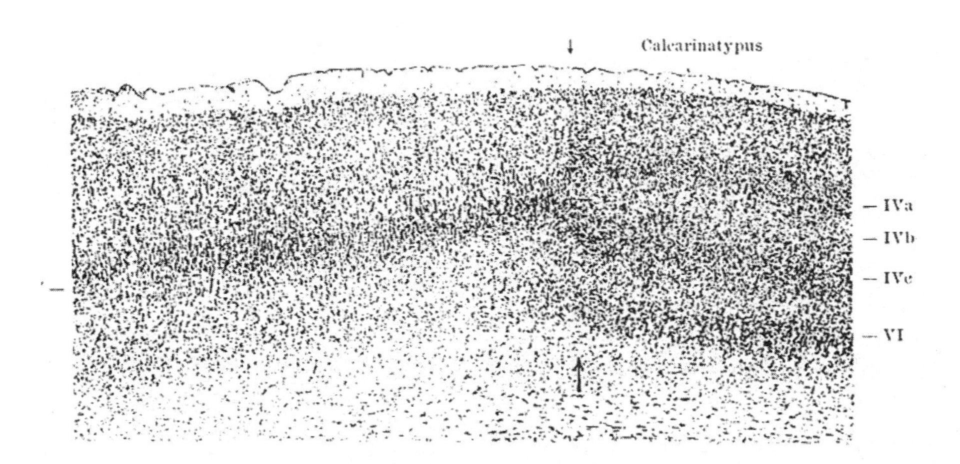

Fig. 21. Black-eared marmoset (*Hapale pennicillata*) (*62). 25:1, 10µm. Sharp transition from calcarine cortex (area 17) to occipital cortex (area 18 of the brain map) as in Fig. 20. There is increased cellularity in all layers throughout the whole cortical depth.

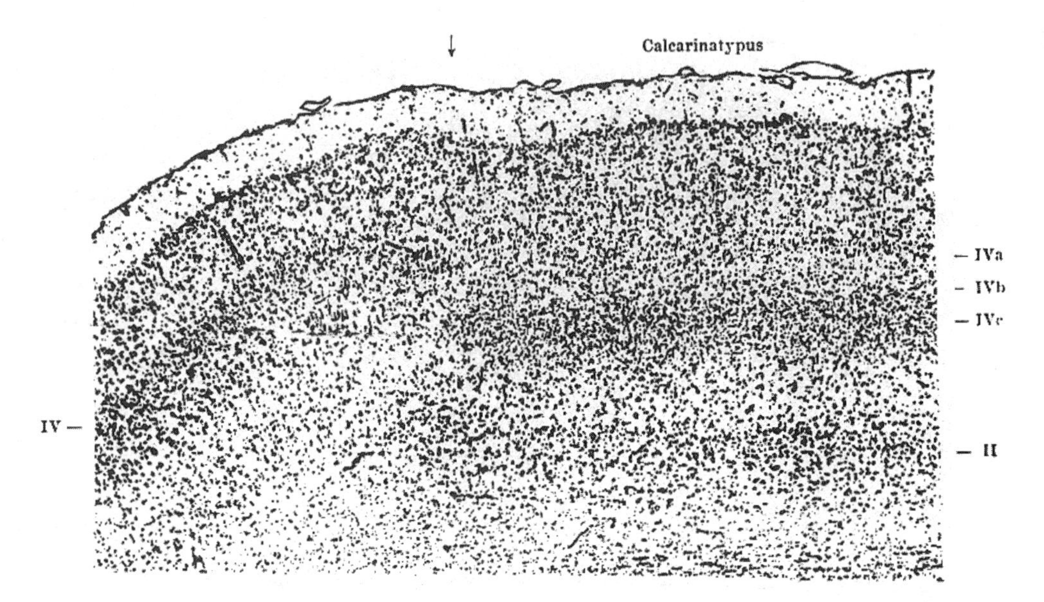

Fig. 22. Adult wallaby (*Macropus dorsalis*). 25:1, 10μm. The same as in the previous figure. Increase in cell density over the whole cortical depth, especially in layers IV and VI. Deepening of the whole cortex. Alteration of thickness of layers.

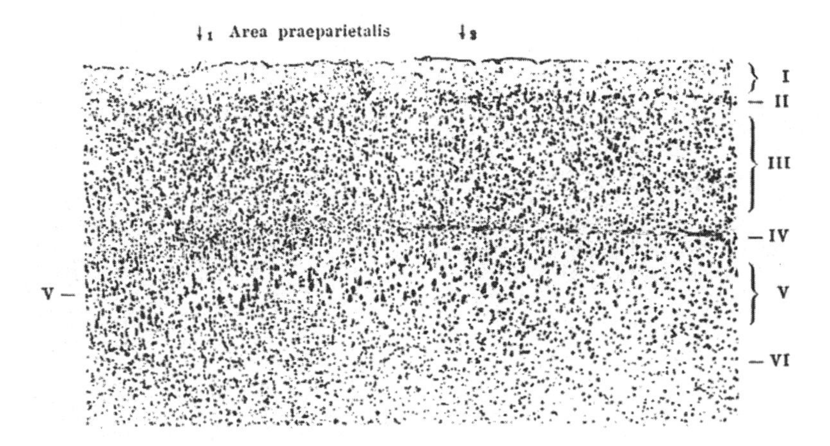

Fig. 23. Prosimian (*Lemur macaco*). 25:1, 10μm. Transition from preparietal cortex to the parietal area on the one hand (↓2) and to the postcentral area on the other hand (↓1). The preparietal cortex is characterised by very large ganglion cells in layer V and an increase in granule cells in layer IV, while the cell density remains the same in the rest of the cortical thickness.

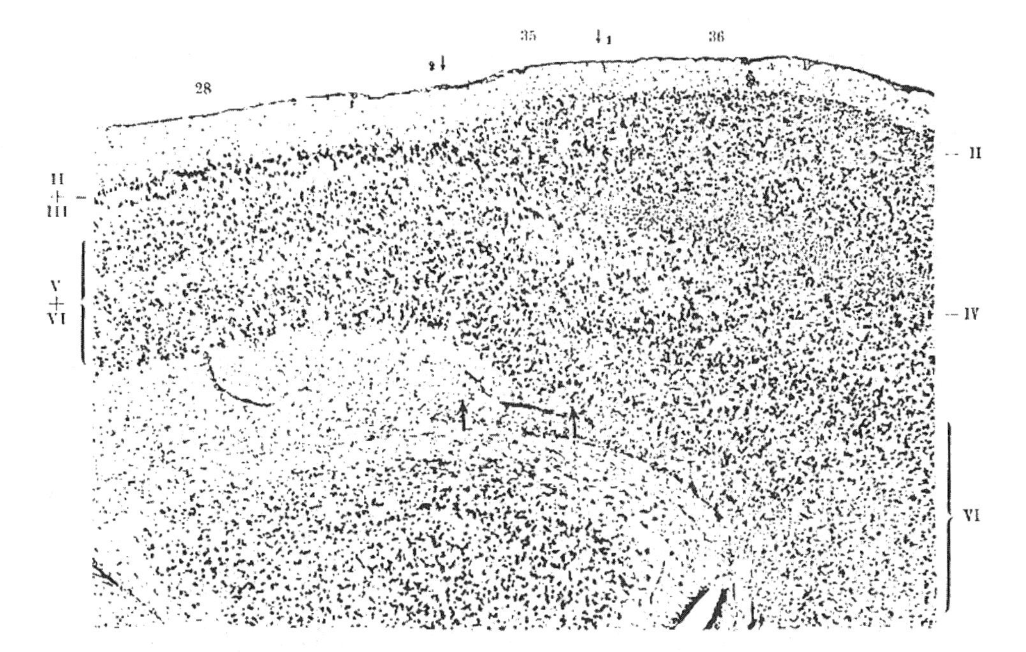

Fig. 24. Marmoset (*Hapale jacchus*). 25:1, 10μm. Transition of the homogenetic cortex of the neopallium (right) to the heterogenetic cortex of the archipallium (left): narrowing of the total cortical depth, widening of layers I and VI with sharp narrowing of layers II and III and loss of layer IV (↓1). Between ↓1 and ↓2 is the *perirhinal cortex* (area 35 of the brain map), to its right the *ectorhinal cortex* (area 36) and to the left the *entorhinal cortex* (area 28). cf. Part II.

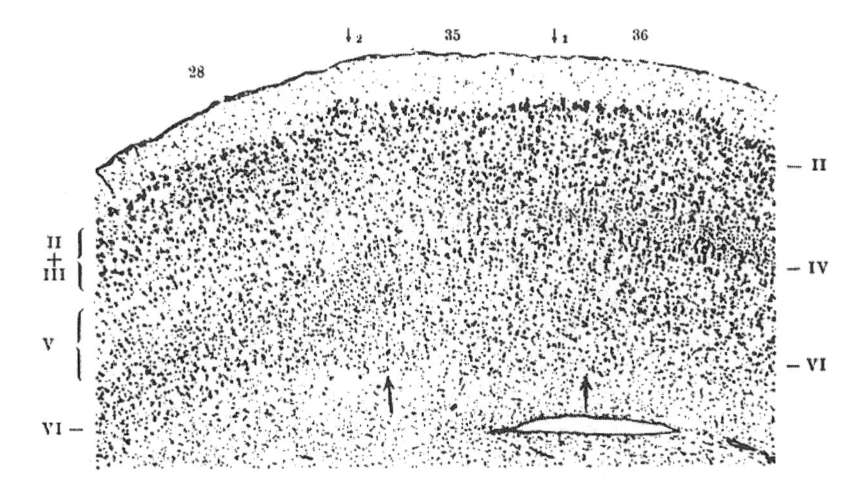

Fig. 25. Prosimian (*Lemur macaco*). 25:1, 10μm. The same as in Fig. 24. The cortical thickness does not change at the site of transition, only layers V and VI widen at the expense of layers II and III. Layer IV stops quite sharply at ↓1. Layer I becomes narrower rather than wider.

a) Through splitting of a basic layer into two or more sublayers. A typical example of this pattern is the calcarine cortex, where the original single inner granular layer (IV) divides into three sublayers, a superficial, an intermediate and a deep (Figures 20, 21 and 22).

The insular cortex also belongs here; as is well known, it is characterised by the formation of the claustrum as a deep cell-rich structure which, as developmental and comparative anatomy demonstrate conclusively, is formed by the breaking off from the deepest cortical layer (VI) of a cellular strip close to the white matter that traverses the extreme capsule and becomes more and more independent (Figures 36 to 37) (*64).

b) Through differentiation and parcellation of new cell types within a principle layer of the basic cortical type. In this way a sublayer separates within the original layer. A typical example is the human occipital cortex (Figure 27) in which there develops a particularly clear division of the pyramidal layer (III) into an outer small-celled layer (*parvopyramidal sublamina*, IIIa) and an inner large-celled layer (*magnopyramidal sublamina*, IIIb). Similar processes are numerous, for instance in the ganglion cell layer (compare, for example, Figures 16, 17 and 23).

2. A reduction in the number of layers can also occur through a double mechanism.

a) One of the original layers of the basic architectonic pattern may disappear completely more or less abruptly. We can observe this process in various parts of the frontal cortex, part of the insula and in the anterior part of the cingulate gyrus, where the prominent inner granular layer (IV) of foetal life later disappears completely, by its granular elements regressing or becoming dispersed within adjacent layers, so that it can no longer be considered to exist. There is also a reduction in the number of layers at the transition from neopallium to archipallium where the inner granular layer suddenly ceases, as shown in Figures 24 to 26.

An especially characteristic and physiologically important example of the sudden interruption of a basic cortical layer, that is to say a reduction in layers, is provided by the transition from the giant pyramidal cortex (area 4) to the postcentral cortex (areas 1 to 3). The same essential architectonic transition can be traced through the whole mammalian class as will be explained elsewhere. In all animals the inner granular layer (IV) stops quite abruptly in this region to be substituted by the appearance of Betz giant cells in the underlying ganglion cell layer (V). In Figures 28 to 31 this transition is illustrated for four different animals.

b) Several layers that were originally separate in the basic pattern may fuse together and form a single layer in the mature brain. This is encountered especially in certain areas of the retrolimbic region that I described in detail for lemurs in my seventh communication. However, the most striking illustration of this category of laminar reorganisation comes from the fact that Meynert's outer granular layer (II), that is clearly expressed as a separate cell layer throughout the immature brain and shown in Figures 1 to 3 and 5 to 11, frequently regresses so much in the adult that it can hardly be distinguished from the underlying pyramidal layer (III), if at all (Figures 12, 13, and 16 to 18). As I have argued repeatedly, this situation is precisely the reason why most authors have not recognised an outer granular layer and have arrived at an erroneous

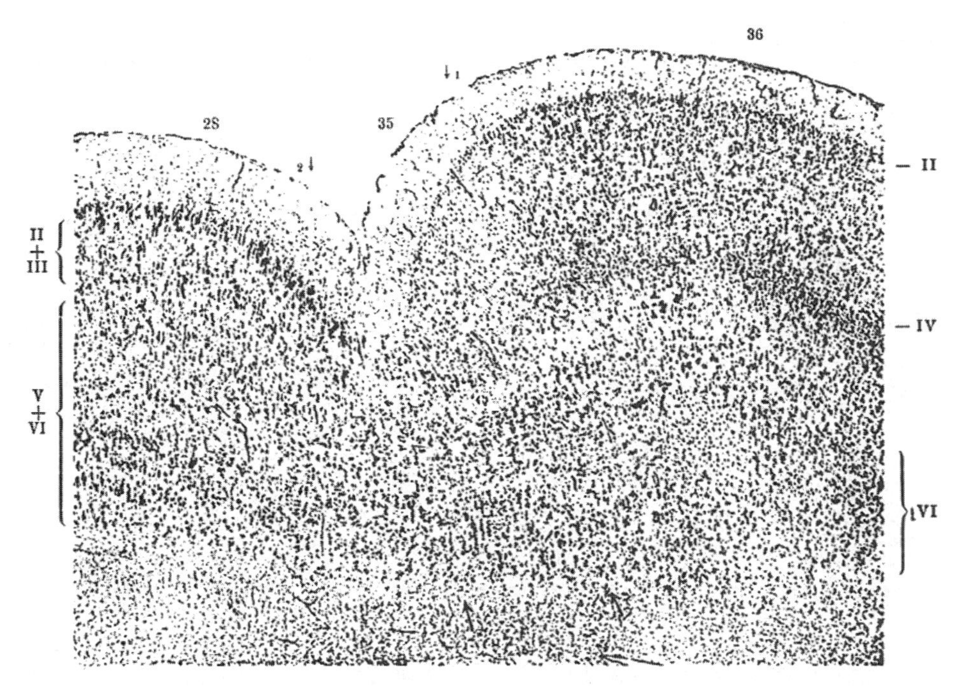

Fig. 26. The same architectonic transition as in Figs. 24 and 25 in the rhesus monkey (*Macacus rhesus*). 25:1, 10μm. The transition site in the archipallium is marked by a sulcus, the posterior rhinal sulcus. The labels are the same as in Figs. 24-26.

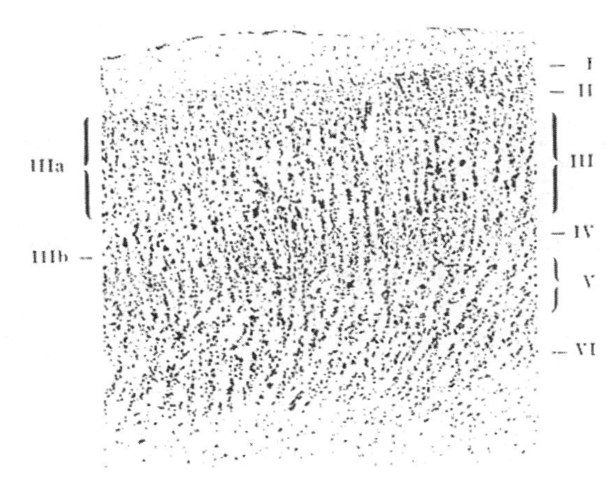

Fig. 27. Human occipital cortex. 25:1, 10μm. Division of the pyramidal layer (III) into two sublayers: the parvopyramidal IIIa and the magnopyramidal IIIb. (see also Fig. 42).

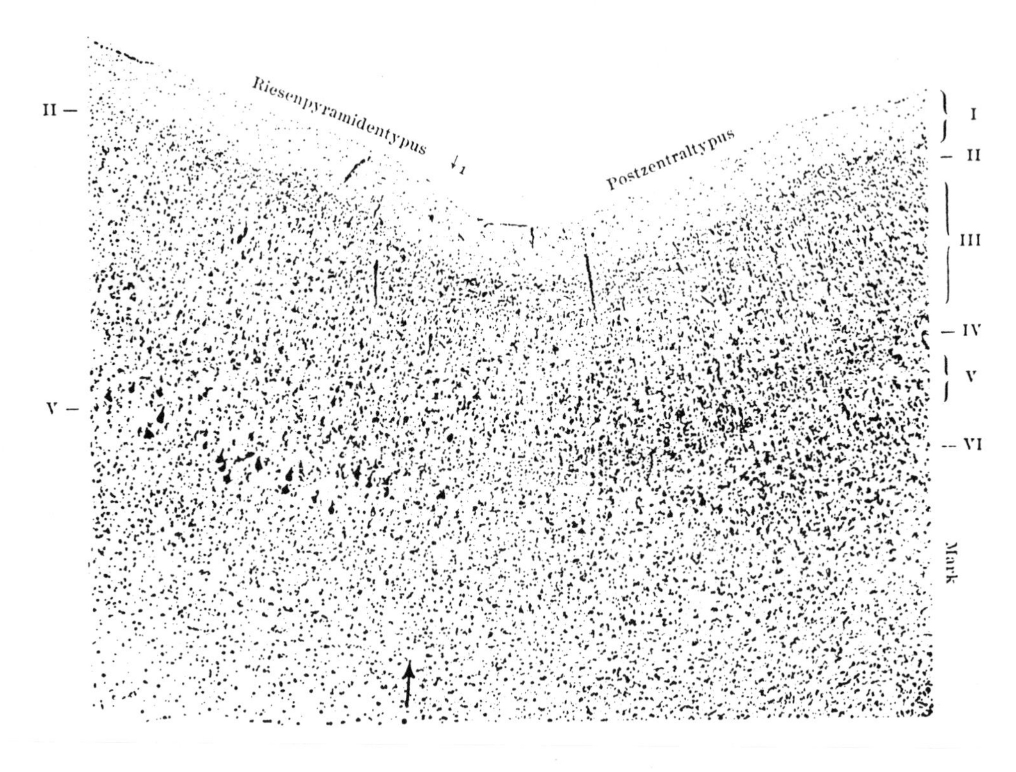

Fig. 28. Adult human. 20:1, 20μm. (For explanation, see Fig. 31).

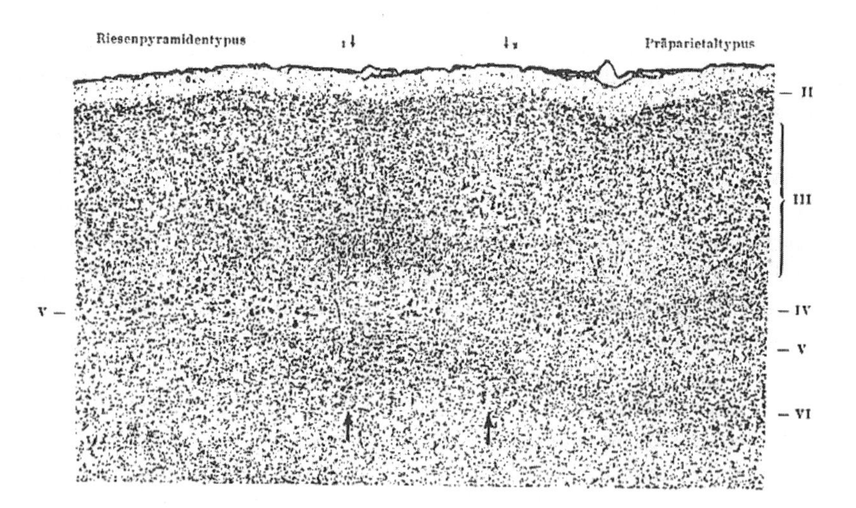

Fig. 29. Marmoset (*Hapale penicilata*). 25:1, 10μm. Transition from the giant pyramidal cortex to the postcentral cortex (↓1) and then to the preparietal cortex (↓2). The borders are very sharp. In the preparietal cortex large ganglion cells appear in the ganglion cell layer (V), similar to the actual giant pyramids.

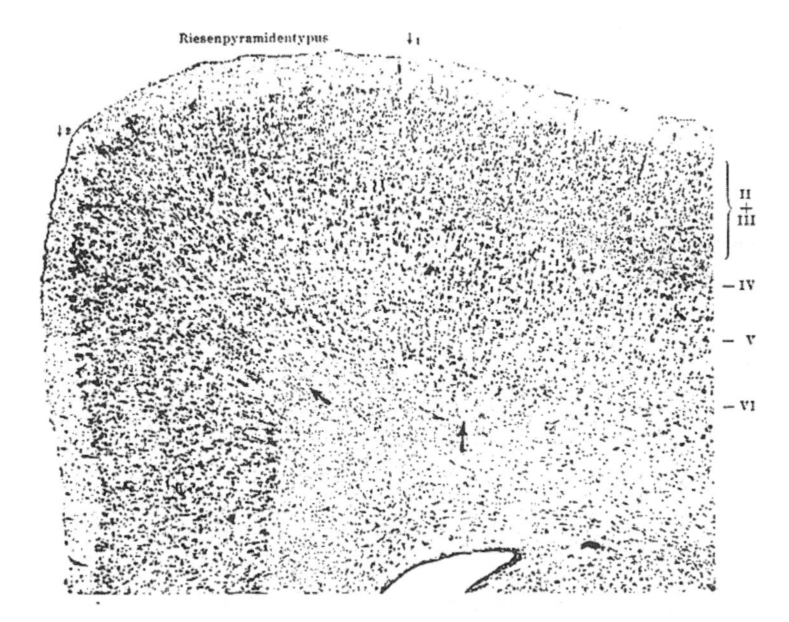

Fig. 30. Rabbit (*Lepus cuniculus*). 25:1, 10μm. Transition from the giant pyramidal cortex to the postcentral cortex (↓1) on one side, and the prelimbic cortex on the other (↓2).

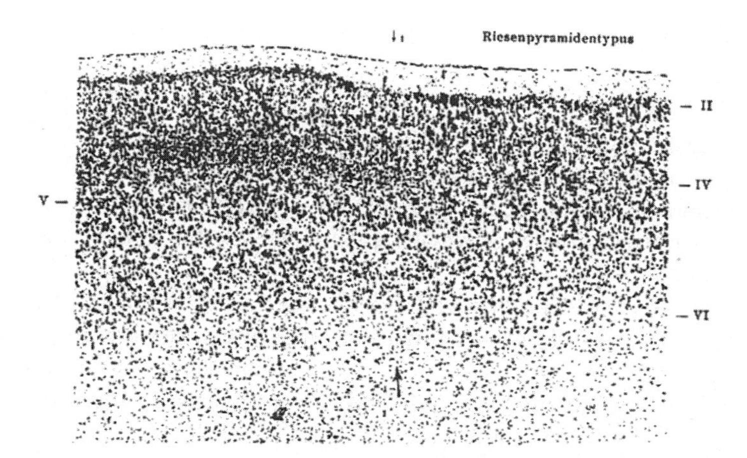

Fig. 31. Foetal wallaby (*Onychogale frenata*). 25:1, 10μm. The same as in the previous figure. There is a sharp end to the inner granular layer (IV) at ↓1.

Figs. 28-31. Transition from the giant pyramidal cortex to the postcentral cortex in man, marmoset, rabbit and wallaby (*65). In the giant pyramidal cortex the inner granular layer (IV) is absent, instead of which the giant pyramids appear in layer V; in the postcentral cortex, on the other hand, a distinct layer IV is formed, whereas the giant pyramids are absent. One can compare the differences in cortical thickness, cell density and the laminar pattern in the different animals.

interpretation of the layers. In marsupials layer II remains clearly separate from the pyramidal layer (III) throughout life in many areas (Figures 15, 19 and 22).

Another form of fusion of two layers is found in those not uncommon cases where the cellular elements of the ganglion cell layer (V) and the fusiform layer (VI) mix in such a way that they appear to form a single layer. Examples in man include the rostral portion of the cingulate gyrus (areas 24, 25 and 33), certain frontal areas (6 and 8), and areas 30 and 35 of my brain map. Illustrations in support of this are not provided here, but can be found in my third and seventh communications.

c) Complete overlap and fusion of all cortical layers can be observed in several regions of the frontal lobe in man and other animals which were clearly laminated and possessed an inner granular layer during foetal development (Figures 32 and 33).

2. Regional characteristics of individual layers. (Constancy and variability.)

We have based our arguments so far on the primitive six-layered pattern as a whole and examined the general principles by which differentiation and reorganisation can take place. It emerges that modifications of cellular lamination respect the same rules in all mammals. There is either regression, for example fusion of individual primitive layers, or duplication and differentiation of sublayers from an elementary layer or, thirdly, there can be less radical changes in the thickness of layers, cellular density and size, and specific cell shape.

For the following comparative anatomical studies it will be helpful to examine each individual layer of the basic cortical pattern separately once again in terms of their regional variations.

This is necessary first of all because the changes in the cortical layers we have discussed are frequently not manifested suddenly and abruptly in a single region, but arise gradually over a broad area. In such cases only the comparison of widely separated regions permits the untrained eye to detect differences in a layer. Thus if one wishes to obtain an accurate picture of the degree of modification of a layer, one must treat each layer as an entity to be examined throughout its whole extent over the cortical surface. From the foregoing arguments the important fact emerges that, in general, certain of the basic layers can be assumed to be very constant and unchangeable and the others highly inconstant and variable. It can further be taken as established that those layers that undergo only slight regional modifications in man, on the whole also change little in other mammals, while on the contrary those layers that undergo marked local changes in man usually show an equally great variability throughout the whole mammalian class.

a) (*66) One can regard layers I and VI - the molecular and spindle cell layers - of the basic pattern as the most constant in this sense. They are not absent in any species or in any cortical area, and also appear in certain abortively developed zones of the cingulate gyrus and hippocampus. Their cellular structure varies within much narrower limits than that of all other basic layers.

The *molecular layer*, the extreme outer cortical layer (I), essentially only

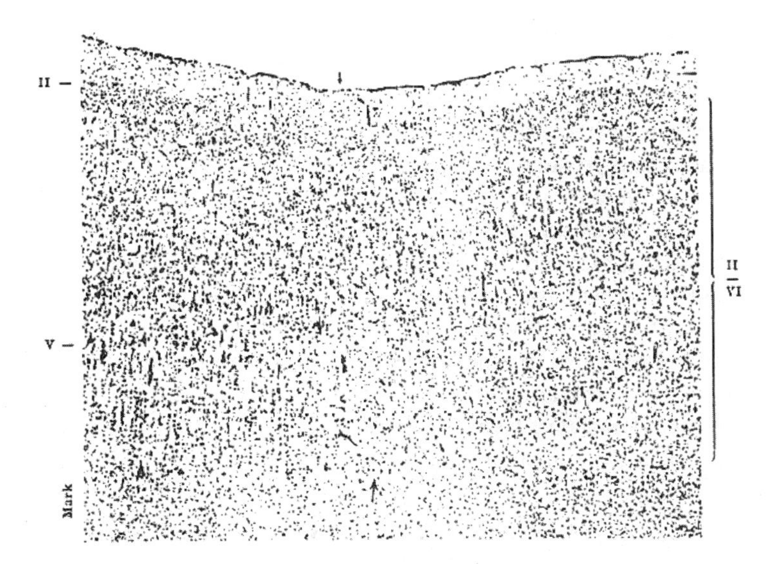

Fig. 32. Agranular frontal cortex in the human adult. 25:1, 10μm. Transition from the giant pyramidal cortex (left) to the agranular frontal cortex, with gradually merging borders. To the right of ↓ the giant pyramids gradually disappear. - Example of heterotypical cortex with fusion of the basic layers and complete regression of the inner granular layer. (Solitary arrangement of the giant cells; see pages 82-86.)

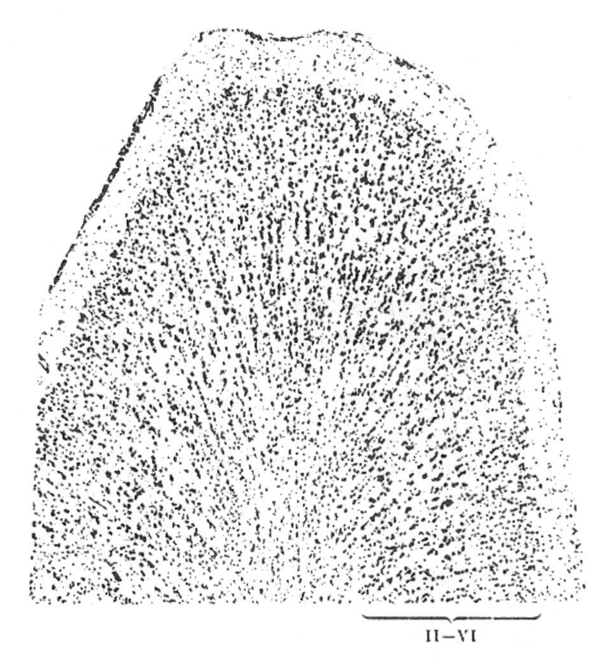

Fig. 33. Agranular frontal cortex in the prosimian. 25:1, 10μm. Fusion of all basic layers and lack of inner granular layer as in Fig. 32.

undergoes changes in its thickness, judged from cell preparations [1]). It is not unusual to find a widening or narrowing occurring quite abruptly in a given region (Figures 24 to 26, 31). Slight differences in the number of cellular elements, most of which are not neuronal, can be neglected for the purpose of cortical localisation. In Figure 20 such an increase in cell number in layer I of the human frontal cortex is visible. On the other hand the thickness of the layer varies noticeably, often more so in animals than in man. For instance, one can compare the occipital cortex of the monkey with the basal parts of the frontal cortex. The latter possesses an almost threefold thicker molecular layer than the former. The same is true of most other mammals. The insula as well as part of the limbic cortex is characterised by an unusually thick layer I. Examples appear in my paper on monkey and prosimian cortex.

The *spindle cell layer*, the extreme inner cortical layer (VI), is equally absolutely constant and is never absent from any cortical area, like layer I, even the so-called "defective" cortex of Meynert. Indeed the latter consists almost entirely of components of layers I and VI of the basic cortical structure. In contrast to former ideas (Schaffer, Cajal) that all typical cortical layers continue in the hippocampus [2]), I must stress that, on comparative anatomical grounds, the hippocampal cortex represents exclusively a continuation of the extreme inner and outer layers of the neocortex and is thus only formed from two elementary layers of the basic cortical pattern. Layers II to V inclusive stop sharply at the subiculum and only the inner and outer layers continue. This organisation can be seen clearly in Figures 34 and 35, the interruption of layers II to V and the enormous widening of layer I being especially striking in Figure 35 of the kangaroo.

Thus the spindle cell layer, like the molecular layer, forms a continuous sheet of tissue covering the whole extent of the cortex, not only in homogenetic cortex but also in all heterogenetic and "abortive" or "rudimentary" structures, unlike the other primitive layers.

Its local structure is variable, in spite of this consistency. In many places there are sudden changes in thickness as well as in cellular density. A typical example of this feature is again the transition from the calcarine to the occipital cortex (Figures 11, 12, 21, 22), where the layer becomes sharply broader and denser. Other fairly abrupt changes in layer VI occur in many other places; in particular, layer VI undergoes a sudden noticeable thickening at the border of the rhinencephalon, as this layer is on average much more developed in almost all heterogenetic structures than in homogenetic ones (Figures 24 to 26, 34 and 35).

However, the thickness of layer VI in the neopallium is subject to large local variations. While in many cortical areas, such as the human occipital cortex and the anterior bank of the postcentral gyms (area 3), it measures only a few microns, in other regions, especially frontal and temporal, it occupies one third of the whole cortical depth, that is ten times more. Cell density and size

---

[1]) It is different for the myeloarchitecture, in which there are substantial differences in layer I in both man and other mammals, according to O. Vogt, Mauss and Zunino.

[2]) Cajal views the hippocampus and the dentate gyrus as "special brain organs" on the basis of their special structure but then writes: "As to the number of layers and their general composition, they fully resemble those of typical cortex, as Schaffer has explained in detail." Zeitschr. f. wiss. Zoolog. Vol. 56. 1893. p.619

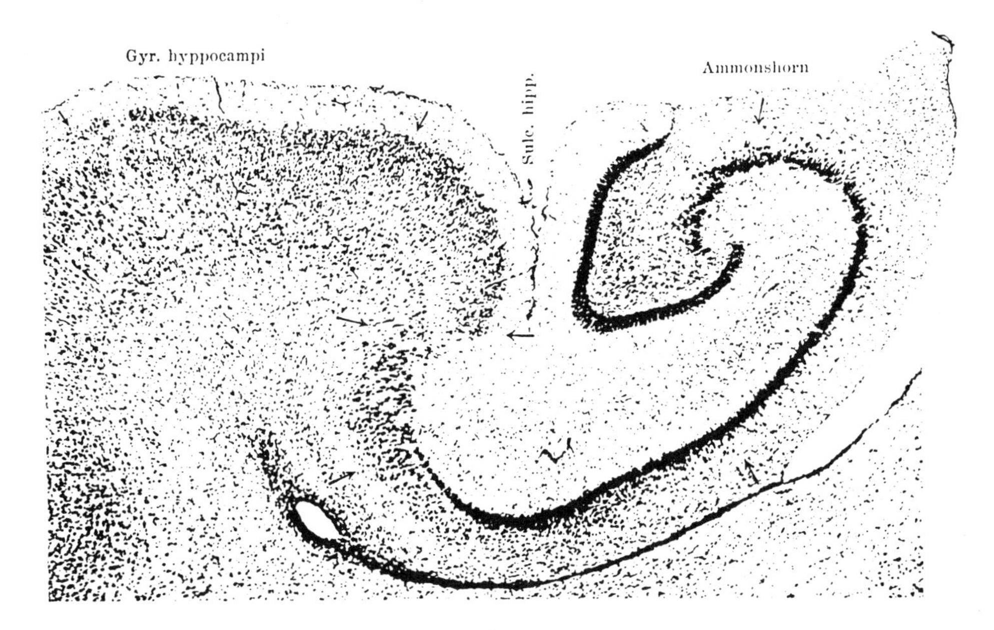

Fig. 34. Young wallaby (*Onychogale frenata*). 25:1, 10μm. Cross-section of the hippocampal gyrus with the transition to Ammon's horn. Abrupt interruption of layers II-V at the beginning of the subiculum (←), with only basic layers I and VI continuing into Ammon's horn in greatly widened form. cf also Figure 35.

Fig. 35. Cross-section of the hippocampal gyrus of the kangaroo at the transition to Ammon's horn. 25:1, 10μm. Fusion of layers II-V in the presubicular cortex (right of ↓1). Abrupt interruption of layers II-V in the subiculum (↓1-↓2), with enormous widening of layers I and VI.

are also variable. The layer undergoes regional modifications in many areas such that it partially splits into two sublayers, an outer cell-dense triangular layer (VIa) (*67) mainly consisting of triangular and stellate cells, and an inner, sparser *fusiform layer* (VIb) (*68) containing mostly spindle cells. Sometimes the layer changes its structure by more or less fusing with layer V, the ganglion cell layer; then the separation of the two layers is often no longer possible and the persistence of the spindle cell layer is only ascertained by the presence of its characteristic spindle cells. Notable cases of fusion of the basic layers V and VI are illustrated in Figures 38 to 41.

The most substantial local modification of layer VI in all mammals is found in the structure of the insula. Here, as we have already seen, the insertion of the extreme capsule splits the layer into two distinctly separate cell groups of which the inner, at least in higher mammals, attains a certain morphological independence as the claustrum (*64). Thus one can recognise three sublayers VIa, VIb and VIc, instead of the usual single basic layer VI (Figures 36 and 37), rather as the primitive single layer IV divides into three sublayers in the calcarine cortex. The process, from the point of view of developmental and comparative anatomy, is the same in both cases, and we should see in this example a confirmation of our concept of the origin of cortical lamination from a primitive tectogenetic pattern.

b) The most inconstant or variable layers are Meynert's two so-called granular layers, layers II and IV of the basic pattern (the outer and inner granular layers). They alter their original cytoarchitectonic features so extensively during ontogeny that it is often only possible to correlate their mature structure with their primitive tectogenetic form by following the whole developmental sequence. Extreme variations, such as the disappearance or doubling of layers that we have already discussed, occur particularly in them, and their specific organisation also varies widely throughout the mammalian class, as we shall see.

The outer granular layer (layer II) is a major layer of the basic tectogenetic pattern, present over the whole extent of the cortex during foetal life and infancy. It stands out clearly as a thick, compact cellular lamina deep to the cell-sparse molecular layer, as can be observed in Figures 1 to 3, 8 to 11 and 13 to 15 from man, cat and wallaby (*69), To understand the genesis of the layer it is significant that it is much distincter in structure during early development than in the mature brain, not only in man but also in lower mammals, such as the marsupials. One only has to glance at the micrographs of the immature cortex mentioned above. As the individual grows older, the layer regresses and its essential transformation consists of its disappearance as an independent structure and its more or less complete fusion with the underlying pyramidal layer in most cortical areas of the mature brain. Thus this is a form of regression or disappearance of a primitive layer. The outer granular layer maintains its independent character as a separate layer only in relatively few regions, where it represents a strip of densely packed, small polymorphic cells deep to the molecular layer, distinguished from the actual pyramidal layer by the density and small size of these cells. This organisation is illustrated in man in Figure 32; a distinct layer II can be seen in the monkey in Figures 21 and 29, and even more clearly in the kangaroo in Figures 15 and 19, whereas it is virtually absent in the rabbit (Figure 18) and the kinkajou (Figure 17).

Layer II possesses a particular organisation in certain heterogenetic areas

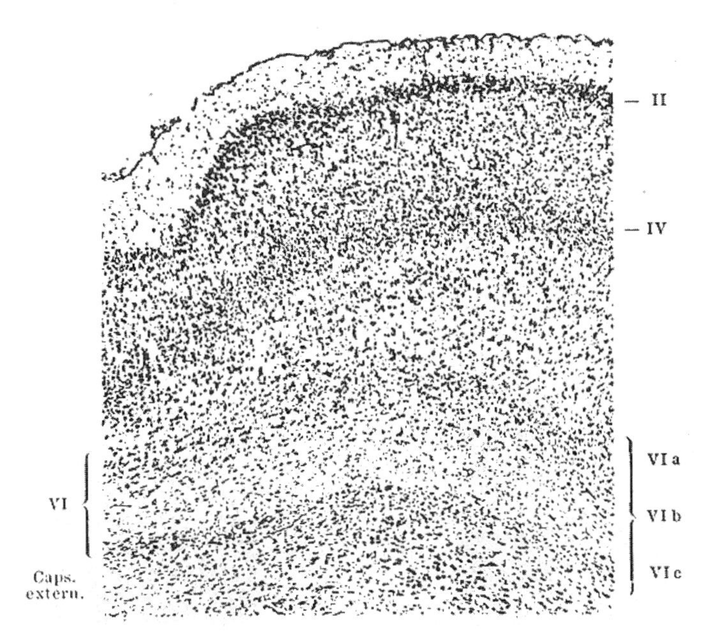

Fig. 36. Wallaby (Macropus dorsalis). 25:1; 10μm. Granular insular cortex. Distinctive lamination of the whole cross-section; division of layer VI into three sublaminae VIa, VIb and VIc. VIb = extreme capsule, VIc = claustrum. The outer (II) and inner granular layer (IV) are well developed.

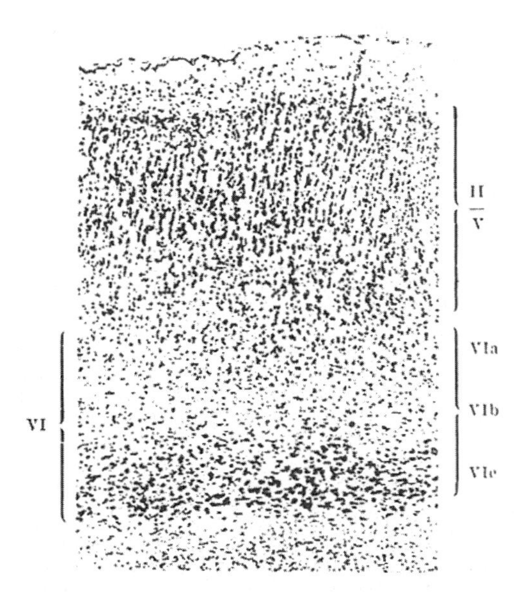

Fig. 37. The same as in Fig. 36. Agranular insular cortex. The inner granular layer (IV) is completely absent, and the layers are fused so that the lamination is severely regressed; only layer VI is organised in sublaminae VIa, VIb and VIc, as in Fig. 36.

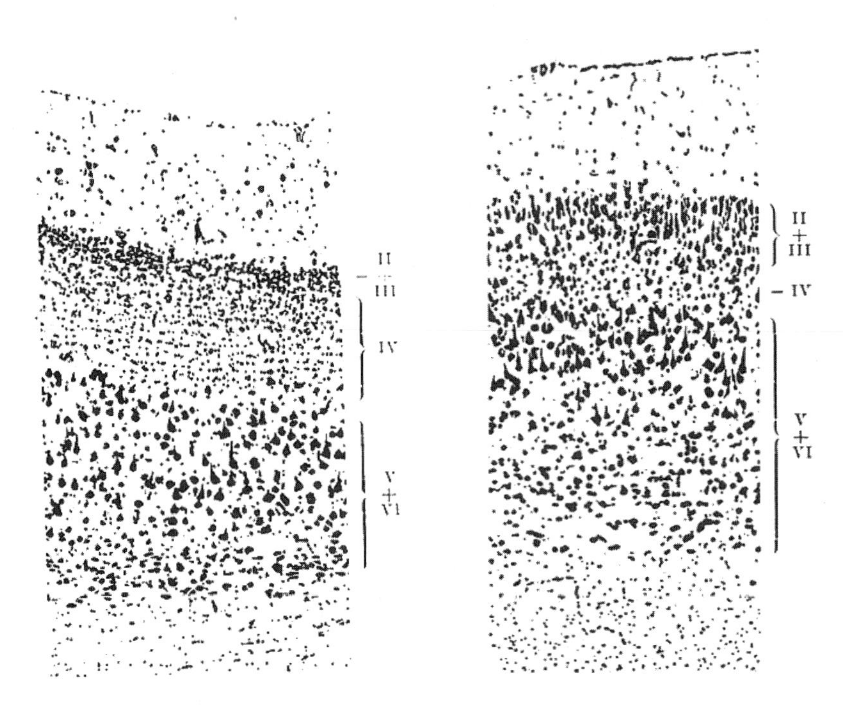

Fig. 38. Rabbit.

Fig. 39. Flying fox.

Fig. 38-41. Retrosplenial cortex, area 29 of the brain map, of 4 different animals as an example of a polymorphic heterotypical cortex. 66:1, 10µm. (see pages 54 and 87). Common features in all 4 animals are:

a) the regression and fusion of layers II and III,

b) the massive development and simultaneous fusion of layers V and VI,

c) the unique development of layer IV,

d) the very thick layer I.

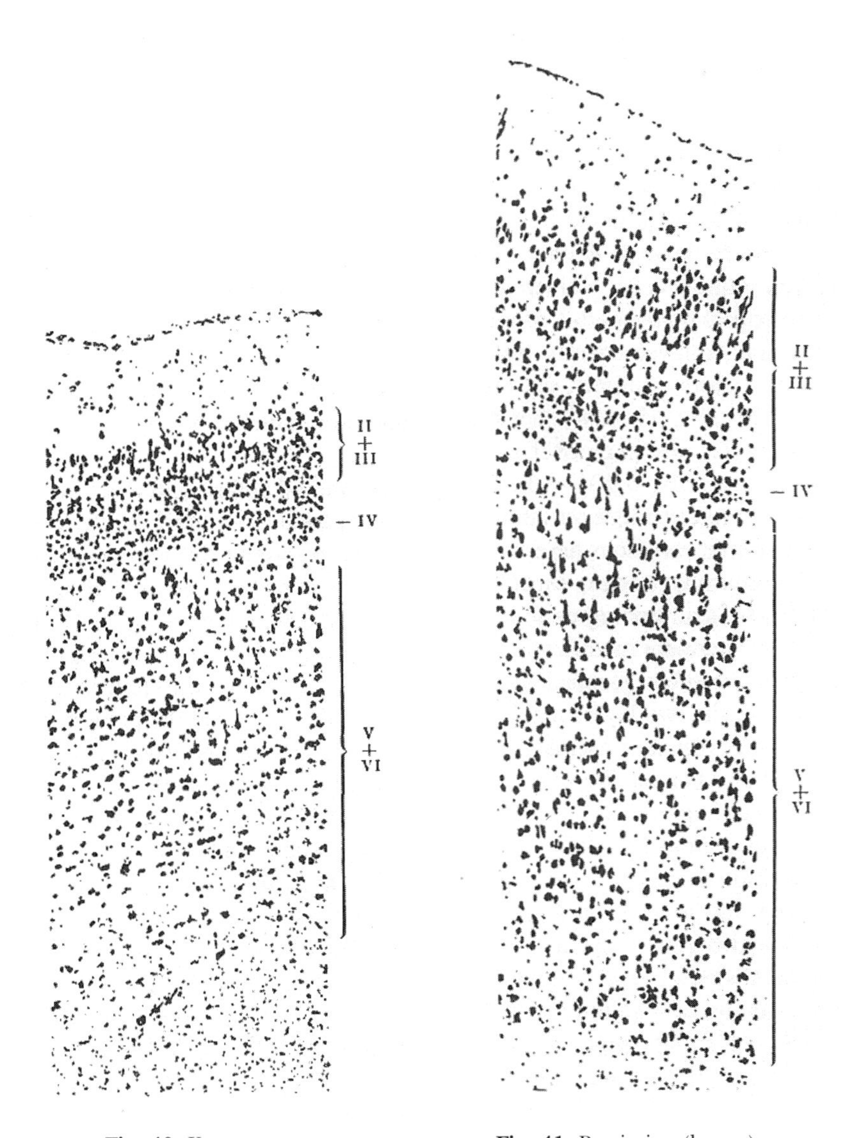

Fig. 40. Kangaroo.

Fig. 41. Prosimian (lemur).

of the rhinencephalon, especially in area 28 of my brain map, in the part of the temporal lobe directly adjacent to the posterior rhinal sulcus, in areas 35 and 36, and finally in the basal parts of the insula. In all these regions the primitive granule-like elements have been modified to have strikingly large and polymorphic shapes, thus producing a clearly demarcated layer II with a continuous compact cellular structure. Examples are found in Figures 24 to 26 and 36 to 37.

An unusual regression of layer II together with extreme atrophy of layer III is found in the *granular retrosplenial cortex*, illustrated in Figures 38 to 41 for four different animals. Here, in addition to the regression of layers II and III, there is fusion of the original layers V and VI and at the same time an isolated massive increase of the inner granular layer, that is particularly prominent in Figures 38 and 39; thus, this pattern represents a typical example of the concurrent appearance of regressive and progressive transformations in the basic laminar structure of a cortical type.

In both man and animals Meynert's *inner granular layer* (IV) undergoes even greater local modifications than the outer granular layer. It is generally the most variable layer and its transformation provides a basis for the profoundest variations in cortical architectonics. The two principle modifications, on the one hand complete regression of the layer and on the other its duplication, have already been discussed in the context of the ontogenetic derivation of the basic cortical pattern. Comparison of the accompanying micrographs shows that these modifications are repeated in essence throughout the whole mammalian class. One should refer to Figures 58 to 63 which show the complete lack of the layer in the agranular giant pyramidal cortex in various orders, and also to Figures 68 to 76 illustrating the splitting of the inner granular layer in the calcarine cortex, likewise in several species. Considerable local differences in the cell density of the layer also occur in granular cortex. There exist certain homotypical cortical areas, that have thus preserved their six-layered pattern, in which the inner granular layer dominates the whole cortical thickness because of its massive development On the other hand in other areas this layer, although present, has regressed so much that it is only recognisable after detailed scrutiny. Figures 16 and 18, of man, should be compared. In the former layer IV constitutes a dense, compact, small-celled lamina; in the latter it is only weakly distinguishable by relatively sparse, diffusely scattered granule cells. Numerous similar examples are illustrated by micrographs in my two papers on the cortex of monkeys and prosimians. The specific cell types of the inner granular layer equally differ in different regions; in some places it consists entirely, or almost entirely, of small darkly-staining round cells with little cytoplasm, the granule cells, while in other regions it is composed of smaller or larger polymorphic cells. (See also page 91 ff.)

c) Layers III and V of the basic pattern, the pyramidal and ganglion cell layers, manifest an intermediate variability compared with the layers discussed above. There are no extreme variations here, in the sense of those just described, but in several areas of the retrolimbic cortex, especially in macrosmatic mammals, there is widespread fusion and concurrent regression of layers III and II, as shown in Figures 38 to 41. The rest of their regional modifications involve predominantly the formation of special cell types that are often grouped in a new sublayer. Next, large local differences in the thickness of the two layers may appear, and finally one may find that the predominant feature of a particular regional cortical type is fusion with adjacent layers to form a single cellular lamina. We shall illustrate these relationships by examples.

Meynert's pyramidal layer (III) varies widely in thickness in different gyri of the same brain. An example of how its thickness can change quite suddenly can be seen at the transition from calcarine to occipital cortex in many mammals, such as man in Figure 68 and monkey in Figures 69 to 71. A similar reduction of layer III in the retrosplenial region is illustrated in Figures 38 to 41. Here, at the same time, there is complete fusion of layers III and II such that the two are represented only by a quite narrow lamina of medium-sized pyramidal cells above the inner granular layer (IV). Examples of similar processes of fusion in different areas, especially in man, were given earlier. It has been known for a long time, and was already described by Meynert, that the pyramidal layer is frequently split into two laminae, an outer with mainly small- and medium-sized cells, the *parvopyramidal sublayer* (IIIa) and a deep, large-celled *magnopyramidal sublayer* (IIIb), as illustrated by the example of human occipital cortex in Figure 42.

The best known form of differentiation in Hammarberg's ganglion cell layer (V) is the appearance of the Betz giant cells in the giant pyramidal cortex; the "motor area" of the literature (Figures 7 and 43). The characteristic of this cortical formation, apart from the loss of the inner granular layer, is the appearance of these gigantic cell types, first described by Betz. Layer V of the preparietal cortex (area 5 of the brain map) is also distinguished by such huge ganglion cells (Figures 16, 17 and 23). In much of the human frontal cortex a special lamina of medium-sized, polymorphic ganglion cells appears within the ganglion cell layer just deep to the inner granular layer so that the whole separates into two sublayers, an outer one with high cell density and an inner cell-sparse one. I have described and illustrated the same process in monkeys and prosimians in my third and seventh communications. In yet other regions the cells of this layer become mixed with those of the subjacent layer VI so that both are entirely fused; this is particularly the case in rostral portions of the cingulate gyrus and in rostromedial parts of the frontal cortex. It is not possible for me to give examples of all the individual variants; it must suffice to point out the principles according to which a layer can be modified from its original form.

Thus there are very varied elementary processes during embryonic and postembryonic development that convert the originally uniform embryonic Anlage of the cortical laminae to a multiplicity of structural divisions, a sort of organ complex, that are the basis of the sophisticated specialisation found in mature cerebral cortex. In man as well as animals, differentiating modifications sometimes occur in narrowly defined locations (as is clear in the accompanying micrographs) such that at a given place a sharp transition between two different structural patterns results. However sometimes spatial architectonic transformations in cortical structure proceed only gradually, so that there is no question of sharp borders. Transitions in the laminar pattern then have a smooth character, similar to or more striking than the examples in Figures 17 and 23. In such cases it takes considerable practice and experience to detect unequivocally differences in structure or borders between adjacent areas. In the

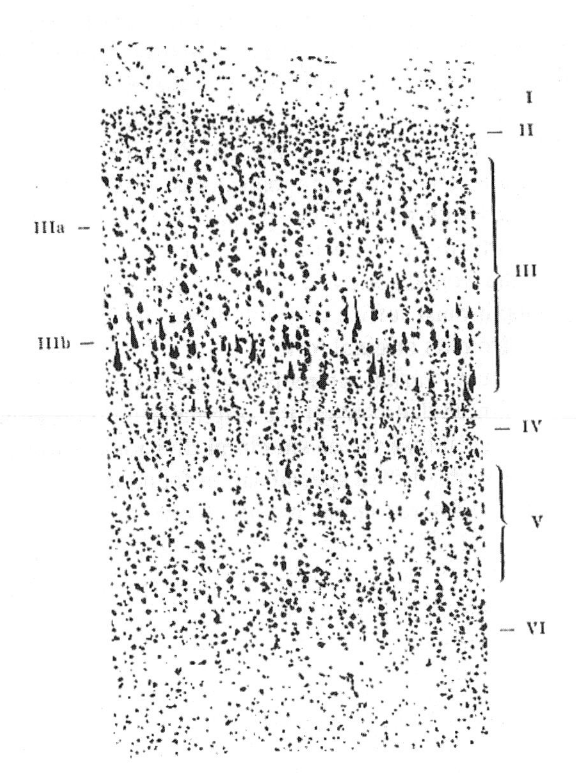

Fig. 42. Adult human occipital cortex. 66:1, 10μm. Division of the pyramidal layer (III) into small and large celled sublaminae IIIa and IIIb.

human brain in particular these form the vast majority; this is especially the case for many areas of the parietal and frontal lobes, and I freely admit that frequently it is only possible to decide with certainty about the presence of such differences from photographic records. In this respect microphotography has rendered me an invaluable service for studying localisation.

As we have seen, modifications of cytoarchitecture consist in general either of disappearance or fusion of individual primitive ontogenetic layers, or of splitting-off and differentiation of sublayers from a primitive layer, or finally in variations in the relative thickness of layers, cell size, cell shape or cell density. It should be noted that none of these processes of modification alone results in the emergence of a new structure; it is always rather through a combination and association of several different factors that the architectonic transformation and differentiation of a cortical area is achieved at a given location.

Thus it is frequently observed, for instance, that a narrowing of the cortex is associated at the same time with an increase in cell density and a reduction in cell size and, on the contrary, an increased cortical depth brings a decrease in cell density and an increase in size. In other words, densely cellular cortices are usually narrow on average, while relatively cell-sparse areas belong in general to thicker types of cortex; in the same way large-celled formations mostly occupy a very substantial depth of cortex and, on the other hand, small-celled types, especially the granular occipital varieties, are among the narrowest. There are however exceptions to this, especially in the cingulate and parahippocampal gyri.

Examples of all these rules are given in my earlier publications on histological localisation and in the accompanying figures. If one compares the micrographs in my third communication on the guenon with those of homologous cortex of lemurs in the seventh communication, one will find, on close inspection, that the reciprocal relationships of cell density, cell size and cortical thickness are generally confirmed. The same is true for man.

Many other similar relationships can be identified. The most important result of our findings to retain is that where individual new cell varieties appear in a particular part of the cerebral cortex, other aspects of the laminar cytoarchitecture are immediately modified in one way or another and, inversely, where the overall architectonic features of the cortex change, one can always expect some sort of variation in the morphological or histological characteristics of the individual cellular elements. However, it is thus immediately apparent that from the point of view of localisation we must turn our attention not to an individual layer or cell but to the whole cortical cross-section, and that the whole depth of a particular part of the cortex represents the localisational unit. The following comparative anatomical observations also support this emphatically. Naturally, it is easier to detect a single new cell type than to identify a complex of histological structural features, and considerable practice and experience are needed in specific cases to recognise all the above-mentioned characteristics correctly and to evaluate possible pathological deviations.

Chapter III

Particularities of the cytoarchitecture in different animals.

So far we have examined the principles underlying the potential of the mammalian cerebral cortex to undergo regional modifications in cellular lamination, and have determined that the process of architectonic transformation is essentially the same in the whole mammalian class. From this emerges the important biological principle that the genesis of mammalian cortex is not only conceived according to a common plan, but it completes its further development according to standard rules.

In spite of similarities in design, development and growth, different mammals display specific structural features throughout the whole cortex, as well as in particular cortical regions. They are a result of heredity and adaptation dependent on the precise nature of cerebral activity during development. These cortical regions are quite characteristic for each species, or at least each major order, and bestow specific characteristics on the brain structure so that in appropriate circumstances it is possible to recognise the species or order of an animal from its particular cortical structure.

The extent of specific variations in animals can be judged by the fact that most research workers dealing with the lower mammals, such as Cajal for rodents and Haller for monotremes, marsupials and microchiropterans, postulate that they possess basically different cortical organisations.

It would be excessive to discuss all individual cases of variation here and would go beyond our aim of describing basic processes. I can once again only sketch the outlines of the ways in which modifications of individual layers and of the whole cortical depth are accomplished, thus merely giving indications of the direction which future comparative cortical research should take.

In general terms there are three possibilities to distinguish:

1. generalised structural modifications throughout the depth of the whole cortex,
2. specific changes in individual layers in certain animals and
3. architectonic particularities of individual cortical areas in a single species or order.

We shall examine these three forms of variation separately, but must bear in mind that in reality they occur together and interact to determine the particular characteristics of the cortex in a given species.

1. General particularities of cortical architectonics in different animals.

If one compares a series of mammalian brains from different orders on the basis of cortical structure, even a superficial examination reveals broad general differences. The variations, apart from the special regional organisation of individual cortical types discussed above, concern general laminar structure, cortical thickness, cell density and mean cell size.

These differences were familiar to previous researchers and, as local variations in cortical structure of a brain were still largely unknown, they sought in them specific structural markers for individual animal groups. Thus there has been no lack of attempts to group brains systematically according to general architectonic characters. In particular, cell density, cell size and cortical thickness have been supposed to provide criteria for ranking a given cortex high or low in the animal kingdom, or for attributing a high or low organisational level to the brain or "psyche" (*70) of the particular animal.

In any case, the views of various authors in this respect are often mutually contradictory.

As far as cortical thickness is concerned it was formerly generally accepted that it diminished with lowering position in the animal kingdom, man having the thickest and lower animals a progressively thinner cortex as one descended systematically. The same applied to cell size, that was also supposed to decrease in lower mammals. In particular Marburg found that in monkeys the size of pyramidal cells diminished and small cells dominated with decreasing taxonomic position. The view that lamination was less developed in lower animals than in higher has recently been defended by Cajal and Haller, as we have seen. In contradiction to this Marburg maintains that the layers are actually better developed in lower monkeys than in higher; the more densely cellular a brain, the more its layers are supposed to be blurred. With regard to cell density, Nissl's view, based on his theory of "nervous grey" (*71), that the more cells per unit volume there are in the cortex the lower is the developmental stage of the animal, has mainly prevailed so far. Accordingly, the lowest animals are supposed to have the highest cortical cell density. The observations of Kaes oppose this view, finding lower (inferior) cortex more sparsely cellular, and Marburg agrees with him for monkeys, noting similarly that the cortex becomes more sparsely cellular as one descends the monkey series.

Thus we have a whole selection of contradictory declarations and findings. What are the real facts? I believe that my illustrations, even when examined rapidly, can give an unequivocal answer to the questions that have been posed [1]).

a) Cortical thickness.

In a recent publication "Über Rindenmessungen" (*72) I demonstrated that in man the thickness of the cerebral cortex is subject to regular local variations, within very broad limits, even under physiological conditions. I determined mean extreme values of 1.5 to 4.5mm, from which it emerges that in man certain cortical areas are normally three times thicker than others.

The same is true of other mammals and it is evident from the accompanying micrographs that those types of cortical structure that are particularly thick in man *ceteris paribus* (*73) are also relatively thick in animals, compared to other cortical types in the same brain, although exceptions do occur. Thus one arrives at the conclusion that when attempting comparative measurements one must always consider only identical regions or, if the comparison is between different animals, only homologous regions. Systematic studies of this are in progress and partially completed; the comprehensive tables cannot be reproduced here and I must limit myself to a summary of the measurements of a few cortical types in various animals. Examples will be given of cortical areas whose homologues in different animals are unequivocal. Table 3 contains the results. The figures are means based on the principles that I have recently defined for cortical measurement [2]).

Table 3 shows first that similar differences in cortical thickness between different regions exist in animals as in man. The giant pyramidal cortex (4) is always much thicker than the calcarine cortex (17), in monkeys easily twice as thick; the entorhinal cortex (28) has a different trend from these two areas, usually taking a middle course. The retrosplenial region (area 29) contains the thinnest cortex.

In many animals the thickest cortex (in areas 4 and 6) represents two to three times that in other regions of the same animal, such as areas 17 and 29.

If one compares individual homologous cortices in different animals it emerges that man has the absolutely thickest cortex as far as the areas under discussion are concerned; but an exception to the above-mentioned rule arises in the lower monkeys, for they possess extraordinarily thin areas 17, 28 and 29,

---

[1]) It should be emphasised that these micrographs and sections are at a standard magnification and thickness. Also the preparation of the brains was standardised. The figures thus permit a direct quantitative comparison of cortical thickness, lamination, cell density and cell size.

[2]) Über Rindenmessungen. Zentralbl. f. Nervenhlk. 1908; see also Neurol. Zentralbl. 1909.

Table 3. Thickness of homologous cortical areas in different animals (averages, in mm).

Area[1])	Man[2])	Guenon	Marmoset	Prosimian (Black lemur)	Flying fox	Hedgehog	Kinkajou	Rabbit	Ground squirrel	Kangaroo
4	3.0-4.5	3.0	2.15	2.3	1.9	1.87	2.17	2.7	2.1	2.8-3.1
6	3.0-3.8	2.5	2.17	2.3	1.6	2.1	2.0	2.33	2.18	-
7	3.08	2.0	1.73	1.67	1.7	1.78	1.7	2.2	1.73	2.2
17	2.3-2.6	1.7	1.26	1.55	1.76	1.5	1.9	1.8	1.37	1.9
28	2.5	1.6	1.14	1.35	1.52	1.6	1.9	1.2	1.13	1.7
29	2.3	1.1	1.07	1.19	1.4-1.76	0.8	1.67	0.8-1.5	0.75	1.2

Table 4. Relationship between body and brain weight and cortical thickness.

Order	Species	Body:Brain weight in grams	Cortical thickness in mm	
			Area 4	Area 7
Primates (*74)	Guenon	2500:85.0	2.8	2.0
	Marmoset	200:8.0	2.1	1.7
Prosimians	Lemur	1800:23.0	2.3	1.67
	Mouse lemur	62:1.9	2.0	1.5
Chiropterans	Flying fox	375:7.0	1.9	1.7
	Bat	23:0.3	0.4	-
Insectivores	Hedgehog	700:3.5	1.87	1.78
	Mole	75:1.3	1.3	1.0
Rodents	Rabbit	2200:10.0	2.7	2.2
	Ground squirrel	200:2.2	2.1	1.8
	Mouse	20:0.4	1.25	0.8
Marsupials	Kangaroo	5000:25.0	2.8 (3.1)	2.2
	Opossum	1100:5.5	1.4	1.1

being situated far below most representatives of the lower orders in this respect. Of particular importance in this question is the fact that the marsupials, a very low systematic order represented here by the kangaroo, have a moderately thick cortex, in excess of all other higher animals and in several areas almost reaching that of man. Even the echidna has a relatively thick cortex.

From this it is incontestable that there is no direct correlation between relative cortical thickness and the taxonomic position of an animal. I have been able to ascertain further that the mean cortical thickness in a brain, within limits, is more related to the body size of its owner, or the brain volume (or weight), than to membership of a particular order.

To prove this I have sought large and small species from various orders (primates, prosimians, chiropterans, insectivores, rodents, marsupials); the results are summarised in Table 4. The figures are again means of several measurements. They demonstrate manifestly a certain correlation, albeit only superficial, between body size, brain weight and cortical thickness.

The above findings imply that, in general, of two species from the same family or order, the smaller of the two, or the one with the smaller brain volume, will also have the smaller absolute cortical thickness. The narrowing of the cortex is, however, not really proportional to the fall in brain weight or body volume. If one compares the relationship between rabbit, ground squirel and mouse, the body weights are in the proportion of 1:10:100; but the cortical thicknesses only not quite 1:2:3. Thus small animals possess a relatively thicker cortex than large animals of the same order.

Apart from this, as I see it, purely superficial correlation between brain size and cortical thickness, no other consistent relationships with cortical thickness are detectable in homologous cortices throughout the mammalian class, and not even within an order. The larger animals of a given group will always possess an absolutely thicker cortex at maturity, without regard to their systematic relationships. Numerous examples are provided in the orders of carnivores and ungulates (*60). To attempt to produce correlations outside a given order or even from the whole mammalian class is contrary to all the evidence. Such general rules as the cortex of lower animals being thinner, or Kaes' even farther-reaching interpretation that the most highly developed cortices are the thinnest, are quite untenable. The inexactitude of this last rule has already been demonstrated by Marburg using comparative measurements in monkeys.

b) Cell size.

Here, relationships are even more complicated; size differences between different cortical cell types are enormous in a given animal. For example, one can compare sections of the frontal, precentral and occipital regions of the same monkey, as in Figures 50 to 52, shown at the same magnification. Hammarberg has made quantitative estimates in a number of cortical areas of man by immensely laborious measurements. No obvious simple relationship between cell sizes of different animals emerges. Of course, strictly speaking only homologous cell types should be compared, but the proof of such cellular homology is entirely lacking. The homologous nature of only one cell type is established with certainty, the giant pyramids of Betz, and so I will limit myself to this example to explain the principles.

In Figures 43 to 49 Betz giant cells of area 4 of our brain map, the giant pyramidal cortex, from various animals are illustrated for comparison at a standard magnification of 66:1. One notices that cells of this type are approximately the same size in man and in the kinkajou. If one compares many sections one is certainly convinced that the largest of these cells in the kinkajou are larger than similar ones in man. In man and, it seems to me, also in animals considerable individual differences exist in this respect, so that one must be very careful when suggesting relationships.

These cells are on average smaller in the lower gyrencephalic monkeys, even smaller in prosimians, and smallest of all, among primates and prosimians, in the lissencephalic marmoset. In other orders (except the ungulates) I could not detect that these cells were large enough to be easily differentiated from the largest cell types of adjacent cortex. This can be seen in Figure 48 of rabbit and Figure 49 of the kangaroo (*76).

Thus man and certain carnivores, such as cats and bears (kinkajou) (*77), possess the absolutely largest giant pyramids [3]) in the homologous cortex of area 4. Then come the gyrencephalic monkeys, then the prosimians. Certain ungulates (medium-sized domestic animals) are quite similar to the prosimians, while just after these in terms of size of the Betz giant cells follow the lissencephalic monkeys. Chiropterans, insectivores and rodents possess even smaller Betz cells. These relationships are quite clearly visible in low magnification micrographs (Figures 58 to 63) [4]).

In this context it should however be noted that cell size in different species of the same order or family can be very different. Thus I have found, for example, giant cells in a mature prosimian, the indris, that easily exceed in size those of all lemurs and are certainly no smaller than those in monkeys and great apes; on the other hand the mouse lemur possesses strikingly small Betz cells [4]). Furthermore, there are small carnivores, especially weasels, in which these cells are very poorly developed, in contrast to other carnivores. It should also be noted that in a given animal these cells are often polymorphic and of very varying sizes. Whereas in man, for example, all giant pyramids have a rather similar average size, their dimensions vary very widely in the kinkajou - see Figure 46. Thus one must always only compare the sizes of the largest cells of a species with the largest of another species and then one can perhaps say that the biggest Betz cells in the kinkajou (or tiger or lion) are on average

---

[3]) Today all sorts of cell types are understood and described as "giant pyramids" in the literature, including sometimes those of the occipital cortex. I use the expression only in the narrower sense of the cells of the ganglion cell layer of the precentral gyms and its homologues, as originally used by Betz.

[4]) I must contest that giant pyramids are larger in lemurs than in monkeys, as Marburg proposes. (cf. Figures 44 and 45.)

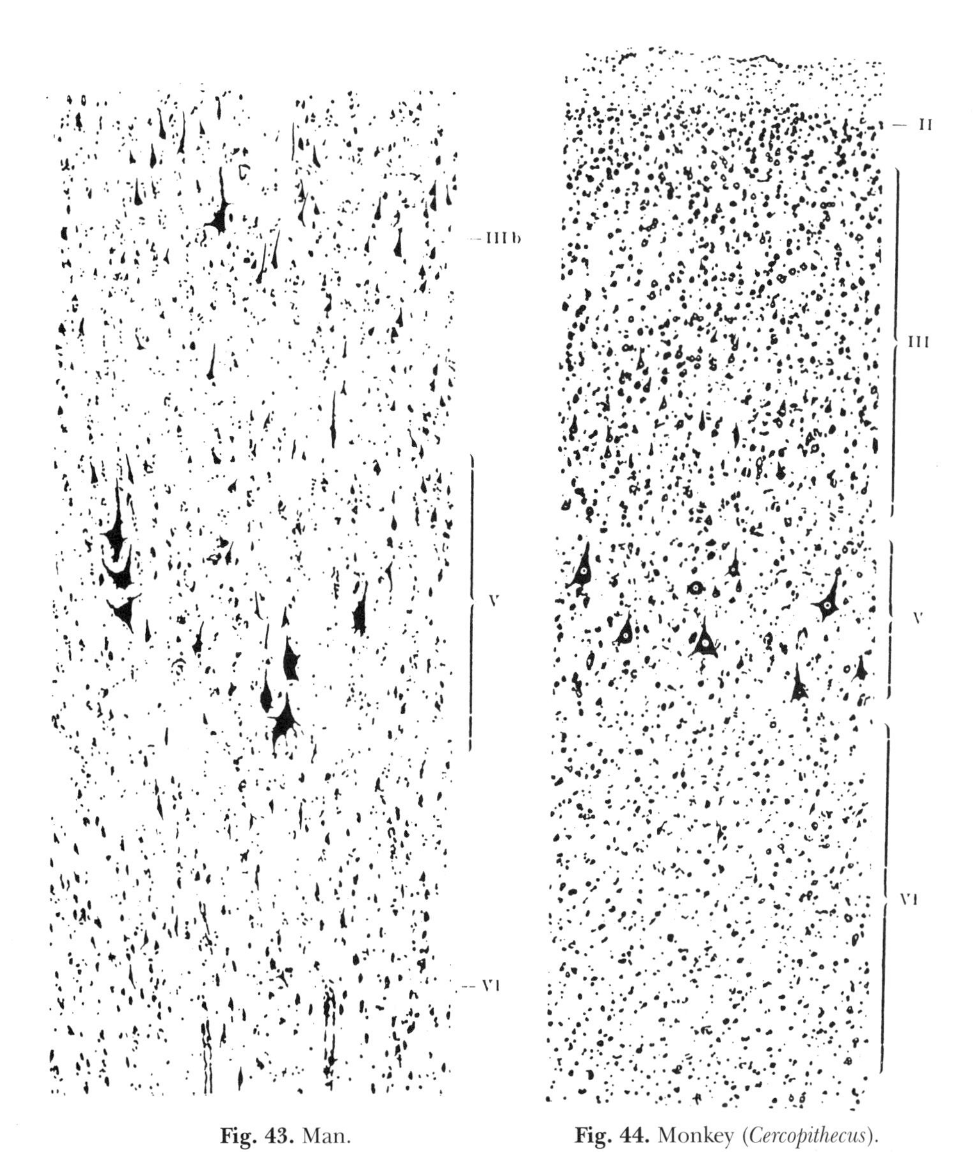

Fig. 43. Man.

Fig. 44. Monkey (*Cercopithecus*).

Fig. 43-49. The same cortex in different mammals. 66:1, 10μm. Giant pyramidal cortex (area 4 of our brain map) of man, monkey, prosimian, kinkajou, flying fox, rabbit and wallaby. Note the differences in cortical thickness, cell size, cell density, and especially the arrangement of the giant pyramids. Because of lack of space, only layers IIIb-VI are shown in man. With respect to the arrangement of the giant cells (*75) one can distinguish:

a) the cumulative type (Fig. 43),
b) the unilaminar type (Fig. 46),
c) the multilaminar type (Figs. 44, 45 and 48),
d) the solitary type. (see also pages 83ff.)

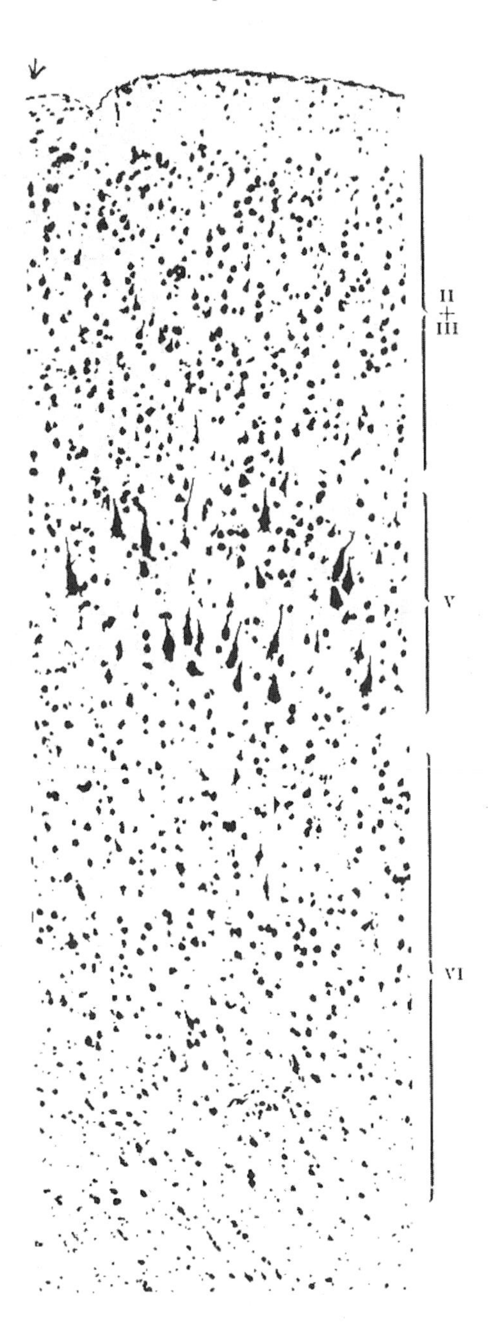

Fig. 45. Prosimian (lemur).

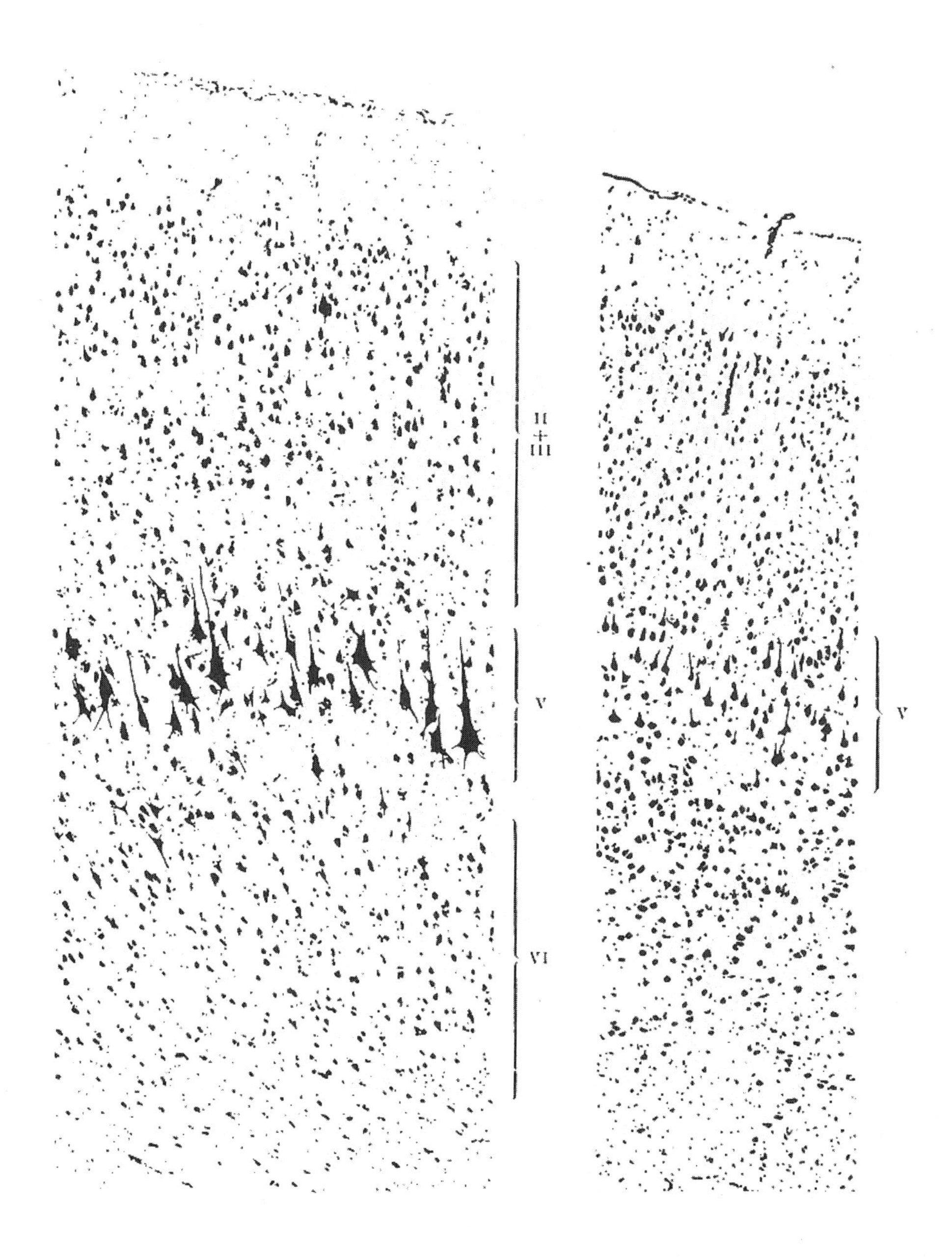

Fig. 46. Kinkajou.

Fig. 47. Flying fox.

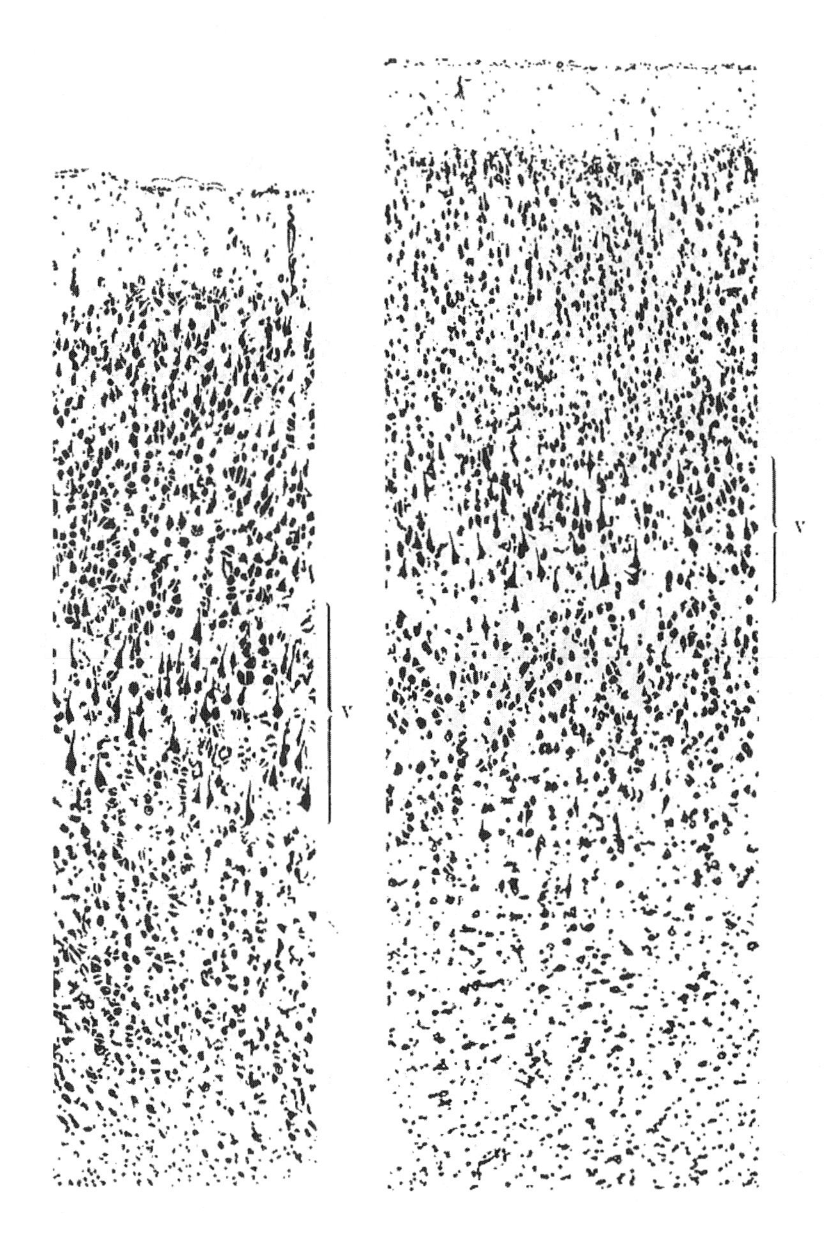

Fig. 48. Rabbit.

Fig. 49. Wallaby (young).

bigger than in man, while all other animals have absolutely smaller giant cells than man.

Therefore, with respect to this single cell type, there exist extraordinarily complex relationships that show absolutely no connection with zoological systematics, to say nothing of there being a general rule of cell size for the whole cortex and for all mammals. With all due consideration it can simply be said that certain groups are characterised by a predominance of, on average, strikingly smaller cortical cells. This is true, for example, of a family of primates, the marmosets, as well as of chiropterans and rodents; it is however equally true for quite lowly orders, like the marsupials such as the kangaroo, that are characterised by a lack of all large cell types. On the other hand one can group together certain species that possess on average many medium and large cells in particular cortical regions, notably in the frontal lobe, and yet others that develop true giant cells within a regional cortical type. Among the latter group we have recognised primates (except marmosets), prosimians, carnivores and ungulates.

Totally different interpretations concerning the absolute size of Betz giant cells in different animals, and its dependence on particular somatic factors, have been published in the literature. We should like to have one more word on this question because it demonstrates clearly how necessary it is to first collect unbiased factual material before one can enter into the proposition of any ingenious theory.

In the past the size of pyramidal "motor" neurons has often simply been related directly to body size, and it was accepted than the bigger an animal was the bigger its motor cortex cells must be. A second theory, proposed by Bevan Lewis and supported by Campbell, states that the size of cells depends on the length of the axon arising from them; the longer the pathway that the motor impulse has to traverse from the cortex the more massive must the cortical cell of origin be. "The greater the distance along which a nerve cell has to transmit its energy the larger will that nerve cell probable be" (Bevan Lewis) (*78). And finally a third theory of Hughlings Jackson (*79) supports a relationship between cell size and the size of the dependent muscle (*80). Cell size would be directly proportional to the size of a muscle or movement; the stronger the muscle or the movement for which it is responsible the bigger would be the central motor neuron providing the impulse.

It seems to me that none of these theories is supported by the facts.

1. Regarding the first, most widely held view I need only refer to my illustrations. Further, I have made measurements of Betz cells in several animals, the results of which are summarised in Table 5.

Table 5. Size of Betz giant pyramidal cells in different animals (width and length of the cell body in microns).

	Maximum [5]	Average
Man	**53 x 106**	27-50 x 66-100
Monkey (mona guenon)	**40 x 72**	23-40 x 56-68
Prosimian (indris)	**44 x 80**	20-44 x 60-70
Prosimian (lemur)	**30 x 70**	10-27 x 50-68
Flying fox	**16 x 36**	7-10 x 20-30
Brown bear	**53 x 100**	30-50 x 65-100
Kinkajou	**50 x 110**	26-50 x 67-108
Lion	**60 x 133**	27-53 x 67-110
Tiger	**60 x 100**	33-55 x 65-90
Rabbit	**18 x 40**	8-16 x 27-36

Table 5. The greatest length and width of the cell body of numerous individual cells were determined and - in addition to the maxima - average values were calculated for individual animals. The figures need no further explanation; they confirm the variations visible in the text figures. One may simply point out again that all carnivores (brown bear, kinkajou, lion and tiger) have relatively large giant cells and that even small ones (kinkajou) have higher maximum values than man. In another small carnivore, the cat, Bevan Lewis also observed unusually large giant pyramids, up to 32xl06μm with an average of 37x83μm - values that I essentially confirm - whereas in, for example, the sheep the largest examples only measured 23x65μm (*81). It thus emerges that carnivores, and especially cats and bears (as far as they have been studied), exceed all other mammals in terms of the size of their giant "motor" neurons.

My figures for man do not entirely agree with those of other authors. Maximum values are:

Bevan Lewis	55xl26μm
Betz	60x120μm
Hammarberg	40x80μm
Brodmann	53xl06μm

These discrepant results need confirmation. Extensive studies are in progress, especially directed toward comparative anatomy.

Concerning the other animals, I would point out that many species manifest mainly long, narrow giant cells (lemur and indris), while others are characterised by wider, rounded types (cats, especially the tiger).

The most important conclusion from these figures for the present question

[5] I have measured maximum values of 35x60μm in an elephant; however, my material was from an animal with a infection and was, in addition, poorly preserved, so that no firm conclusions can be drawn. It is known that anterior horn cells in the spinal cord of the elephant are also small.

is that the size of the giant pyramids cannot be said to be exclusively or even predominantly dependent on body size or brain volume. The delicate kinkajou of some 2kg body weight possesses much larger cells than the similarly sized black lemur (2kg body weight) and, on average, as large or, in isolated cases, even larger giant cells than man with a mean body weight of 75kg. Even within a given order body volume is not a decisive factor. In the three great carnivores, lion, bear and tiger, with body weights of several hundred kilograms the cells in question are only insignificantly larger than in the tiny kinkajou. On the other hand the ungulates, often with massive body weights and correspondingly heavy brains, possess without exception smaller giant "motor" neurons than these carnivores.

2. The second theory of Lewis and Campbell, that in its general formulation is based on that of Pierret and Schwalbe, cannot be reconciled with the fact that the representation of the trunk at least contains no larger giant pyramids than the cortical region devoted to the upper limb, although the pathway from the cortex to the trunk is longer than that to the proximal part of the arm, that is the shoulder and upper arm. Campbell answers this objection by claiming that the whole mechanism controlling the trunk muscles is less highly specialised than that to the limbs. Even accepting this view, the objection to the theory that larger motor neurons correspond to a longer pathway still remains, and this theory cannot be accepted in its general form [6]).

3. The same is true of the, in itself plausible, hypothesis of Jackson (*79). Trunk movements are undoubtedly "large movements", just as the trunk muscles are very massive structures compared with those of the hands and fingers; nevertheless we do not find correspondingly larger giant cells in the relevant motor cortex [7]).

One can therefore simply state that none of the three hypotheses agree entirely with the facts. One must be satisfied provisionally with the assumption that other unknown factors determine the size of giant pyramids (the motor cortical neurons). The notion may be proposed that the number of intracortical connections of these cells or of the whole giant cell layer all play a role, in physiological terms, in the sophistication of a function, that is to say the finesse of motor control and coordination, as well as muscle strength. It is then

---

[6]) A similar constant relationship between cell body size and axon length (or fibre diameter) was often accepted formerly at subcortical levels of the central nervous system. Pierret (*82) first made the general statement that neurons were larger the longer the centripetal and centrifugal nerve fibres related to them. Later workers agreed with him, Schwalbe in particular showing in frog and man that the diameter of a nerve was directly proportional to its length and was thus related to its cell body. However, later extensive precise measurements showed that this rule was not free from exceptions and was not of general validity, being subject to numerous discrepancies and inconsistencies, particularly in the cerebral hemispheres and notably in the motor pathways. Gaskell obtained similar results for certain medullary and spinal centres, and Fürbringer demonstrated by comparative measurements on spinal nerves in birds that correlations of the thickness-to-length ratio of nerve fibres were far more complicated than expressed in the theories just cited.

[7]) It is possible to find a certain confirmation of this theory in many cranial nerves. The fact has simply to be recalled (Kohnstamm) that Deiters' nucleus, the coordination nucleus for body musculature, is exclusively large-celled, whereas the angular nucleus (*83), the coordination nucleus for eye movements, consists of quite small cells.

acceptable to concede that, in addition to these, the body size of the animal and its muscle volume, or in other words the extent of the territory of innervation and the pathway length, also exert an influence. All these factors probably contribute, together with yet other unknown ones, and their interaction governs the size (and number) of giant pyramids in individual cases, as well as the dimensions of the giant cell layer. Thus one arrives at a satisfactory interpretation by uniting Merkel's morphological hypothesis with Schwalbe's physiological theory, as in Furbringer's proposal for spinal innervation. We can only obtain more information on this by the most minutely detailed study of more material in which all architectonic data for the relevant cortical field (our giant pyramidal area 4) are taken into consideration. One will have to pay particular attention to those animals and individuals with any form of exceptional motor performance, such as primitive races of mankind or other individuals such as athletes and acrobats that stand out by their muscular strength, endurance and dexterity.

c) Cell number. (*84)

Interrelationships of relative cell number or density in different animals are at least as complicated. In order to obtain an unequivocal basis for specific comparative studies, detailed cell counts should be made [8]) and indeed each individual layer or, even better, the whole thickness of every cortical field of each animal should be examined independently to determine the number of elements per unit area. However such an undertaking exceeds the capacity of any individual and requires collaborative work by many people. Thus I could not promise initially that our study would be very useful. Firstly it only attempts to establish general rules as to whether the suggested constant relationship between cell number in the cerebral cortex and the taxonomic position of a given animal or its level of organisation must really be accepted. My illustrations suffice to arrive at a conclusion about this.

First let us compare different regional types from the same brain for cell density, using sections from the frontal lobe, the precentral gyrus and the calcarine cortex of the same animal. I have illustrated the relevant cortical sections of a monkey (guenon) in Figures 50 to 52. In these, the agranular frontal cortex is the least cell-dense, the immediately caudally situated giant pyramidal precentral cortex is somewhat denser, while the calcarine cortex is characterised by an extraordinarily high cell density. This last area contains at least three times more cells per unit volume through the total cortical depth than the other two types. There is a similar, if not so massive, difference in man (Figures 43 and 53) and also a rather smaller one in the kinkajou (Figures 46 and 55).

---

[8]) It is again Hammarberg who first employed such methods in various regions of man. H. Berger made cell counts in the cat cortex using other, simpler techniques, especially in animals that had been blinded immediately after birth (*85).

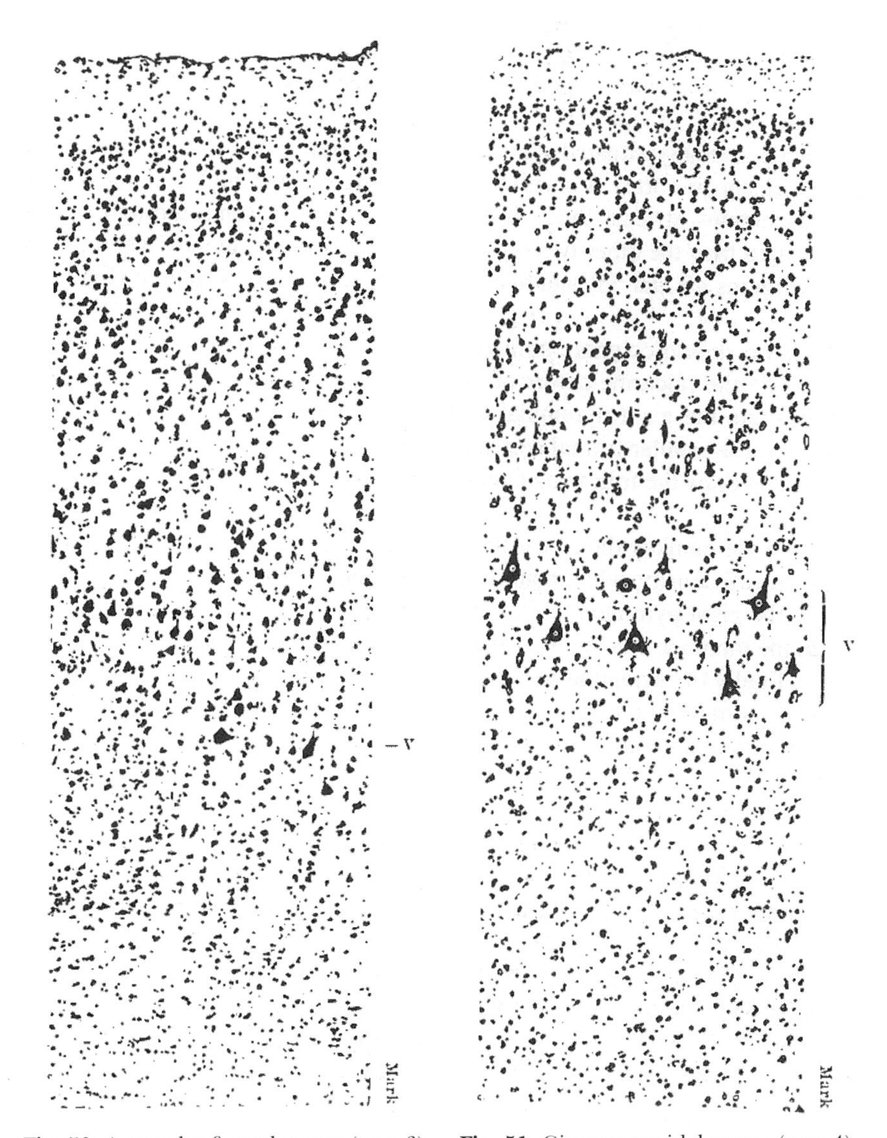

Fig. 50. Agranular frontal cortex (area 6). **Fig. 51.** Giant pyramidal cortex (area 4).

Fig. 50-52. Three different regional cortical types from the same monkey brain (*Cercocebus fulginosus*) (*86). The photographs are all from the same series of sections. 66:1, 10μm. Note the differences in cortical thickness, cell density, cell size, and cell arrangement in the three different regions.

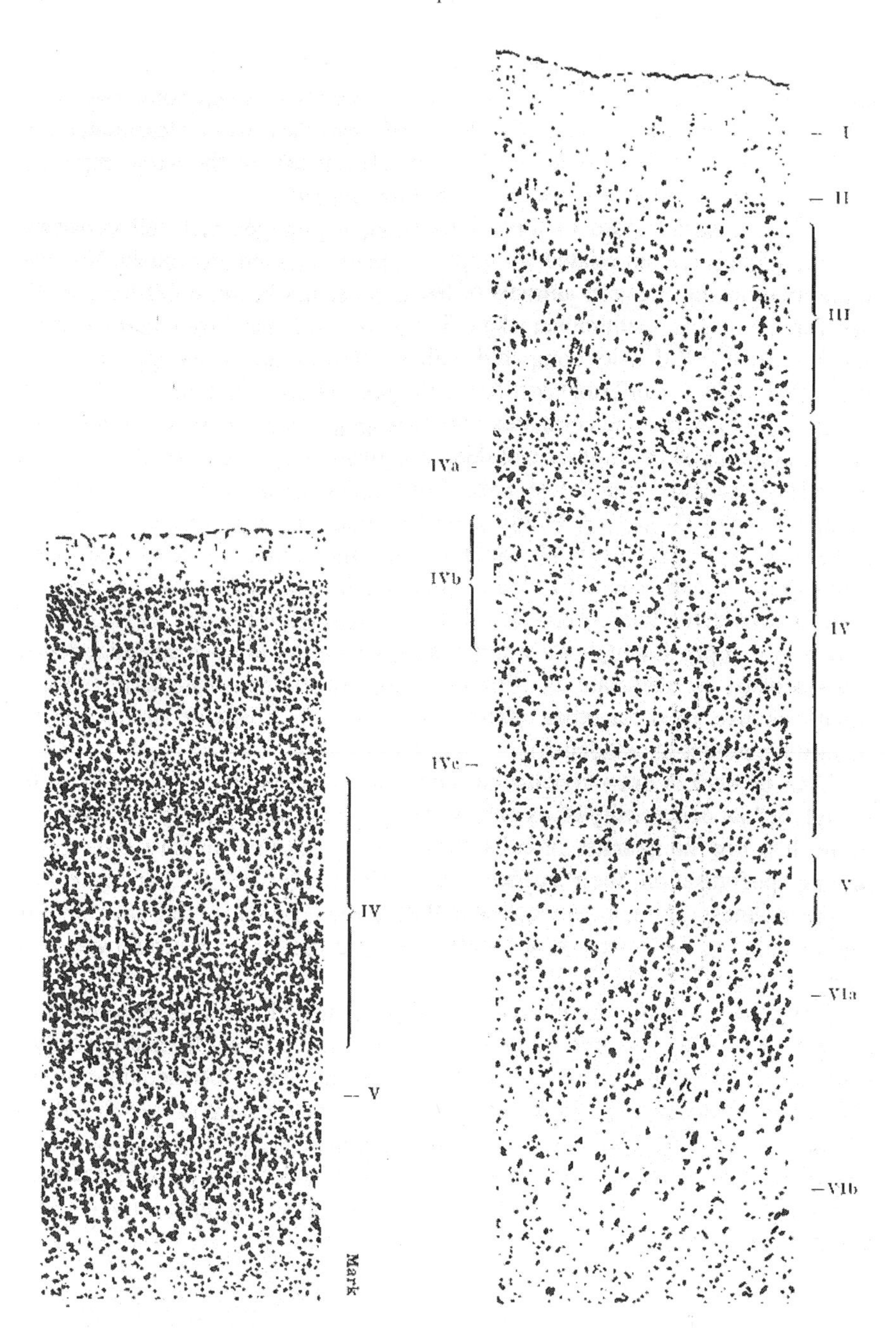

Fig. 52. Calcarine cortex (area 17).

Fig. 53. Man.

Now we should compare homologous cortical types in different animals. In Figures 53 to 55 the calcarine cortex of three different mammals is compared. Figures 43 to 49 illustrate the giant pyramidal cortex of several animals and Figures 56 and 57 reproduce the agranular frontal cortex of two animals (monkey and kinkajou).

One can draw the following conclusions directly from comparison of these micrographs:

a) in the monkey the calcarine cortex manifests the highest, and in the kinkajou the lowest, cell density, while man occupies an intermediate position;

b) the giant pyramidal cortex has a similar trend, being on the whole most cell-dense in the monkey, less so in prosimians, while man is again intermediate;

c) finally the agranular frontal cortex also displays higher cell density in the two illustrated monkey brains with lower density in the kinkajou. On the other hand the homologous human cortex seems to have greater cell density than that of the monkey (Figure 32).

The comparison of heterologous types in different animals produces an even more interesting result. Our examples show that the calcarine cortex of the guenon or the capuchin monkey (Figures 52 and 54) possesses the absolutely greatest cell density, and the giant pyramidal cortex of prosimians (Figure 45) the least. The frontal and giant pyramidal cortices of all brains without exception are much poorer in cells than the calcarine cortex of any animal, whether of high or low taxonomic position.

Other cortical areas show similar trends. For instance, one can compare a number of sections of parietal cortex from different orders, say from man, monkey, kinkajou, rabbit and kangaroo. The kinkajou has the lowest cell count of these animals, in the kangaroo it is slightly higher, even higher in man, while the rabbit has the highest density.

These examples should suffice; anyone can obtain further evidence from a critical comparison of the micrographs; the heterogenetic cortices are particularly instructive in this respect.

Thus we observe:

1. Regional differences in cell density of the cerebral cortex of individual animals are usually significantly greater than the differences between homologous cortical types of different species regardless of how far apart they may be taxonomically.

2. Therefore one should only compare cell density in the same regions, or homologous areas in the strictest sense, of different animals.

3. Also, with regard to topical localisation, it emerges that cell densities of different animals are not related as simply as had been believed previously. Not even a simple relationship between animals of the same order has been established, let alone rules for the whole mammalian class. It cannot be said that lower animals have a greater cortical cell density compared with higher (Nissl), nor is the opposite opinion correct that lower (or inferior) cortices are cell-sparse (Kaes, Marburg). Such relations are quite absent when one compares

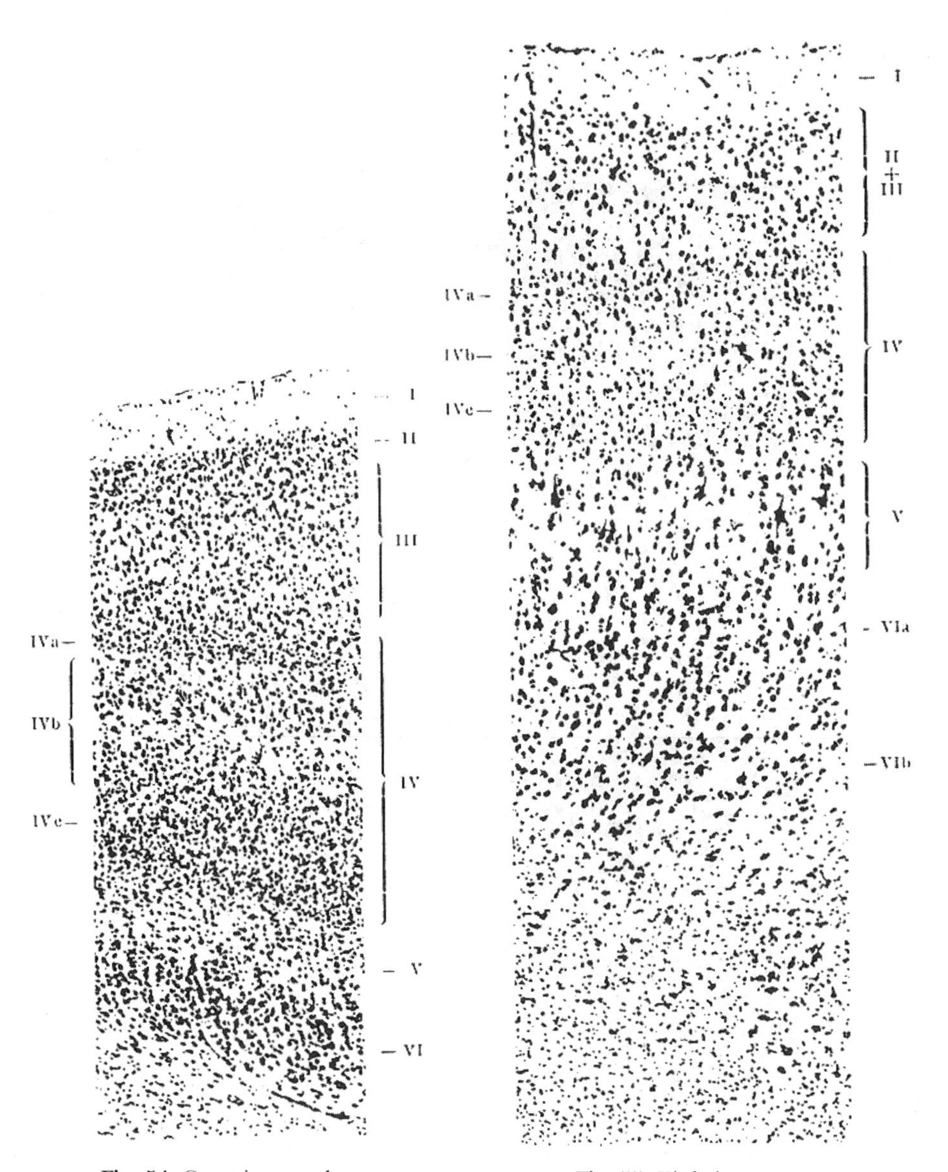

Fig. 54. Capucine monkey.

Fig. 55. Kinkajou.

Fig. 53-55. The same cortex in three different animals (calcarine cortex, area 17, of man, capucine monkey and kinkajou). 66:1, 10μm. Note the differences in cortical thickness, cell density and cell size. The lamination also shows some differences, but is in principle the same.

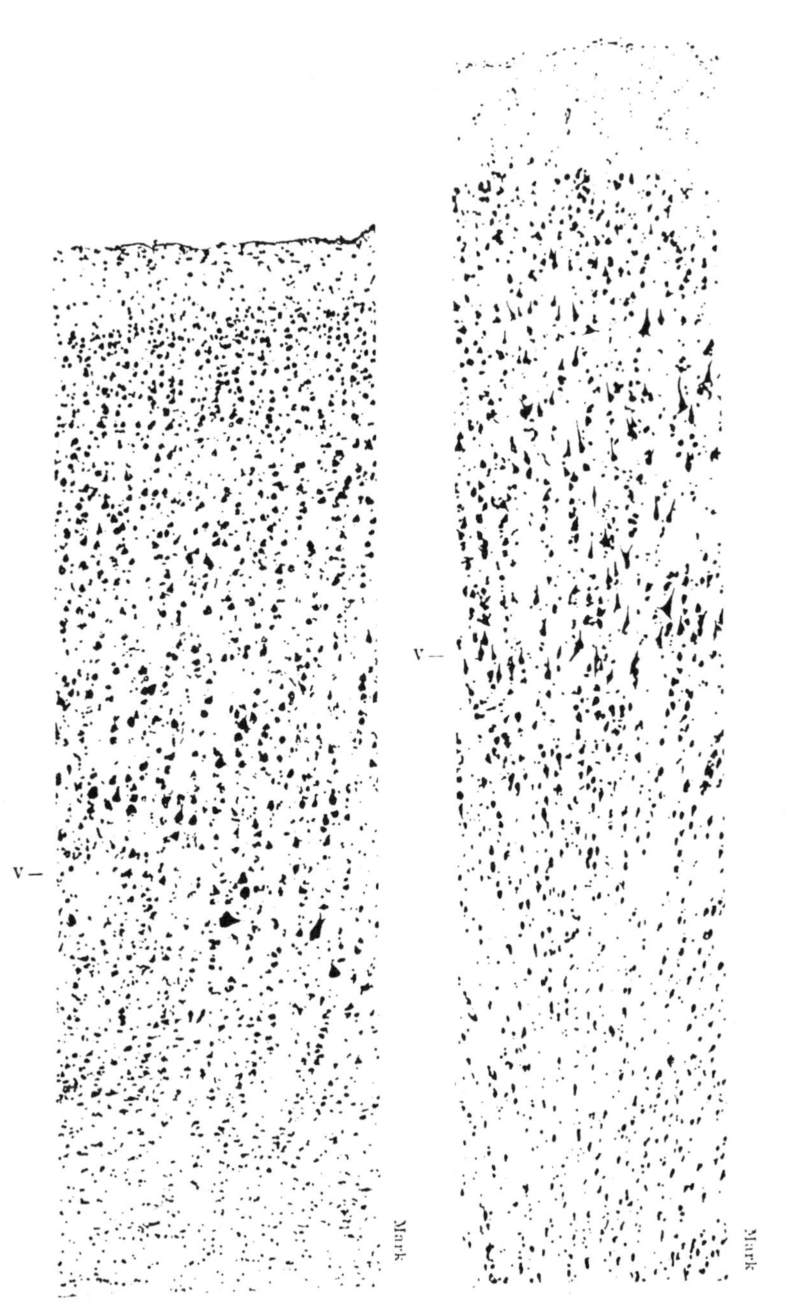

Fig. 56 and 57. Agranular frontal cortex in monkey and kinkajou (area 6 of our brain map). 66:1, 10μm. Indistinct lamination as a result of absence of the inner granular layer; deep, cell-poor cortex consisting mainly of medium-sized cells.

homologous areas; in one case a primate may be in the lead as far as the highest cell density of a given area is concerned, and in other cases a lowly rodent or marsupial. Even within an order consistent relations are usually lacking. It is not correct, as Marburg proposes, "that the lower one descends in the monkey series the more sparsely cellular becomes the cortex". I rather find that one could support the opposite view; the marmosets, the lowest family, have uncontestably the most densely cellular cortex of all monkeys [9]) (compare Figures 69 and 71). But this does not imply that higher positions in the primate order involve a regular decrease in cell density, for there are all sorts of exceptions to the rule. And what is true for primates is also valid for other orders. As a rule, there is a lack of any recognisable regularity concerning cell density.

Thus to wish to draw conclusions about the level of organisation of a brain from the high or low cell density of its cortex must be considered in principle as doomed to failure. It is not the quantity of cells per unit volume, but their quality, their detailed intrinsic specialisation, their surface area as manifested by the number of dendrites, and the richness of their connectivity, that all form a yardstick for the functional sophistication of the cortex or of a particular cortical region, as Bielschowsky [10]) has described in another context.

Nevertheless many morphological data support the idea that physiological performance is reflected to some extent by cell density. Thus we see - to mention only one example of motor innervation - that the oculomotor nuclei of the medulla oblongata (*88) possess an extraordinarily high cell density, corresponding to their highly specialised function, while the phrenic nuclei are comparatively very cell-sparse. Precisely these two examples also demonstrate again that it is not principally the muscle volume or the work load that influence cell density, the massive diaphragmatic muscles having a much smaller number of motor cells than the weak eye muscles. In this respect very different complementary factors interact. Among these, as we have seen above with respect to cell size, the most important roles must be played by the level of sophistication of the overall function of a territory of innervation, together with the extent of the territory, the scope of the movement and, above all, the intracerebral connectivity.

There is however as yet no objective index to evaluate all these factors as far as the cerebral cortex is concerned.

2. Modifications of individual basic layers in mammals.

In spite of the six basic tectogenetic or primitive layers described in Chapter II being constant formations in all mammals, there are such important modifications to individual layers within the class that in certain cases it is only

---

[9]) Marburg's error is partly explicable in that he did not investigate this lowest monkey family.

[10]) Bielschowsky, Die histologische Seite der Neuronlehre. Journal f. Psychol. u. Neurolog. V, 1905. (*87)

Table 6. Thickness of layer I in mm and its ratio to the total thickness of the respective cortical area.

Area 12)	Man	Monkey		Prosimian	Flying fox	Hedgehog	Kinkajou	Rabbit	Ground squirrel	Kangaroo
		Macaque	Marmoset							
4	0.35 *(1:12)*	0.18 *(1:18)*	0.13 *(1:15)*	0.15 *(1:13)*	0.25 *(1:8)*	0.35 *(1:5)*	0.23 *(1:9)*	0.26 *(1:10)*	0.18 *(1:12)*	0.22 *(1:13)*
6	0.37 *(1:10)*	0.16 *(1:15)*	0.16 *(1:13)*	0.19 *(1:12)*	0.19 *(1:8)*	0.32 *(1:6)*	0.2 *(1:10)*	0.22 *(1:10)*	0.17 *(1:12)*	-
17	0.25 *(1:10)*	0.11 *(1:14)*	0.10 *(1:13)*	0.13 *(1:11)*	0.16 *(1:11)*	0.33 *(1:6)*	0.17 *(1:13)*	0.15 *(1:12)*	0.12 *(1:11)*	0.18 *(1:10)*
28	0.47 *(1:5.5)*	0.35 *(1:4.5)*	0.23 *(1:5)*	0.18 *(1:7)*	0.32 *(1:5)*	0.35 *(1:4)*	0.38 *(1:5)*	0.3 *(1:4)*	0.4 *(1:3)*	0.36 *(1:4.5)*

possible for an experienced observer with abundant comparative material to recognise given layers, such as the so-called granular layers, by studying intermediate forms. Only brief general outlines can be indicated here of how the different layers vary in their importance and histological specialisation within the mammalian class.

Layer I - *the molecular layer*. Cytologically it only varies in different animals with respect to its thickness, that is its vertical depth [11]).

The older authors were already aware of these differences in thickness and naturally attempted quite early to draw conclusions about the physiological value of layer I from its varying size, although the results were usually quite contradictory. Thus Meynert was of the opinion that layer I was relatively much thicker, compared to the rest of the cortex, in all other animals than in man and thus represented a "neurologically worthless layer", while on the contrary Stilling invokes precisely its strong development in man and speaks of it as a "formation of considerable sophistication", a plexus containing the finest neuronal processes. What is the present status of this correlation? I think a comparison of our illustrations provides a satisfactory answer to that. In order to obtain an unequivocal standard for comparison I have made additional quantitative studies in a number of animals. Of course it is once again only valid to compare homologous areas in different animals.

In Table 6 the values for four homologous cortices from several animals are summarised. The figures do not need detailed explanation. I wish only to point out the wide regional variations in thickness of layer I in all the animals; they are relatively modest in man and hedgehog, but strikingly large in the ground squirrel. Man has, on average, the absolutely thickest layer I of all animals, while the monkeys (macaque and marmoset) are remarkable for the extraordinary narrowness of the layer. Other lower mammals are intermediate between man and monkey. In certain areas of lower species (kangaroo and rabbit) layer I approaches or even easily exceeds the thickness of the narrowest types of man. Comparison of the relevant text figures also demonstrates this.

One sees here also that there is no set of rules based on the taxonomic position of an animal, and that the physiological sophistication of the outermost cortical layer cannot be judged from its thickness.

Layer II - *the outer granular layer*. Its extensive regression in most animals has already been discussed above. It remains relatively well developed in many cortical areas of lower monkeys (Figures 21 and 54) and layer II appears distinctly cell-dense in the kangaroo. However, it is almost entirely absent over the whole cortical surface of the adult rabbit, as indeed in most rodents and insectivores, a fact that explains the above-mentioned erroneous homologising of layers. It is also generally weakly developed in chiropterans. On the contrary, we find the outer granular layer developing as a clearly differentiated thick cell

---

[11]) I cannot here go into the extensive basic variations in fibre structure of layer I. They form major criteria for the myeloarchitectonic parcellation of the cerebral cortex, as emerges from the neurobiological research of O. Vogt, Mauss and Zunino on man, guenon and rabbit.

lamina almost everywhere in the kangaroo and the phalangers, which again speaks for the independent genesis of this layer. However, its cells are profoundly modified in these animals. They are no longer true granular elements, the layer being rather composed of relatively large, mainly multipolar and stellate cells that have differentiated from the original granules. Compare Figures 15 and 31 of the immature kangaroo, where the primitive granule stage is still present, with Figures 19 and 36 of the adult animal. The situation is rather similar in man; true round granular cells with sparse cytoplasm can hardly be found in layer II of the adult brain; more frequently the "granules" are largely modified to small pyramids by secondary transformation and the layer is often counted with the true pyramidal layer (my layer III) as the layer of "small pyramidal cells". In contrast the granular character of layer II is often preserved in certain regions in various mammals. This is true of the giant pyramidal cortex of carnivores and many ungulates, but only where it lies in the depths of the cruciate sulcus, the "granules" of the free surface having differentiated into polymorphic cells. Also, the primitive neuroblast-like granular appearance persists in the cells of layer II in the retrolimbic cortex of certain rodents.

Layer III - *the pyramidal layer*. Many authors have emphasised the difference in thickness of the pyramidal layer in different animals and attempts have been made to draw conclusions about the high or low level of cortical organisation from the extent of its development. Again, directly contradictory views emerge in this respect. While most authors since Meynert tend to assume that a brain is at a higher level the better developed the pyramidal layer is, that is the thicker it is, Kaes proposed a rule that the level of cortical development directly paralleled the evolution of the major inner layers (that is layers V and VI). Marburg has demonstrated in monkeys that in itself this rule is not generally valid for this single restricted group. From my own research I can only say that on average the human cortex possesses both the absolutely and the relatively thickest pyramidal layer, but that otherwise no firm rules can be laid down in the animal kingdom to support the above thesis, even in homologous areas. The layer varies considerably in thickness and in addition behaves extremely variably even in homologous areas of different animals; thus for example, in area 5 (the preparietal area) it is narrowest in prosimians (Figure 23), a little thicker in the carnivorous kinkajou (Figure 17), thicker still in the marmoset (Figure 29) and thickest in man (Figure 16).

With regard to size and number of pyramidal cells in this layer, no consistent relations emerge.

Layer IV - *the inner granular layer*. The essential alterations undergone by this layer in specific cortical types, such as in the calcarine area of various animals, will be discussed in the next section. First we shall consider the common transformations to which the mammalian inner granular layer is often subjected, particularly in individual species. These modifications are so vast in many animals that the layer has escaped identification by many authors, as we have seen, like the outer granular layer.

In man it is characterised by great polymorphism resulting in multipolar, stellate, triangular and similar variously shaped small cells, whereas in occipital regions of lower monkeys and many prosimians it is composed almost exclusively of small, darkly staining, round cells with little cytoplasm, true "granules", interspersed with just a few large multiangular and stellate cells of neuroblast-like appearance (as also found at immature human stages). In carnivores and ungulates the layer differentiates progressively in the same way as in man; that is it does not consist of round, granular cells but mainly of polymorphic elements whose average size is quite considerable, comparable with cells in adjacent layers. Thus in the mature cortex the layer is hardly distinguishable structurally from other layers and one can only recognise the original granular character of these cells at immature stages, and thus identify the layer. In contrast to the above-mentioned groups, in many lissencephalic animals, especially small rodents and insectivores, but also in many marsupials (such as the opossum), the layer is composed of larger, pale staining vesicular (*89) cells, that are poorly differentiated and rather sparse. Thus the whole layer regresses in these animals and can easily be overlooked, as has often happened.

So we can distinguish at least three possibilities with respect to cellular differentiation of the inner granular layer in different animal groups:

1. the cells maintain a persistent immature form, that is they have a distinct round shape as at early developmental stages (lower monkeys and many prosimians);

2. the cells undergo an extensive progressive differentiation, completely lose their neuroblast-like character and adopt larger polymorphic shapes, hardly distinguishable from the cells of other layers (man, carnivores, ungulates);

3. the layer undergoes a form of direct regressive transformation; whereas its elements form a well demarcated layer of dark, round cells at immature stages, they later lose their coloration and take on a uniform vesicular shape, and the layer seems to regress in sections.

Layer V - *the ganglion cell layer*. Its modifications in mammals consist as much in changes in thickness as in variation in shape of its characteristic "ganglion cells".

Concerning its thickness, the same comments apply as made above for the pyramidal layer. The size of the ganglion cells and their special organisation in individual animals and groups are subject to gross variations. Naturally, only homologous cortical areas are comparable.

We wish to limit ourselves to one cortical type, the giant pyramidal cortex, that illustrates these relationships well, for a proportion of the ganglion cells has undergone specific development to form the Betz giant pyramids.

Three variants in the organisation of the these cells are distinguishable, as comparison of low-power micrographs shows (Figures 58 to 63). This is even more clearly seen at higher magnification in Figures 43 to 49.

a) The cells may aggregate in several widely separated groups or clusters, the so-called "*nests*", as Betz has already described in man: this is *cumulative organisation* of Betz cells. It is found in man in the upper part of area 4, especially in its caudal portions, and to some extent in monkeys and prosimians.

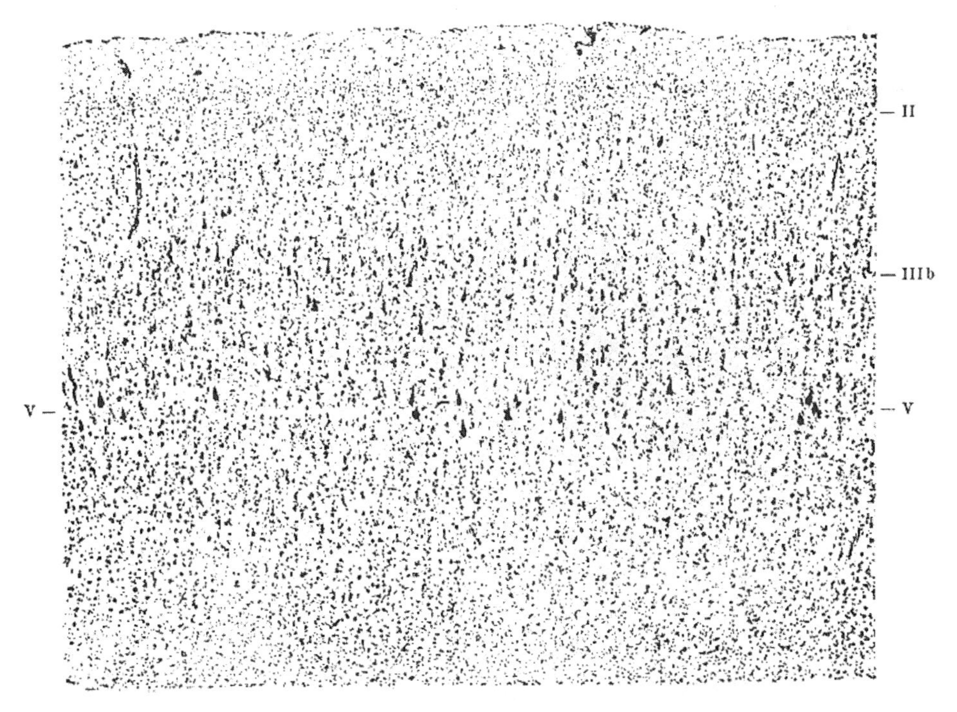

Fig. 58. Giant pyramidal cortex of man. Posterior bank of the precentral gyrus. Mixed arrangement of the Betz giant pyramids, partly cumulative, partly laminar, partly solitary. cf. the purely cumulative type in Fig. 7 from the apex of this gyrus.

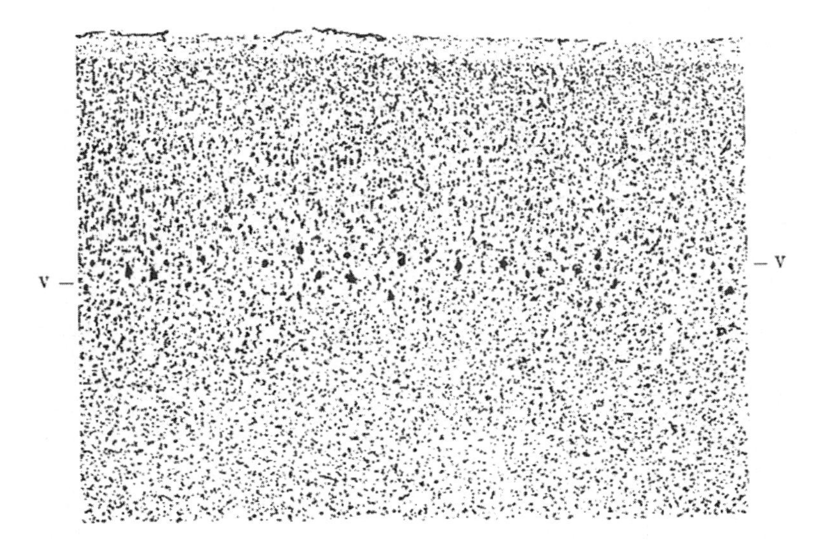

Fig. 59. Giant pyramidal cortex of the monkey (*Cercopithecus fulginosus*) (*86). Posterior bank of the precentral gyrus. 25:1, 10μm. Multilaminar arrangement of the giant cells (cf. Fig. 44).

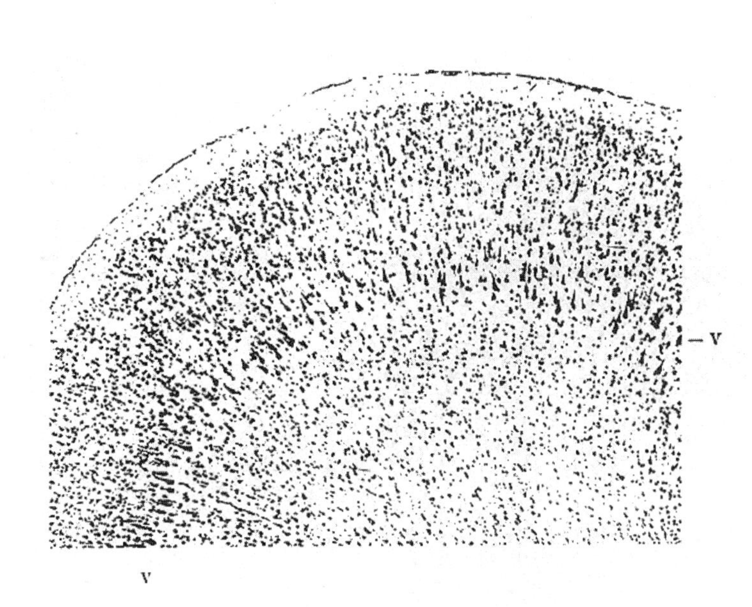

Fig. 60. Giant pyramidal cortex of the prosimian (*Lemur macaco*). 25:1, 10μm. Multilaminar arrangement of the giant pyramids. These cells are relatively small, but very dense and in several rows above each other. (cf. Fig. 45.)

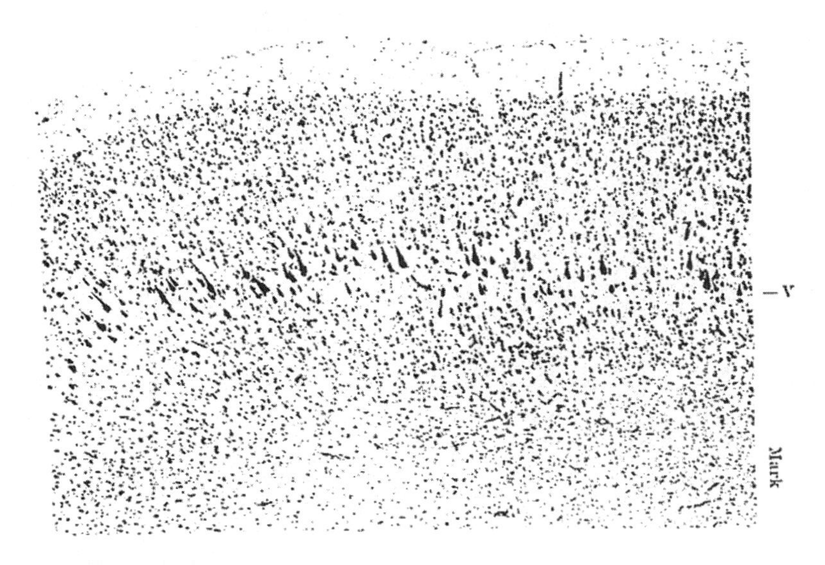

Fig. 61. Giant pyramidal cortex of the kinkajou (*Cercoleptes caudivolvulus*). Posterior sigmoid gyrus. 25:1, 10μm. Unilaminar arrangement of the giant cells. Layer III is strikingly narrow and cell-poor compared with that of man (Fig. 58). (cf. Fig. 46.)

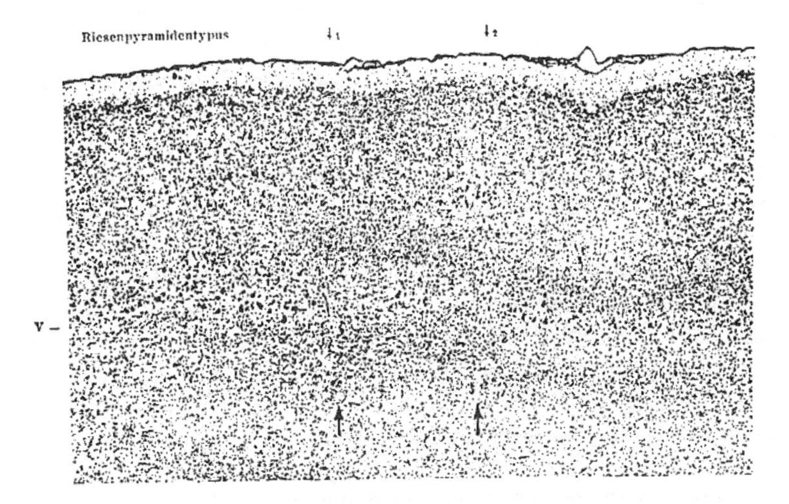

Fig. 62. Giant pyramidal cortex of the marmoset (*Hapale jacchus*). 25:1, 10μm. Multilaminar arrangement of the giant cells. To the right of ↓1 transition to the post-central area: abrupt appearance of the inner granular layer.

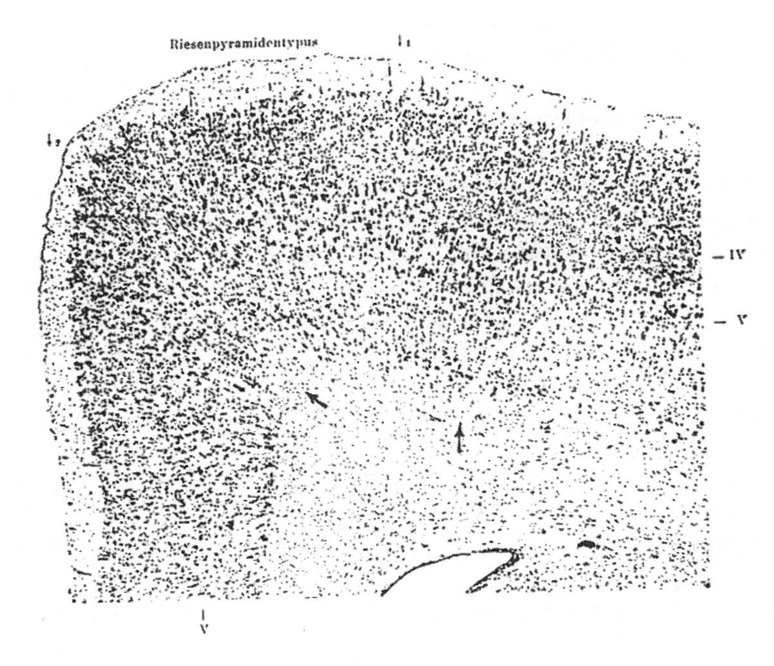

Fig. 63. Giant pyramidal cortex of the rabbit (*Lepus cuniculus*). 25:1, 10μm. Multilaminar arrangement of the giant cells. On the right (↓1) transition to the post-central area, and on the left (↓2) to the anterior limbic area (cf. Fig. 48).

b) The giant cells may form a continuous densely-packed layer: *laminar organisation* (lower monkeys, to some extent lemurs, carnivores including cats, dogs and bears, and, as far as research has gone, ungulates and pinnipeds).

In this variant, cell arrangement can still undergo further modifications. In many animals the giant cells almost all lie at the same depth in the cortical thickness and thus form a single dense continuous row of cells within the otherwise relatively cell-sparse layer V - *the unilaminar type* - (kinkajou. Figures 46 and 61). In other species the giant cells are also arranged in a continuous layer, but they lie at different relative levels and so form a sort of multiple layer - the *multilaminar type* - (guenon. Figures 44 and 59; lemur. Figures 45 and 60). The composition of the layer can vary in its cell size in the former, unilaminar, as in the latter, multilaminar distribution. There are animals in which the giant cells are all of similar cross-sectional area, such as in the guenon, the marmoset and other lower monkeys, while in other animals the giant cell layer is composed of very differently sized elements, as Figure 46 illustrates particularly strikingly for the kinkajou.

c) The third major type of giant pyramid organisation is the *solitary organisation*, in which the cells are isolated and distributed sparsely over the whole ganglion cell layer. It is found in man in the inferior sections of area 4 (Figure 32), in many monkeys, and sporadically in carnivores and marsupials.

The different distributions of Betz cells within layer V implies that in many animals the giant pyramidal cortex, and in particular the giant cell layer itself, is very conspicuous, while in other animals it is more obscure often making its identification quite difficult. Naturally, there is a wide variety of transitional forms involving the various distributions of Betz cells in the ganglion cell layer described above.

As far as the size of these cells in different animals is concerned reference should be made to what was said above (this Chapter, page 63ff).

Layer VI - *the spindle cell layer*. It is true that many lower mammals, such as small rodents and insectivores, possess a strikingly thick spindle cell layer, as can be seen, for instance, in Figure 18 of rabbit. But there are cortical types in higher mammals, even in primates including man, that also have a very thick innermost layer relative to the total cortical depth. Apart from the insular cortex, one might mention various temporal areas. On the other hand there are several areas in lower species that contain a very thin layer VI (compare Figures 19 and 22 of the kangaroo).

Thus it cannot be accepted as generally true that the thickness of the innermost cortical layer increases in lower animals, as has mostly been proposed.

Indeed, it is not appropriate to consider an individual layer in such comparative assessments, however, the whole cortical depth is also not an adequate measure. As I have already stated elsewhere in relation to cortical measurement, it is more accurate to divide the cortical cross-section into two main zones,somewhat similar to Kaes' principal layers, an "outer main zone" encompassing layers I to III and an "inner main zone" consisting of layers IV to VI. If one compares a large series of animals of different orders and levels of organisation from this standpoint, it appears to me possible to arrive at a tangible, generally applicable conclusion. Thus, one sees that the inner main zone of many - but not all - lower animals possesses a relatively greater average thickness than that of higher animals, and it can be further determined that the inner main zone in lower species is more frequently of greater thickness than the outer main zone than is the case in higher species, especially in man. However, for the moment nothing further can be said.

Whether any conclusions can be drawn concerning the functional significance of these two cortical main zones must await further investigation. What has been written so far about this question, however, does not agree with the comparative anatomical facts. In particular, the view that in general a greater thickness of the inner layers represents a higher organisation of the corresponding cortex must be considered erroneous.

3. Specific differentiation of individual homologous cortices in different animals.

Just as the whole cortical depth or individual basic layers can display special features in a given animal group, one also frequently sees that a cortical structure as a whole undergoes particular architectonic modifications throughout the mammalian class or in a given species or family. This does not usually go so far as to make it possible to recognise the species to which the brain belongs from the peculiarities of cell arrangement that characterise a particular single cortical area.

In this respect one can firstly distinguish two types of homologous cortices in mammals:

1. monomorphic types, that are those that keep the same basic characteristics throughout the whole class; the essential specialisation that determines their cytoarchitecture remains largely the same from the highest to the most primitive species;

2. polymorphic types, that are those in which the characteristic modifications of the basic lamination of the group themselves undergo such major transformations that, in spite of homologies between different orders, completely new patterns can arise.

Monomorphic types include most of what were described above as homotypical formations, that is cortical types with six persistent layers. In these cases the six-layered pattern remains essentially the same in all orders, with certain structural features that are characteristic for the particular cortex; an example that could be mentioned would be the cortex of the parietal lobe.

However, even heterotypical formations can develop monomorphically, notably many limbic and retrolimbic regions whose specific structure can be traced in its basic fundamentals through the whole mammalian class, even if here again certain inherent species and family differences are encountered.

In Figures 64 to 67 the granular retrosplenial cortices of the flying fox,

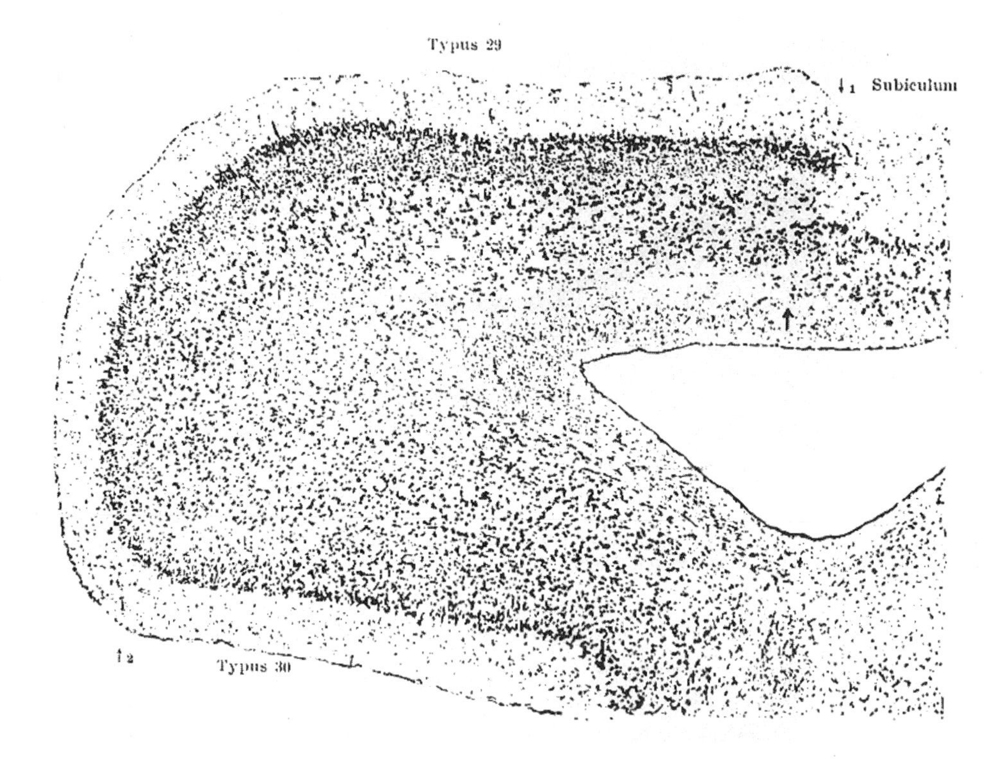

Fig. 64. Flying fox (*Pteropus edwardsi*). 25:1, 10μm. Retrosplenial region (area 29 of the brain map) with transition to the subiculum of Ammon's horn on one side ↓1 and to the agranular retrospenial cortex on the other ↓2 (area 30). (cf. also the higher power view in Fig. 39).

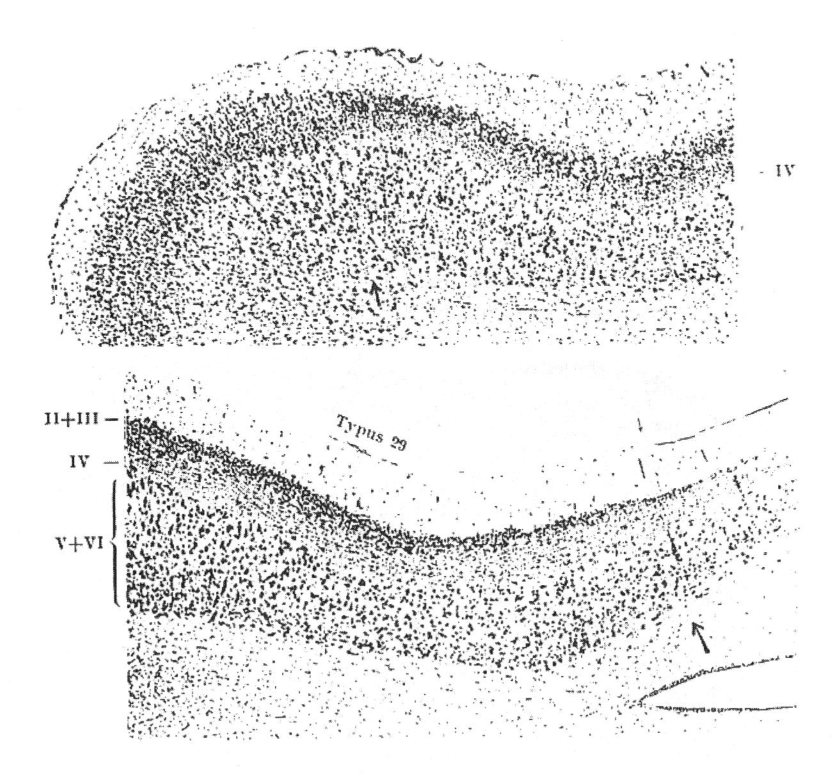

Fig. 65 and 66. Adult rabbit 25:1, 10μm. (Area 29b+c.) The same as in the previous figure. Transition on one side to area 29d and on the other to area 48. (cf. also Fig. 38.)

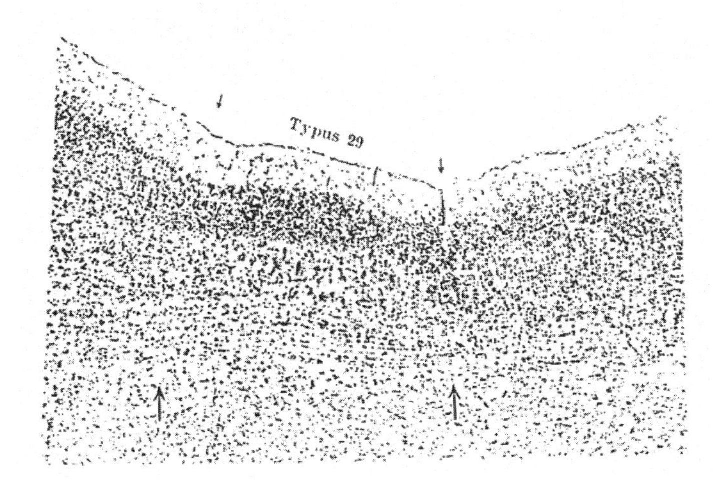

Fig. 67. Wallaby (*Macropus dorsalis*). 25:1, 10μm. The same as in Figs. 64-66. Layer III is wider and layer IV more weakly developed than in the previous figures. (cf. Fig. 40.)

rabbit and kangaroo are illustrated. In all three animals the major distinctive features of this cortical formation are the same, namely on the one hand an extreme regressive transformation (involution) of layers II and III and a corresponding progressive development of layer IV, the inner granular layer, with a concurrent thickening of layers I, V and VI. Thus the inner main zone (layers IV to VI) comes to strongly dominate the outer main zone (layers I to III). These relationships are more clearly visible at the higher magnification of Figures 38 to 41. The same principle of laminar transformation recurrs in all animals in which I have studied this cortex and, although only weakly expressed in some, it is still sufficiently characteristic to make identification possible.

An especially typical example of a monomorphic heterotypical cortex is represented by the giant pyramidal type (area 4 of our brain map). The two main recognition features that distinguish it from the basic tectogenetic type, the disappearance of the inner granular layer and the formation of a particular cell type, the Betz giant cells, in the original ganglion cell layer, are found everywhere in essentially the same form. This can be judged by comparing Figures 43 to 49 and 58 to 63, The same is true of the agranular frontal cortex, area 6 of the brain map; here again one feature has persisted throughout, namely the regression of the inner granular layer. In spite of the monomorphic development of a cortical type, that is despite its architectonic differentiation in all animals being essentially the same and remaining clearly visible throughout the mammalian class, its detailed organisation, one could even say its decoration, can develop so many individual features in a given species that it can easily be distinguished from the homologous cortex of another animal group. The giant pyramidal cortex, as mentioned above in the description of the ganglion cell layer of different animals, is again an instructive example in this respect.

In carnivores, and specifically in certain cats and bears, the giant pyramidal cortex undergoes a quite characteristic development Figure 61 illustrates a section of kinkajou at low magnification and Figure 46 at higher magnification. The basic architectonic pattern is indisputably the same as in the homologous cortex of primates and prosimians (Figures 43 to 45), but nevertheless it is immediately distinguishable from them thanks to particular features that were already discussed above for the individual layers. Layer V stands out more clearly than in any other animal as a compact cellular stripe; its giant cells are very large and numerous relative to the thickness of the cortex, as we have already seen, and indeed they are bigger and more numerous than in any other species studied so far. In addition layer III is unusually narrow, containing plump, poorly differentiated pyramidal cells, and layer VI is compact with sharp borders. Differences in the particular organisation of giant cells in man, monkeys and prosimians are also extensively discussed above (pages 64ff and 82).

The distinct features of human giant pyramidal cortex, apart of course from its great thickness, consist of the massive development of the supraganglionic layers corresponding to layers II and III, the latter of which is distinctive in its well differentiated, narrow pyramids (Figure 43, IIIb), and also especially of the polymorphism and density of medium and small cells in all layers. From these alone one can distinguish this cortical formation in the human from that of all closely-related groups.

In other orders such as chiropterans, rodents, insectivores and marsupials this cortical type is distinguished by its giant pyramids not attaining any particularly striking size, in contrast to that of the species mentioned above. Nevertheless this cortex can be easily identified by the lack of an inner granular layer (IV) and the dense alignment of relatively large cells (homologues of the giant cells) in the ganglion cell layer (Figures 47 to 49).

Of the **polymorphic types**, the calcarine cortex provides one of the most striking examples being, like the giant pyramidal cortex, a heterotypical modification of the basic type with an altered number of layers. We are therefore here dealing with a polymorphic heterotypical cortex, whereas the former is to be considered an example of a monomorphic heterotypical structure. The principle by which this type of cortex has differentiated cytoarchitecturally is by a sort of hypertrophy of the inner granular layer (in myeloarchitectonic terms the formation of the stria of Gennari or Vicq d'Azyr). This principle is indeed discernable everywhere in all mammalian brains, but it is modified in individual groups in such a way that a quite different cytoarchitectonic type appears to emerge. Proof of homology is then only possible by comparative topological localisation or by developmental study.

With respect to the organisation of the inner granular layer, one can distinguish three major structural variants of the calcarine cortex (compare Figures 68 to 77):

a) A complete splitting of the inner granular layer into two separate granular zones with the formation of an intermediate lamina, as described several times above: the *tristriate calcarine pattern*. This form is very obvious in man, all monkeys (including the marmoset) and in lemurs, and less obvious in many carnivores, ungulates and, among marsupials, in kangaroos. Figures 68 to 74 provide illustrations for these orders.

The tristriate form can itself differentiate in various specific directions in different families of the same order. This polymorphism is particularly marked in monkeys. As examples, compare this cortex in a guenon, a marmoset and a capuchin monkey (Figures 69, 70 and 71) (*92). The last example possesses a far more elaborately laminated calcarine cortex than the first two, for within the intermediate granular layer (IVb) further splitting into several sublayers has occured. As Figure 70 (*93) shows, a distinct dense cellular stripe traverses the middle of the intermediate layer, thus splitting this sublayer into three. In Figures 78 and 79 this type of cortex is again illustrated for the capuchin monkey and guenon at higher magnification.

b) The inner granular layer may only divide into two sublayers, a cell-poor outer layer and a cell-rich inner layer, such that the stria of Gennari (IVb) lies within the former: the *bistriate form* of calcarine cortex (eg. cat and rabbit, Figures 75 and 76).

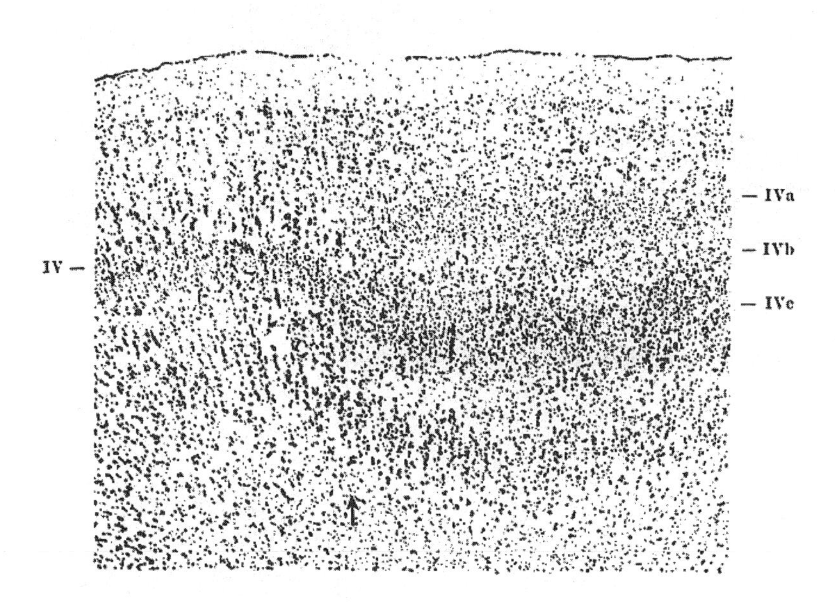

Fig. 68. Calcarine cortex: man. 25:1, 10μm.

Fig. 69. Calcarine cortex: rhesus monkey (*Macacus rhesus*). 25:1, 10μm (*90).

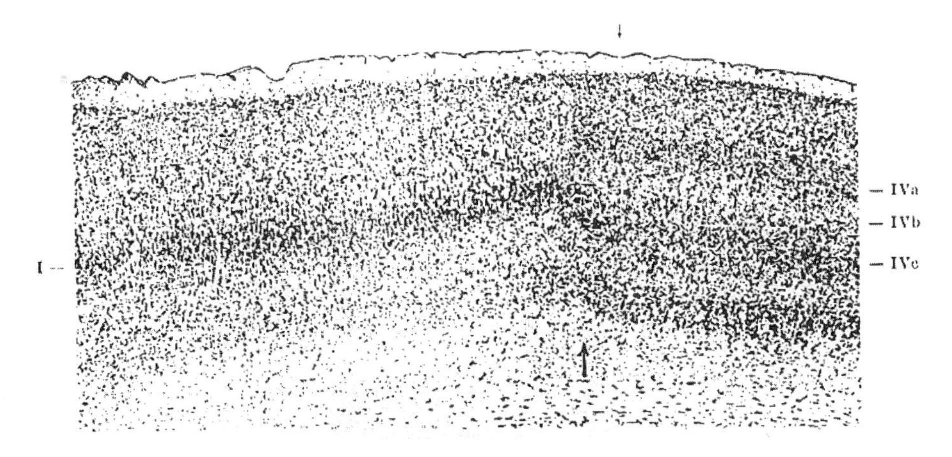

Fig. 70. Calcarine cortex: marmoset (*Hapale jacchus*). 25:1, 10μm.

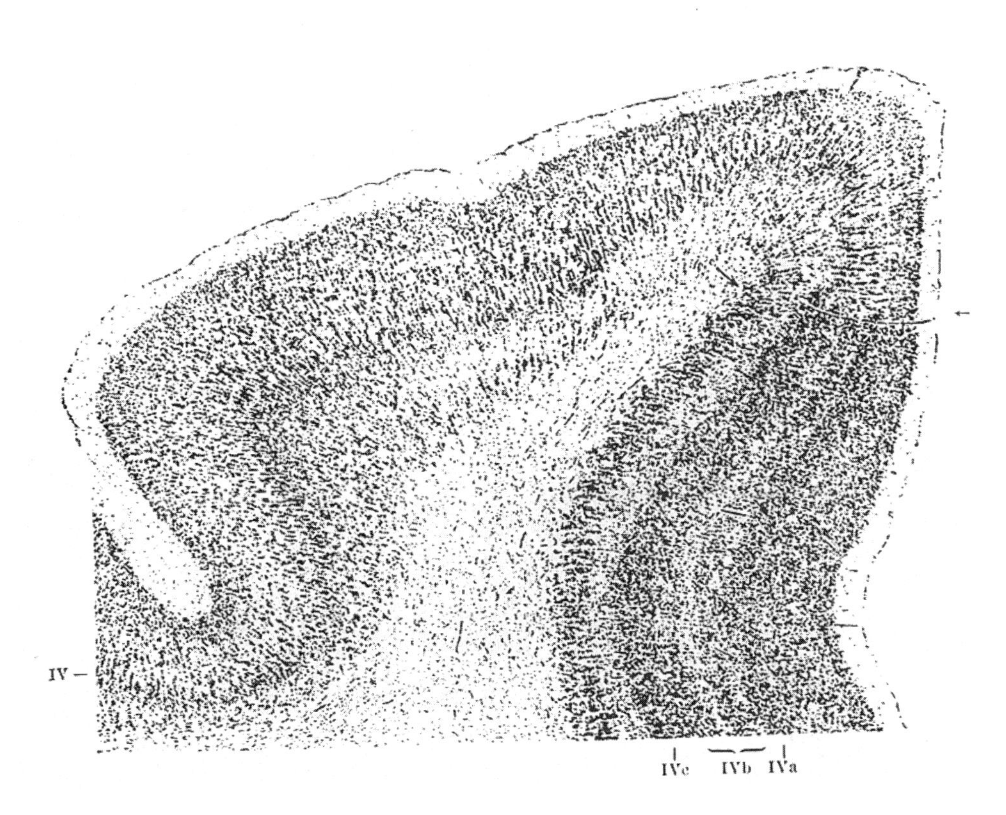

Fig. 71. Calcarine cortex: capucine monkey (*Cebus capucinus*). 25:1, 10μm. The intermediate granular layer (IVb), in contrast to other monkeys (Figs. 69 and 70), shows a further division into three sublaminae. Multistriate calcarine cortex.

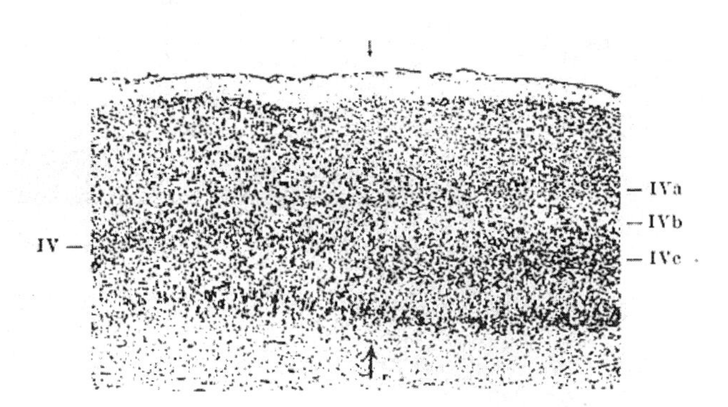

Fig. 72. Calcarine cortex: prosimian (*Lemur macaco*). 25:1, 10μm.

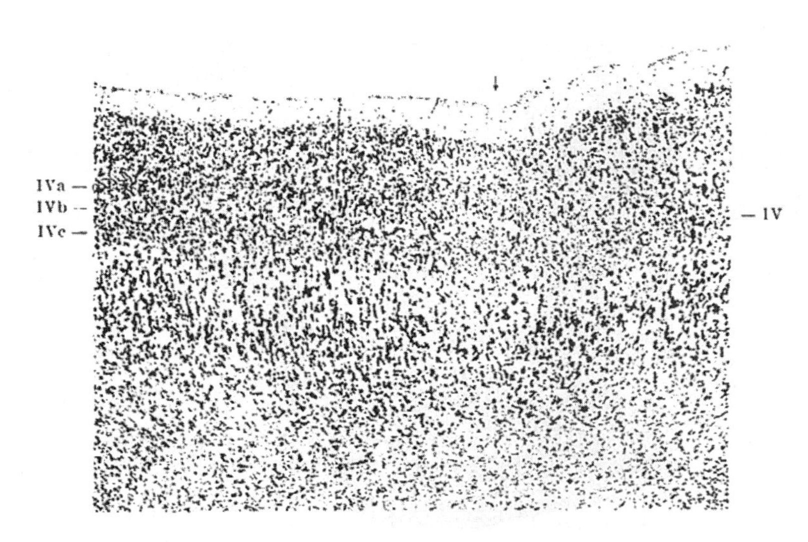

Fig. 73. Calcarine cortex: kinkajou (*Cercoleptes caudivolvulus*). 25:1, 10μm. The tristriate character of the calcarine cortex is less clear than in primates.

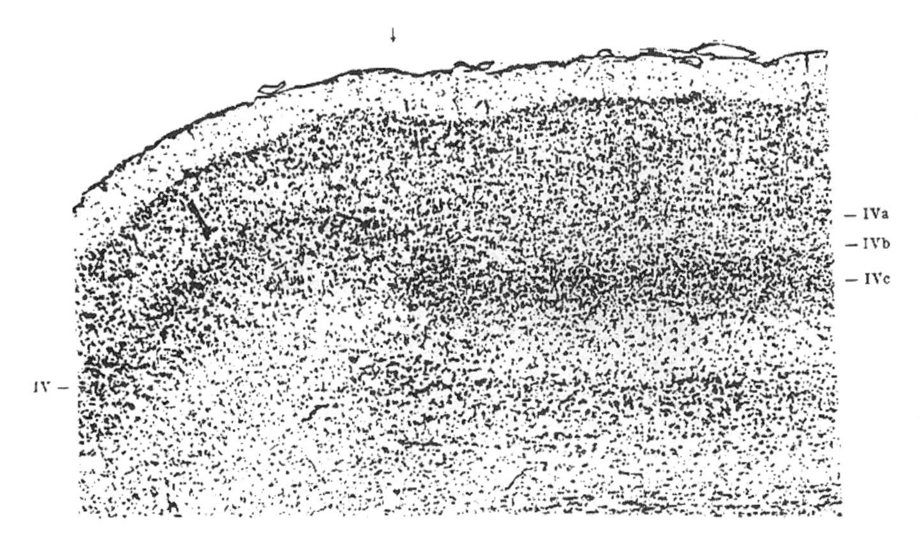

Fig. 74. Calcarine cortex: wallaby (*Macropus dorsalis*). 25:1, 10μm. (Compare Fig. 11 of the immature kangaroo, where the tristriate character of the calcarine cortex, ie. the splitting of the inner granular layer, is expressed more distinctly.)
Figs. 68-74 show the site of transition to the calcarine cortex in primates, prosimians, carnivores and marsupials, with a true splitting of the inner granular layer and the formation of three sublaminae IVa, IVb and IVc (tristriate form).

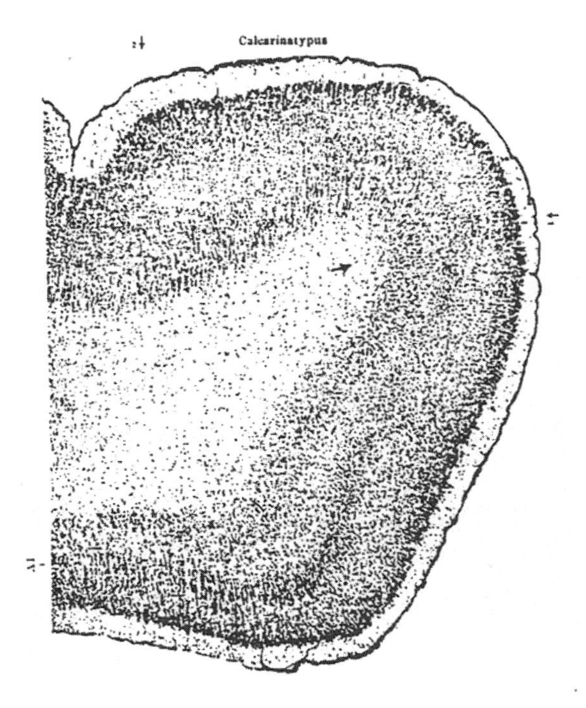

Fig. 75. Cat (*Felis domestica*). 25:1 10μm. Bistriate form of the calcarine cortex between ↓1 and ↓2. To the right of ↓1, transition to the six-layered basic cortex.

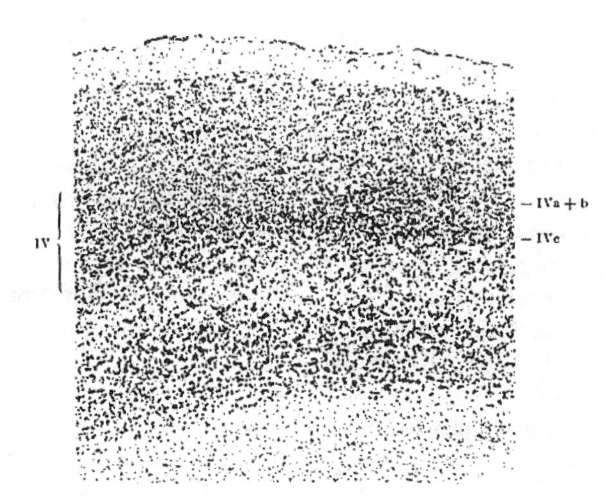

Fig. 76. Rabbit (*Lepus cuniculus*). 25:1, 10μm. Bistriate form of the calcarine cortex.

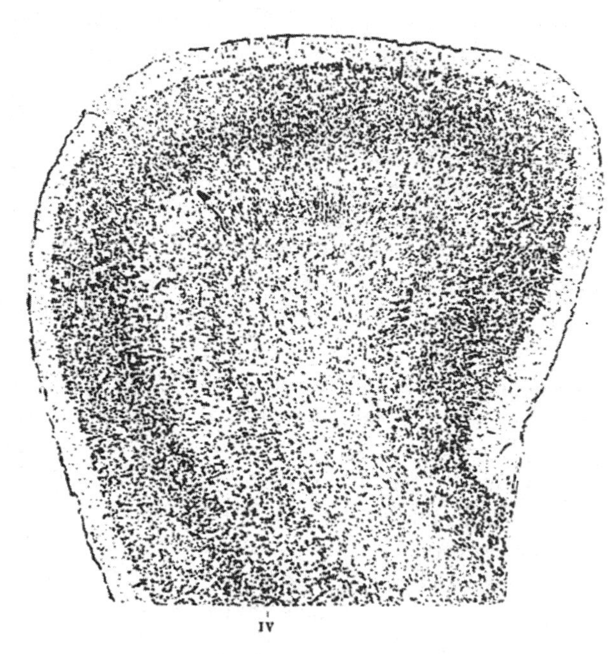

Fig. 77. Flying fox (*Pteropus edwardsi*) (*91). 25:1, 10μm. Unistriate form of the calcarine cortex.

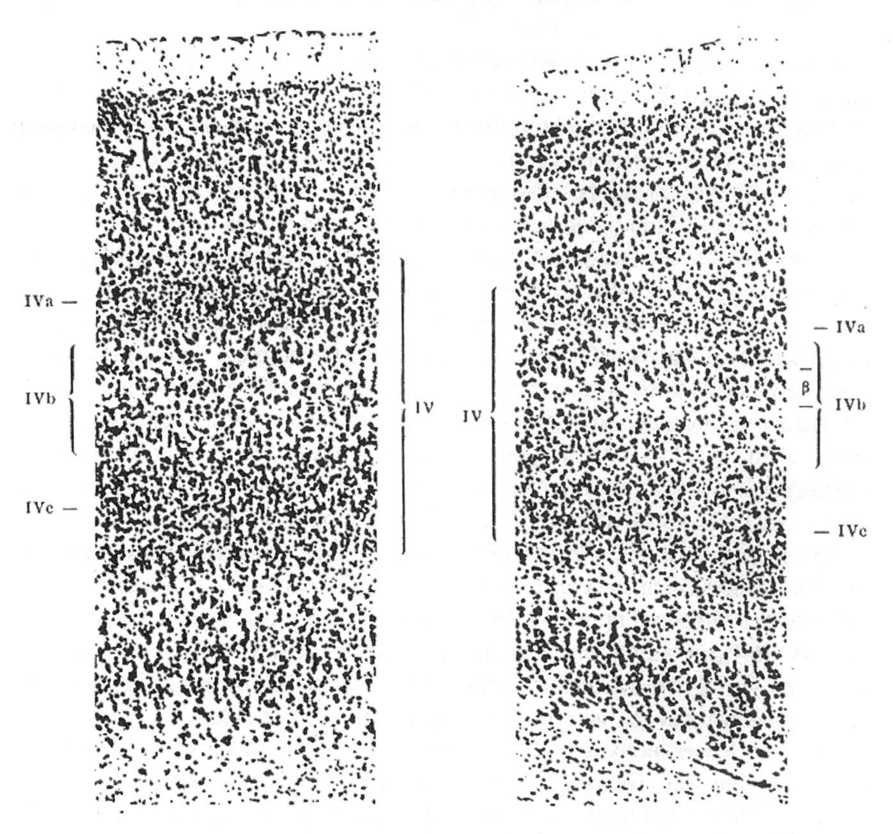

Fig. 78. Guenon (*Cercopithecus*) (*94).

Fig. 79. Capucine monkey (*Cebus*).

Fig. 78 and 79. Calcarine cortex of two different monkeys. 66:1, 10μm. Polymorphism of a heterotypical homologous cortex. The polymorphism is demonstrated in that the inner granular layer (IV) in the capucine monkey is architectonically more richly organised and divided into more sublaminae. In Fig. 79, a compact cellular band of densely-packed large elements (IVβ) stands out within layer IVb such that this sublamina itself contains three separate cell laminae, that are absent in Fig. 78. In the same way, layer IVc clearly divides in Fig. 79 into a cell-poor outer and a cell-rich inner half. Thus the calcarine cortex in the capucine monkey has 11 layers, while in the guenon only 8 layers can be recognised. (cf. also Figs. 69-71.)

c) The inner granular layer may simply represent a thicker and denser variant of the granular layer of the neighbouring cortex: the *unistriate form* of calcarine cortex (eg. small rodents, flying fox, Figure 77).

Other examples of a polymorphic heterotypical cortex are provided by certain areas of the retrolimbic region, especially in rodents, chiropterans and insectivores.

One should also just mention that even heterogenetic cortices of various animals can specialise in the direction of monomorphic as well as polymorphic differentiation. Figures 24, 25 and 26 can be cited as examples. In the hedgehog, the olfactory region in particular has differentiated polymorphically to a high degree.

From these few examples, the great variability of homologous mammalian cortices should be clear enough. To summarise the essentials once again:

In spite of consistency in the basic pattern of cortical structure, the specialised architectonics of individual types, and especially heterotypical formations, manifest particular features in most mammals that are characteristic for each species.

The carnivores, and in particular certain cats and bears, show such specific development of the giant pyramidal cortex that it proves easy for the experienced observer to distinguish it from that of other species. The same is true of the giant pyramidal cortex of man and especially the large prosimians (lemurs and indris).

On the other hand, the monkeys possess a quite systematic and standardised structure of their calcarine cortex that is unique to them and that is a sure recognition feature of the pithecoid brain. Among them there are families, especially the capuchin monkeys (*95), in which this cortical type has further differentiated in a quite unique way. Even in carnivores and ungulates one can often find original variants of the calcarine pattern that characterise them compared with other orders.

In many lower monkeys and prosimians, the preparietal cortex develops differentially in a similar way. The rabbit, and less so smaller rodents, but also macrochiropterans, carnivores represented by the weasel and ungulates by the pig, goat and chevrotain (as far as I have had the opportunity of studying them), all possess such a characteristic differentiation of the retrolimbic area that this alone suffices to identify their brain as belonging to one of these groups. And finally the cortical cytoarchitecture of the marsupial kangaroos and phalangers displays, both as a whole and in individual architectonic divisions, equally specific peculiarities that allow easy differentiation from other species and even from more simply organised marsupials (opossum). Even the monotremes (echidna) are on the whole distinguished from other orders by particular features of their cortical lamination.

Part II.

The principles of comparative field organisation in the cerebral cortex.

As a result of the observations described in Part I we are now in a position to divide the cerebral cortex into structurally circumscribed regions, in other words to construct a histological topographic map of the surface of the hemispheres.

We demonstrated above that modifications of cortical architecture often involve the whole depth of the cortex at a given locus on the surface and that such modifications frequently take place abruptly and quickly, resulting in relatively sharp borders between neighbouring structural regions. As an example, Figure 80 represents a coronal section of the hemisphere of a cat through the middle of the coronal sulcus and reveals several cortical structural types, and their borders, lying in a single plane. Figure 81 shows similar, even sharper, transitions at the borders of the calcarine cortex in a coronal section of the occipital lobe of a monkey. Finally, in Figure 82 the transitions from the hippocampus to the subiculum, presubiculum, entorhinal area and perirhinal area in that order are shown for a marsupial, the wallaby.

We likewise learned that sudden transitions between different cortical types exist in all mammalian orders, often just as dramatically in lower species (such as the kangaroo and rabbit) as in man, and that certain forms of regional modification of lamination that are found in man are also identifiable in essence throughout the whole mammalian class, although more or less modified in individual species (Figures 20 to 22, 24 to 26, 58 to 63, 64 to 67, 68 to 77).

These facts form the point of departure for the establishment and spatial demarcation of homologous cortical fields, that is those with similar histological cellular structure - the cytoarchitectonic areas - in different mammalian orders and so provide the foundations of a **comparative surface parcellation of the cerebral cortex**.

It will thus be our task in Part II to determine the specific forms of field structure in individual animals and so establish common features and variations in the arrangement of fields in different animals. Only in this way will common aspects of comparative localisation emerge. The problem of comparative cortical topography can thus be reduced to the following questions:

1. Is there a homology or merely a certain similarity in the topographic field organisation in different mammalian orders; in other words is the cortex made up of homologous structural regions based on common rules throughout the whole mammalian class or must one assume a special topographical structural principle for each species or at least each major animal group?

2. What are in general the common features and variations in the cortical parcellation of different species?

3. Are there constant and inconstant fields and how do the former vary in form, size and place in individual families or species?

Before we enter into a major discussion of the questions thus posed we should make a few remarks about the material and the brain maps.

As to the material, it is evident that a study that is only intended as an initial general basis for a discipline cannot extend to the whole zoological system nor even to all orders of a class, let alone to the majority of families or species.

To obtain an exhaustive topographic division of the cortex in a single species necessitates the production of several uninterrupted series of sections through the whole cerebral hemisphere and their comparative study, work that, apart from its interpretive difficulties, is technically extraordinarily time-consuming and tedious. Such a study must be limited initially to a few major animal groups, or a small number of representatives of each, for the purpose of orientation. As a preliminary I have determined the field distribution in the cortex of representatives of seven orders, namely

1. primates by man, several cercopithicids and marmosets,
2. prosimians (*96) by the lemur and mouse lemur,
3. chiropterans by the flying fox (*97)
4. carnivores by the kinkajou (and partially the dog),
5. rodents by the rabbit (*98), the ground squirrel and partially the mouse,
6. insectivores by the hedgehog,
7. marsupials by the kangaroo and possum.

I can present finished brain maps of man, guenon, marmoset, flying fox, kinkajou, rabbit, ground squirrel and hedgehog. In addition I have taken limited samples from many other animals and animal groups (orang-utan, capuchin monkey, indris, mouse lemur, pipistrelle, cat, weasel, opossum, echidna and others) but have not yet completed a definitive cortical localisational study of the whole hemisphere.

This material is certainly not exhaustive, but it permits the establishment of the principal common features in the overall organisation of the cortex of

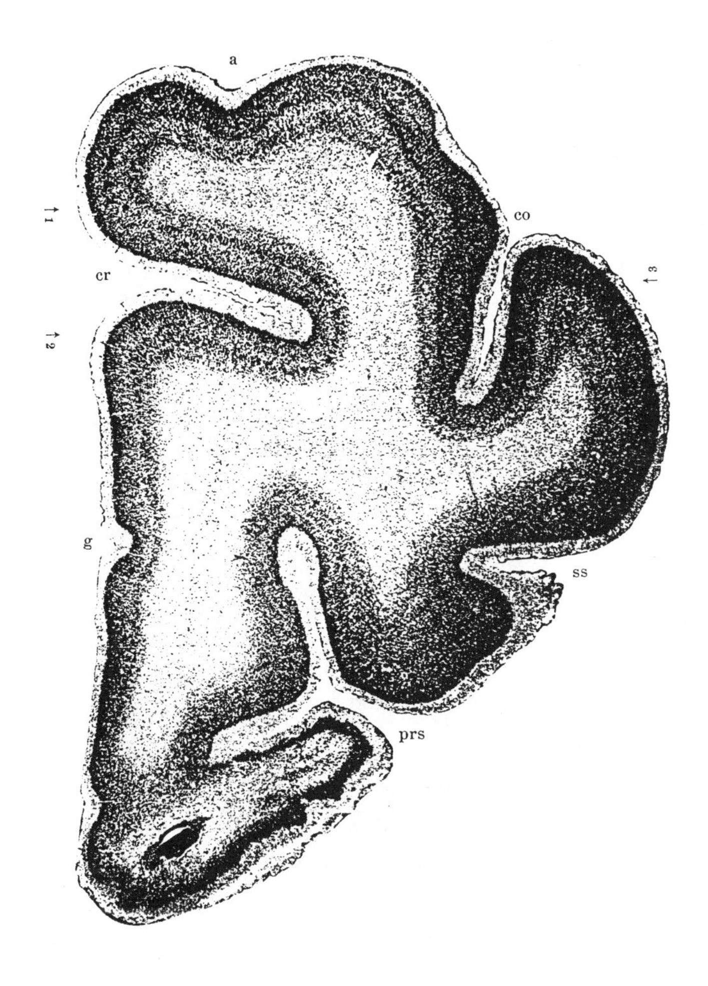

Fig. 80. Coronal section through the hemisphere of a 14-day-old cat (*Felis domestica*). Transition of the basic six-layered cortex to the agranular giant pyramidal cortex (at ↑1) on one side and to the granular frontal cortex (↑3) on the other. On the medial side is the precingulate cortex (↑2). a = ansate sulcus, cr = cruciate sulcus, co = coronal sulcus, ss = suprasylvian sulcus, prs = presylvian sulcus, g = genual sulcus.

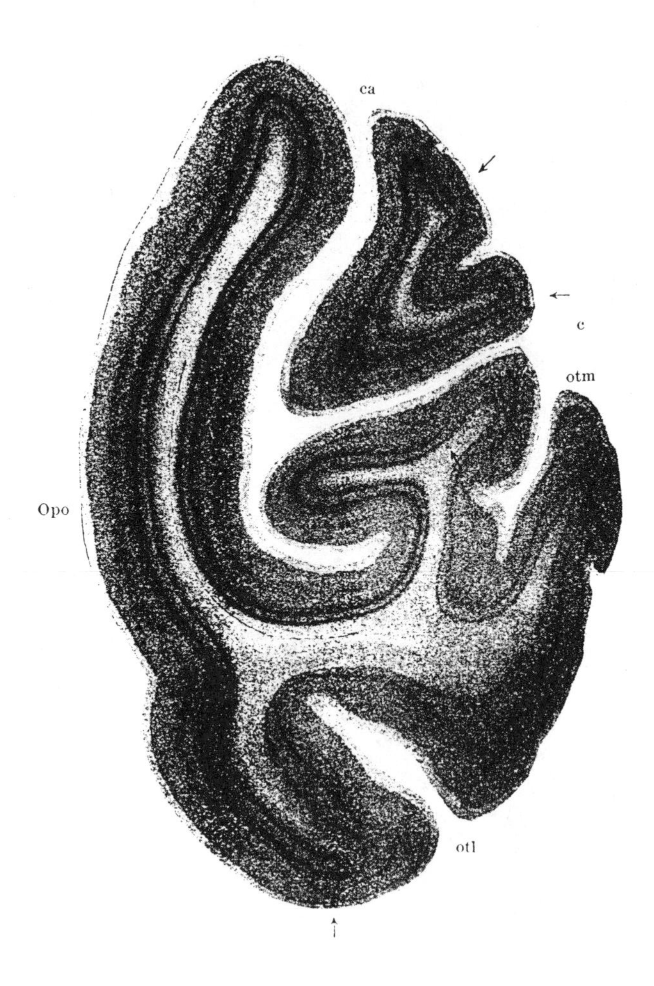

Fig. 81. Coronal section through the occipital region of a monkey (*Macacus rhesus*). There are four transitions from calcarine cortex to adjacent cortex. A diminution in the number of layers is clearly visible at the arrows. c = main stem of the calcarine sulcus. ca = ascending ramus of this sulcus. otm = medial occipitotemporal sulcus. otl = lateral occipitotemporal sulcus. Opo = occipital operculum.

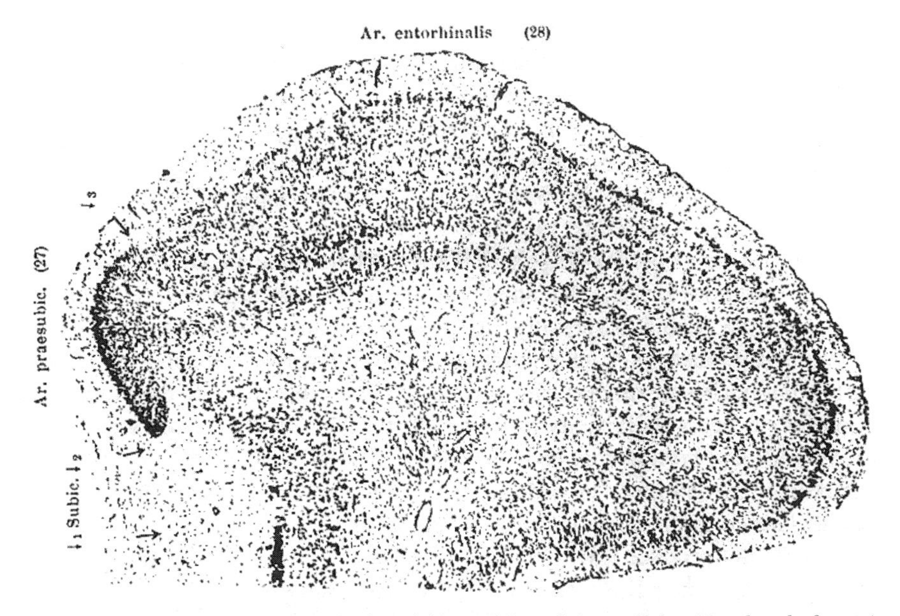

Fig. 82. Coronal section through the piriform lobe of the wallaby (*Onychogale frenata*) with several abrupt architectonic transitions.

various mammals as well as the detection of certain special particularities in the cortical organisation of individual orders or families. This suffices for the object of the present investigation that simply aims to elucidate the basic principles of the comparative subdivision of the cortex. The description of special peculiarities for each individual species must remain the task of later detailed study.

With regard to the brain map, one should note that the individual structural fields are marked with various diacritical symbols on the lateral and medial surfaces of the hemispheres of each brain studied. Homologous cortical types or areas have, in general, received identical symbols in the different animals, but the same symbols do not always indicate absolute homologies, for often one can only speak of partial or relative homologies. When there is superimposition or fusion of several adjacent fields in a given animal, which are separate in others, this is indicated in the brain maps by mixing and superimposition of the corresponding symbols. The reproach has been raised from various quarters, although it seems to me not always with sufficient knowledge of the facts and also often without adequate grasp of the pertinent work, that such brain maps give a false picture of the true relationships and inspire erroneous localisational concepts by drawing sharp borders where there are none and by spatially segregating structural regions that should not be absolutely separated. One may reply that a brain map, being essentially like any schematic representation, necessarily implies certain distortions. I have pointed this out whenever the opportunity arose in my earlier publications, emphasising particularly the difficulties of the graphic representation of surface subdivisions in a convoluted hemisphere. A surface view must represent fields and borders that lie in the depths of sulci or on buried gyri; furthermore it must project curved areas on a flat surface which will result both in distortions of perspective and in spatial displacements. In addition, transitions between neighbouring fields often do not produce truly sharp borders, but are gradual, whereas the map must always draw a boundary if it is to represent topical relationships. Thus certain inaccuracies must be admitted, as in any diagram. In spite of this, the brain maps present the true position and mutual relations of fields and anyone can use them beneficially in comparative studies as long as they do not seek more from them than is intended, that is to serve as aids to orientation.

We shall proceed with the following descriptions by first considering the features of cortical field subdivision in a series of individual animals by means of the brain maps, and then debate more general questions in the following chapters. We shall begin with man and continue progressively down the mammalian class.

Chapter IV.

Description of individual brain maps.

Unlike the preceding descriptions of details of individual cortical fields according to lobes or gyri, the next section will deal with the division of the surface of the hemispheres of different animals into major, structurally homogeneous zones that only partially coincide with the morphological formations of earlier nomenclatures, such as lobes, lobules and gyri, each of which may encompass several architectonic areas. The basis for this is essentially-comparative anatomy and depends on the following considerations. One can indeed roughly subdivide the hemispheres of man and related gyrencephalic animals into morphologically homologous lobes. Nevertheless what corresponds in lower orders, such as small rodents and insectivores, to the frontal or temporal lobes is unfortunately impossible to determine by external inspection. However, it is quite possible to identify histological structures, and a number of such structural areas can be demonstrated in all mammals.

We therefore group together large zones of similar composition as individual structural entities, the so-called "regions" as opposed to the individual fields or "areas". So from now on we shall no longer merely differentiate the fields of the frontal, temporal, occipital lobes etc. but shall take regions as our point of departure, within which the individual areas are delimited according to their histological homogeneity.

A considerable number of such homologous regions can be distinguished in man and the other mammals (*99). They are the:

1. postcentral region
2. precentral region
3. frontal region
4. insular region
5. parietal region
6. temporal region
7. occipital region
8. cingulate region
9. retrosplenial region
10. hippocampal region
11. olfactory region

Many of these regions are massively developed in the higher orders and have a rich variety of individual areas, while demonstrating a simple organisation in lower orders. Other regions show an opposite trend, with more differentiation in lower, more primitive species than in more highly organised animals. Certain zones, like the olfactory region, are extremely reduced in certain animal groups and only developed in a rudimentary way so that they cannot be represented on the brain map, while in other orders, like the macrosmatic animals, they occupy a considerable portion of the cortical surface.

In Figures 83 and 84 the regions of the human cerebral cortex (except the insular region that is shown in Figure 89) (*100) are represented schematically from the medial and lateral aspects of the hemispheres. As can be seen, they only partially coincide with the subdivisions habitually used so far; it should be especially noted that the morphologically homogeneous "Rolandic region" is structurally divided into two separate regions, precentral and postcentral, each of which in turn contains several areas. Also, to avoid erroneous interpretations it should again be stated that not all these regions are demarcated from each other by sharp borders but may undergo gradual transitions as, for example, in the temporal and parietal regions.

I. The human brain map (Figures 85 and 86).

I first gave a brief description of the human cortical pattern in 1907 and at the same time drew up the accompanying surface map of the subdivision of the whole cortex into areas. In general I have nothing to remove from it, nor anything essential to add. I could thus content myself with a reference to that description. Nevertheless, in view of later comparative studies I shall give a detailed description here of the whole extent of the cortex and a precise delineation of the more physiologically or clinically important fields in relation to their position and their topographic relations to sulci and gyri. There will also be a general discussion of the sulci.

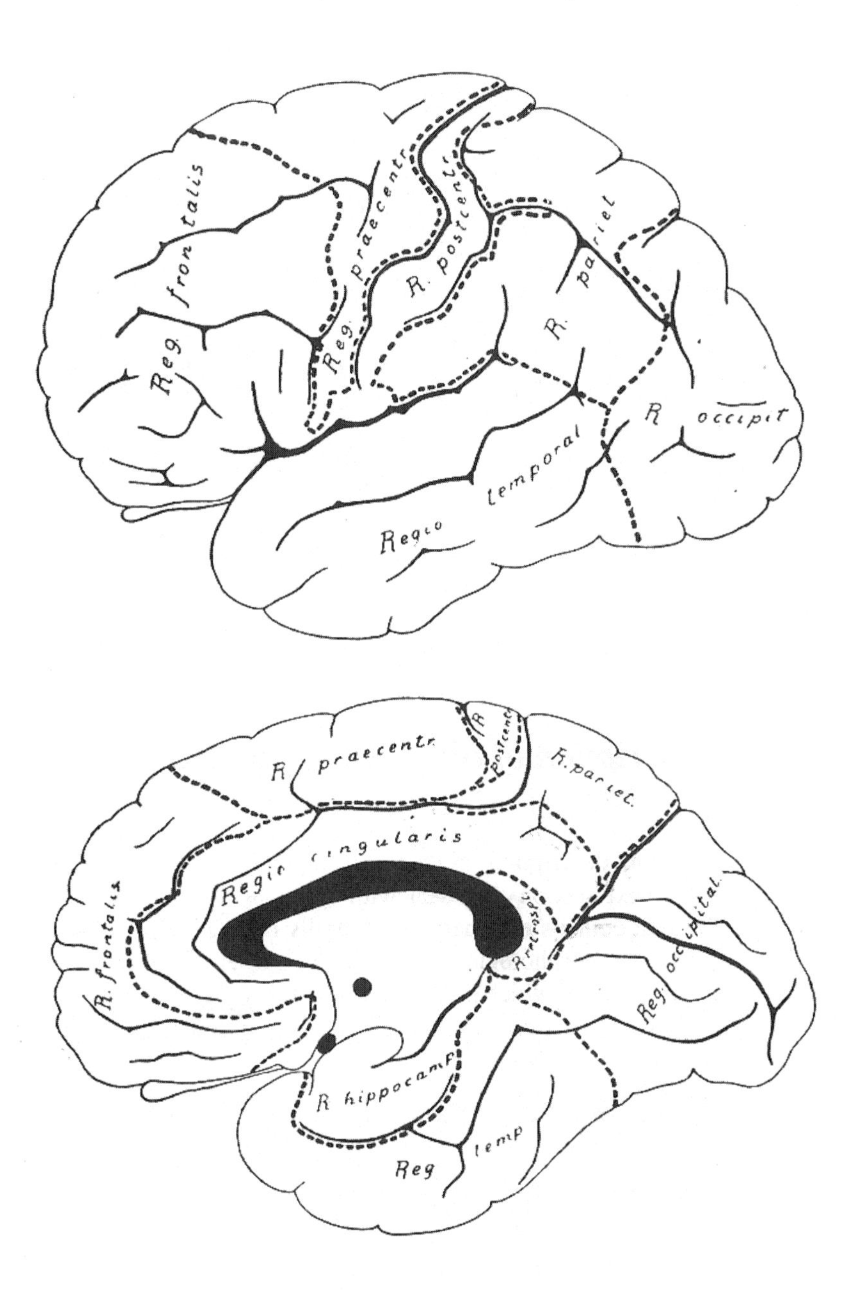

Fig. 83 and 84. The cytoarchitectonic regions of man. The olfactory region is not indicated.

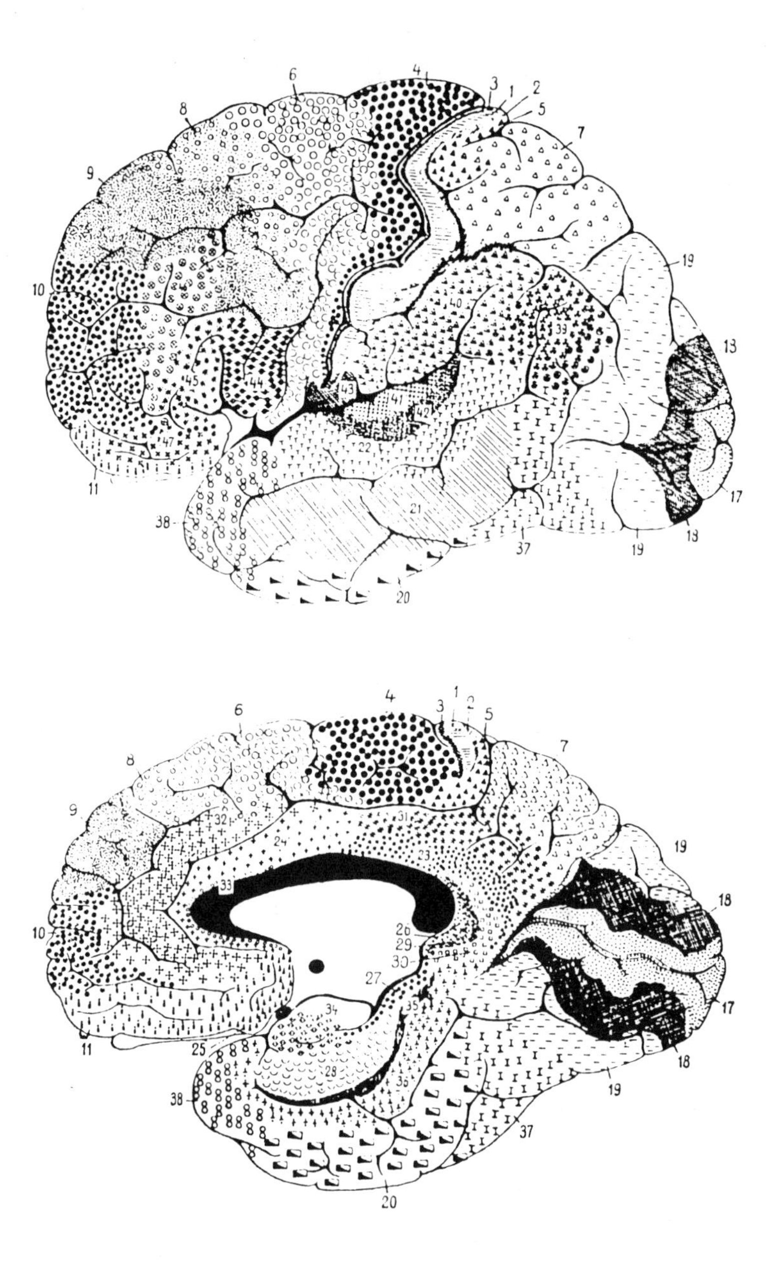

Fig. 85 and 86. The cortical areas of the lateral and medial surfaces of the human cerebral hemispheres. (Sixth communication, 1907.)

Postcentral region.

Concerning the cortical areas of the pericentral gyri, I wrote in 1902/03: "The Rolandic region of man is split by the central sulcus through its whole dorsoventral extent into two anatomical zones that are totally different in their cytoarchitectonic structure; the anterior zone is characterised by the appearance of giant pyramids and the lack of an inner granular layer, the posterior one by the presence of a distinct granular layer and a lack of giant pyramids. The border between the two zones is largely formed by the bottom of the sulcus apart from a short transition zone possessing a mixture of both structural types. At the dorsomedial end of the central sulcus this border continues on the paracentral lobule such that it forms a linear prolongation of the central sulcus as far as the junction with the callosomarginal sulcus, thus separating the paracentral lobule into two histologically different fields, an anterior one of which the structure corresponds in all respects to that of the giant pyramidal cortex, and a posterior one that represents the prolongation of the postcentral gyrus and has the same cytoarchitecture" (my first communication, Journal f. Psychol. und Neurol. Vol. 2, p.93/94).

This already contains the description of the splitting of the Rolandic region into two histological zones, that I now distinguish as the postcentral and precentral regions.

The postcentral region lies directly behind the central sulcus and comprises essentially the postcentral gyrus as well as its medial prolongation on the caudal third of the paracentral lobule (except the most posterior part of this gyrus that is occupied by area 5) and the greater part of the Rolandic operculum. Topographically it is further subdivided into four architectonically related, but substantially different, structural types: areas 1, 2, 3 and 43.

Area 1 - *intermediate postcentral area.* - This area lies in the middle of the granular postcentral region, that is between areas 2 and 3, separated from them by a quite distinct, but certainly not absolutely sharp, border and occupies a narrow band approximately along the whole length of the apex of the postcentral gyrus. At the upper margin of the hemisphere it follows the curve of the postcentral gyrus onto the medial surface as a quite narrow strip on the caudal (retrocentral) part of the paracentral lobule. Its main expanse on the lateral convexity varies in width, being extremely narrow and constricted at the upper margin of the hemisphere and extending more widely in the middle of the gyrus. In places the area also encroaches somewhat on the cortex of the depths of the central sulcus and the postcentral sulcus, compressing the adjacent areas 2 and 3 into the fundus of these sulci. At the lower end of the postcentral gyrus it narrows markedly, its structure changes somewhat and its borders with areas 2 and 3 become less distinct so that a sort of mixed cortical type appears. This transitional form is demarcated quite sharply from the subcentral cortex (area 43) on the Rolandic operculum.

Area 2 - *the caudal postcentral area* - forms, like area 1, a narrow stripe-like zone, that includes chiefly the posterior aspect of the postcentral gyrus and, therefore, the anterior bank of the postcentral sulcus. Its borders are not everywhere sharp and constant; sometimes it does not extend forwards entirely to the apex (*101) of the gyrus, while more often it crosses the fundus of the postcentral sulcus posteriorly and encroaches on the superior parietal lobule (*102). In the course of the intraparietal sulcus (*103), where it is continued as the postcentral sulcus, a narrow strip extends fairly far caudally. Elliot Smith equally described this stripe-like zone and tends to regard it as a special field, the "*sensory band β*" (*104) although he has to concede that it is not possible to separate it from the caudal postcentral area or from the inferior parietal area. There are undoubtedly considerable individual differences concerning this, as there are in the sulcal pattern, that require further special study. Cytoarchitectonically speaking, it is not very feasible in my opinion to separate this strip round the intraparietal sulcus from area 2 without arbitrariness.

Area 3 - *rostral postcentral area* (*105). - This area covers the anterior extent of the postcentral gyrus, thus forming the posterior bank of the central sulcus along its whole length. Its borders are sharper than those of area 2; it is especially sharply demarcated from the agranular area 4 (giant pyramidal area) anteriorly. The transition to the giant pyramidal area is not always precisely in the deepest part (fundus) of the central sulcus, but is sometimes markedly anterior to it and in other places posterior. Thus the area has a variable width at different locations along the central sulcus, a feature that is also determined by the varying depth of the sulcus. At the upper (medial) and lower (lateral) ends, area 3 encroaches on the precentral gyrus around the central sulcus thus, as it were, pushing area 4 anteriorly. At these places, as well as in the retrocentral part of the paracentral lobule and in the posterior section of the Rolandic operculum, there is a noticeable obscuring of the borders such that the adjacent fields seem partially fused, forming composite areas and making parcellation very difficult in some brains.

Instead of my areas 1 to 3, Campbell differentiates only two fields on the postcentral gyrus, a "*postcentral area*" and an "*intermediate postcentral area*", while Elliot Smith leaves the question open as to whether two or three different areas should be recognised.

Area 43 - *the subcentral area* - is formed by the union of the pre- and postcentral gyri at the inferior end of the central sulcus and thus lies on the Rolandic operculum. From its architecture, this area belongs to the postcentral cortex. Its anterior border is quite sharp and coincides approximately with the anterior subcentral sulcus; posteriorly it disappears gradually around the posterior subcentral sulcus in the retrocentral transition zone and in the anterior portion of the supramarginal area (40). It extends widely over the inner surface of the operculum, that is to say in the depths of the Sylvian fissure; in this region it has a distinct boundary with the insular cortex.

In his Plate 1 (*106) Campbell also pointed out a small region on the Rolandic operculum but did not describe it as a special area but rather as a mixed zone. Elliot Smith equally delimits a narrow strip (z) (*107) at the same place but he takes it to be a continuation of his "*area postcentralis A*" (my rostral postcentral area) that he extends round the inferior end of the central sulcus as far as the vertical ramus of the Sylvian fissure anteriorly, in spite of a change in structure, and erroneously in my opinion.

Precentral region.

This consists of the territory lying directly anterior to the central sulcus and is chiefly characterised by the lack of an inner granular layer. It extends rostrally beyond the Rolandic region as it is usually understood, in that the dorsal half of its anterior border crosses the precentral gyrus and encroaches significantly on the superior and middle frontal gyri. Its anterior borders are rather clear but vary between individuals, while the posterior boundary is everywhere sharply demarcated from the postcentral region (*108), and particularly from area 3, by the depths of the central sulcus, although, as mentioned above, the border does not always correspond to the deepest part of this sulcus.

Within the precentral region two distinctly different fields - areas 4 and 6 - can be identified, both characterised cytoarchitectonically by the lack of an inner granular layer, and area 4 further distinguished by the presence of Betz giant cells that do not appear in area 6. (See Figures 94 and 95, page 130)

Area 4 - *the giant pyramidal area* - is one of the most strikingly differentiated and cytoarchitectonically delimitable structural regions of the whole human cerebral cortex.

It consists of a wedge-shaped cortical segment along the course of the central sulcus, narrowing from superior to inferior and enclosed entirely within the precentral gyrus and the adjacent part of about the middle third of the paracentral lobule. On the medial aspect of the hemisphere it covers approximately the middle third of the paracentral lobule. Laterally it only includes the whole width of the precentral gyrus near the superior edge of the cortex - often encroaching somewhat on the base of the superior frontal gyrus - and then, more ventrally, becomes restricted to the posterior half of this gyrus, narrowing progressively (with individual variation) and withdrawing to the buried cortex of the posterior bank of the precentral gyrus, where its sharp border ceases quite a distance above the lower end of the central sulcus as it fuses with area 6.

I have already described its borders elsewhere, in complete agreement with Campbell. They are very variable, especially in the paracentral lobule; it is not uncommon for area 4 not to extend as far medially as the callosomarginal sulcus, but to only include the dorsal half of the paracentral lobule, sometimes precisely to the level of an unimportant secondary sulcus that has been appropriately named the medial subcentral sulcus [1]). Rostrally the borders give way gradually to area 6, caudally they lie in the central sulcus, sometimes somewhat anterior or posterior to its deepest point. Ventrolaterally area 4 does not quite

---

[1]) Contrary to the usual nomenclature, Elliot Smith calls this sulcus the sulcus paracentralis.

reach the lower end of the central sulcus; in many cases it stops 2 to 3cm higher in the depths of the sulcus.

I must definitely classify as erroneous the idea, proposed by Elliot Smith, that the anterior subcentral sulcus is "a limiting furrow" for area 4, especially as Elliot Smith himself admits that the precentral fields vary considerably in their boundaries and that macroscopic analysis in unstained preparations is often unsatisfactory [2]).

Campbell's suggestion that the myeloarchitectonic border of area 4 (his *precentral or motor area*) extends 1 to 2mm further anteriorly than the "cell area" takes on a special importance [3]).

The rostral border of the giant pyramidal area on the convexity of the gyrus is rather unclear and variable, for areas 4 and 6 undergo a gradual transition and because isolated Betz giant cells occur in "solitary" fashion more or less extensively rostrally, such that the identification of the line of transition is purely subjective and can only be determined from numerous individual brains. On the upper edge of the cortex the border usually lies just ahead of the superior part of the precentral sulcus but then soon runs backwards and downwards on the superior frontal gyrus to continue approximately down the crest of this gyrus. At the superior frontal sulcus the field bulges out, widening anteriorly once again, after which it becomes pushed towards the posterior edge of the gyrus, and in the lower half of the gyrus, or even somewhat higher, retreats onto its posterior bank so that from here down the the field is restricted to the cortex deep in the central sulcus, only being visible as a narrow strip.

Within this circumscribed zone, as has been known since the work of Lewis and Clarke, there emerge considerable local differences in the number, size and distribution of the giant pyramids, in addition to individual variations. Lewis and Clarke claimed to observe column-like accumulations of these cells corresponding to the physiological centres for the legs, the trunk, the arm and the face, but later their views did not enjoy universal confirmation and also require verification by physiological experiments. It is however clear that the size and number of the Betz giant cells decrease on average from superior to inferior, that is from the paracentral lobule laterally, and also that the dense cell clusters gradually disappear toward the ventral end of the central sulcus, making way for a more scattered arrangement of these cells. It can further be noted that the distribution of the giant pyramids in the upper third of the field and at the summit of the gyrus is essentially "cumulative", while it becomes almost entirely solitary or laminar more ventrally. Equally, the total cortical thickness decreases ventrally. However, the cytoarchitecture is not sufficiently

---

[2]) "The naked-eye appearances of the praecentral areas is subject to a wide range of variation" (Elliot Smith, 1907, p.246).

[3]) "A discrepancy which must be mentioned, however, is that the fibre area is one or two millimetres more extensive than the cell area; to understand this difference we have only to take note of the size and extensive ramifications of the enormous dendrons possessed by these cells, as well as the numerous collaterals given off by their axis cylinder processes, and also remember that the existence of cells of great size has a marked influence on the fibre wealth of the part and apparently makes its presence felt at a considerable distance" (Campbell, 1905, p.35).

characteristic to enable a subdivision of the giant pyramidal area into spatially circumscribed subfields, in spite of the above-mentioned regional differences.

Area 6 - the *agranular frontal area* - could be considered part of area 4 on account of its lack of a granular structure, and for convenience is included with the latter in a major regional division, separate from other frontal types that possess an inner granular layer. It is very similar to area 4 in shape and extent. Area 6 consists of an upper very broad zone, becoming increasingly narrow inferiorly and laterally, and covering the whole vertical extent of the frontal lobe from the callosomarginal sulcus to the upper bank of the Sylvian fissure. The following gyri contribute to it: medially, the anterior part of the paracentral lobule with the neighbouring parts of the superior frontal gyrus, and in many cases also almost the whole dorsal bank of the callosomarginal sulcus except its posterior third; laterally, the bases of the superior and middle frontal gyri, and further inferiorly the whole precentral gyrus except where it is occupied by area 4.

Campbell also includes in this area (his intermediate precentral area) the whole inferior frontal gyrus, but according to my studies this is undoubtedly to be separated as a special region, the opercular, triangular and orbital areas (44, 45 and 47), as it possesses a distinct inner granular layer, a feature that Campbell overlooked.

Elliot Smith divides my area 6 into a dorsal *area frontalis superior* and an *area frontalis intermedia*. I admit that area 6 gradually changes its structure in a dorsoventral direction (this also applies to the myeloarchitecture). However, the cell structure gives no conclusive indication for a division into two specific fields, and even Elliot Smith concedes that the difference is not always clearly manifested ("This contrast has not been sufficiently clearly" - p.249) (*109).

Frontal region.

The frontal region is by far the most extensive region of the human cerebral cortex in terms of area; it includes the whole of the frontal lobe anterior to the central sulcus, with the exception of the precentral region, and the precingulate region on the medial surface. This constitutes, as a surface estimate, around 20% of the total cortical area of a hemisphere. It should be treated as a single frontal structural region because, in contrast to the agranular precentral region, all its subdivisions again contain a compact inner granular layer. We shall study the major importance of this architectonic feature in more detail below when we make comparisons with other brain maps. Its limits are easy to enumerate: caudally it gives way to the agranular frontal area at well-marked boundaries, rostrally it extends round the frontal pole, and on the medial surface to near the callosomarginal or superior rostral sulci. However, as can be seen from the map, the borders do not correspond exactly to these sulci.

I distinguish eight individual fields in the frontal region of man, namely areas 8, 9, 10, 11, 44, 45, 46 and 47. Of these, areas 44, 45 and 47 on the inferior frontal gyrus form a particularly homogeneous subgroup on the grounds of major cytoarchitectonic similarities, that can be termed the *subfrontal region*. The exact parcellation of these areas (with the exception of the subfrontal zones) is often fraught with great difficulties, for the architectonic differences in cell preparations are sometimes not at all striking. Elliot Smith also draws attention to this with the words: "The accurate mapping out of this area (frontal) presents great difficulties, because the contrasts between adjoining areas are often exceedingly slight and at times quite impossible to detect." (*110) Elliot Smith divides the frontal lobe into eight fields, similar to my divisions, even though in many respects differences exist in relation to individual areas.

Campbell only differentiates two subfields within the whole region, a "frontal area" and a "prefrontal area". He includes the inferior frontal gyrus in his "intermediate precentral area", and thus, mistakenly, in an agranular structure.

Area 8 - the *intermediate frontal area* - consists of a strip-like zone, wide superiorly and narrowing laterally which, like the agranular frontal area (6), crosses from the callosomarginal sulcus on the medial surface over the upper edge of the hemisphere onto the lateral surface; but there it only reaches to about the middle frontal gyrus before gradually vanishing without distinct borders. Especially on the lateral convexity of the hemisphere it is much less extensive than area 6.

Area 9 - the *granular frontal area* - is a field of similar shape and position to the preceding area, but much more extensive. On the medial surface its only approximate morphological boundary is provided by the callosomarginal sulcus, and on the lateral surface it stops ventrally in the region of the inferior frontal sulcus.

Area 10 - the *frontopolar area* - covers the frontal pole, that is approximately the anterior quarter of the superior and middle frontal gyri on the convexity of the hemisphere, but does not extend medially quite as far as the callosomarginal gyrus. Inferomedially it is fairly precisely demarcated by the superior rostral sulcus. It corresponds approximately to the frontal area of Elliot Smith.

Area 11 - the *prefrontal area* (*111) - forms the rostroventral part of the frontal lobe on its orbital and medial surfaces, thus including most of the straight gyrus (*112), the rostral gyrus and the extreme anterior end of the superior frontal gyrus. The borders are: medially the superior rostral sulcus, laterally approximately the frontomarginal sulcus of Wernicke, and on the orbital surface the medial orbital sulcus.

It is possible to detect fine architectonic differences within this area and with some arbitrariness it can be subdivided. Thus one could separate the zone between the superior rostral sulcus and the inferior rostral sulcus from area 11 as a specific rostral area; equally the straight gyrus and the medial orbital gyrus that lies medial (*113) to it demonstrate certain structural differences, which in principle permit a division (into an *area recta* and a *medial orbital area*).

For reasons of clarity and because this whole area forms a histologically circumscribed zone, I have tentatively only included one area in the brain map, coinciding fairly precisely with the prefrontal area of Elliot Smith. (According to O. Vogt this region can be subdivided into a much larger number of individual areas myeloarchitectonically).

Area 44 - the *opercular area* - is a well-differentiated and sharply circumscribed structural region that on the whole corresponds quite well to the opercular part of the inferior frontal gyrus - Broca's area. Its boundaries are, posteriorly, approximately the inferior precentral sulcus, superiorly the inferior frontal sulcus and anteriorly the ascending ramus of the Sylvian fissure. Inferiorly or medially it encroaches on the frontal operculum and borders on the insular cortex. The area then stretches around the diagonal sulcus, and there are again minor structural differences between the cortex in front of and behind this sulcus to justify the separation of an *anterior opercular area* from a *posterior opercular area* by the diagonal sulcus. As there is much variability and inconstancy of these sulci one will find rather mixed topographical relationships of these structural areas in individual cases.

Area 45 - the *triangular area* - is cytoarchitectonically closely related to the previous area that corresponds approximately to the triangular part of the inferior frontal gyrus. Consequently its caudal border lies in the ascending ramus of the Sylvian fissure, its dorsal border in the inferior frontal sulcus and its rostral border near the radiate sulcus of Eberstaller, although it may extend in places beyond this last sulcus as far forward as the frontomarginal sulcus of Wernicke, and this area may also encroach partially on the orbital part; on the inferior surface of the inferior frontal gyrus it borders the insular cortex.

Concerning the exact morphological borders of the last two areas, that are so extrememly important on account of their relationship to the motor speech area, I should like once again to expressly point out the great individual variations of the sulci in this region. As emerges from Retzius' great monograph "Das Menschenhirn" (*114), the diagonal sulcus is not infrequently fused with the inferior precentral sulcus or communicates with the ascending ramus, is often very strongly developed, but sometimes is entirely absent. The radiate sulcus and the ascending ramus vary widely in shape and structure so that naturally the relations of areas 44 and 45 to these sulci must be subject to major individual variations. Elliot Smith also recognised this with the words: "it must be admitted that its relations to these morphological boundaries is rarely, if ever, preserved with mathematical exactness" (Elliot Smith, 1907, p.249).

Area 47 - the *orbital area* - shares certain architectonic affinities with areas 44 and 45 such that it can be combined with them to form a *subfrontal subregion*. It lies essentially around the posterior branches of the orbital sulcus, generally well differentiated from area 11, but without constant morphological borders. Laterally it crosses the orbital part of the inferior frontal gyrus.

Area 46 - the *middle frontal area* - is not clearly distinguishable from neighbouring areas by its cell structure and can thus only be delimited with uncertainty. It includes about the middle third of the middle and the most anterior part of the inferior frontal gyri at the transition to the orbital surface. There are no constant topographic relations to particular sulci.

Parietal region.

The parietal region coincides essentially with the parietal lobe, but the most posterior segment of the paracentral lobule with the ascending branch of the callosomarginal sulcus also belongs to it. In the inferior part of the parietal lobe it is especially difficult to differentiate it histologically and morphologically with certainty from temporal and even from occipital cortex; it is better distinguishable from the postcentral region for which the postcentral sulcus forms the approximate boundary. On the medial surface the subparietal sulcus and the parieto-occipital sulcus form approximate, but not precise, borders. Within its boundaries four or five individual areas can be distinguished.

Area 5 - the *preparietal area* - is a cytoarchitectonically well characterised area, clearly delimited from neighbouring areas, for which the major distinguishing feature is the presence in layer V of extraordinarily large pyramidal cells that sometimes attain the size of Betz giant cells, and in addition a thick inner granular layer (Figure 16). The cortical thickness noticeably exceeds that of the postcentral cortex. Although the architectonics of this area, especially the size of the pyramidal cells, varies considerably in individual cases, its position is essentially rather constant. The area begins in the most caudal portion of the paracentral lobule, and narrows markedly in the depths of the terminal branch of the callosomarginal sulcus on its rostral bank, extending over the edge of the hemisphere to the lateral surface; it forms a rather wider zone posterior to the superior postcentral sulcus that spreads out between the fork-like diverging terminal branches of the superior postcentral sulcus in the cases that I have examined. Thus overall area 5 has a sack-like shape. The characteristic lateral part of the cortex included in this area appears to be very constant and, from its histological structure, to be of great importance, corresponding in the literature to the anterior portion of the anterior arcuate parietal gyrus (Retzius). In spite of its conspicuous structure, the preparietal cortex has been neglected by all authors. Judging from comparative studies, it has a great biological importance as it can be traced down through much of the mammalian class.

Area 7 - the *superior parietal area* - corresponds essentially to the superior parietal lobule laterally, where this is not occupied by the preparietal area, and medially with the precuneus. The approximate boundaries are, medially the subparietal sulcus, laterally the intraparietal sulcus, posteriorly the parieto-occipital sulcus, and anteriorly the superior postcentral sulcus, with the limitations mentioned earlier. Its structure changes gradually from anterior to posterior so that one can distinguish a division at the superior parietal sulcus into an anterior and a posterior half, or an anterior and posterior superior parietal area (in Figures 84 and 85 this is shown by different densities of the symbols) (*115). This difference also struck Elliot Smith, although he did not find it clearly expressed in all brains [4]).

Area 40 - the *supramarginal area* - lies ventral to the intraparietal sulcus around the terminal branch of the posterior ramus of the Sylvian fissure, thus corresponding to the supramarginal gyrus. Anteriorly the supramarginal area abuts against the postcentral regions, notably areas 2 and 43, separated from it by the inferior postcentral sulcus and the posterior subcentral sulcus. Caudally it gradually gives way to the angular area with the sulcus of Jensen forming the approximate border. It has no sharp boundary with the temporal region (area 22).

Area 39 - the *angular area* - corresponds broadly to the angular gyrus, widening around the posterior end of the superior temporal sulcus, especially caudal to it. Its boundaries with the occipital and temporal regions (areas 19 and 37) are ill-defined; the border with the parietal area is formed approximately by the intraparietal sulcus.

Occipital region.

The occipital region includes the whole occipital lobe, that is laterally the superior, middle and inferior occipital gyri, medially the whole cuneus and the posterior portions of the lingual and fusiform gyri. Its borders are morphologically poorly defined on all sides and also architectonically indistinct. It is divided into three structurally very markedly different areas, the striate area, the occipital area and the preoccipital area.

Area 17 - the *striate area* - as we have seen above, is characterised by the most strikingly differentiated of all homogenetic cortical types, the so-called calcarine type (Figures 12 and 53). As a result of its remarkable structure this area is so easily recognisable, even with the naked eye, either in stained sections or in fresh specimens, that the precise delimitation of its extent can usually be determined macroscopically. This makes it all the more extraordinary that the topical localisation of this area was only established a few years ago and that even today many erroneous interpretations are still disseminated. I have described its situation and exact boundaries in detail in various places (my second, fifth and sixth communications); we shall come back to this below in the comparative discussion and the consideration of individual and species variations. In general the striate area corresponds to the cortex of the calcarine sulcus and closely neighbouring zones. At the posterior end of the calcarine sulcus it extends round the occipital pole onto the lateral surface of the hemisphere, but only very little in Europeans, at most no more than about 1cm; the bulk of the area lies on the medial side and includes a wider cortical field than would appear from the surface of the hemisphere, for the calcarine sulcus is very deep and often forms a true "calcarine fossa". The real extent of its deep spread can be judged from coronal sections of the region (Figures 87 and 88).

---

[4]) "In most specimens I have found it quite impossible to distinguish the cortex of the area in front of these furrows from that placed behind them" (Elliot Smith, 1907, p.245).

The borders of this area, especially laterally, are extraordinarily variable, which is particularly important for pathology. But even medially there are no regular and constant relationships to any "limiting sulci", and in particular the cuneate and lingual sulci cannot be taken as true upper and lower boundaries of the striate area (the *sulcus limitans superior* and *inferior* according to Elliot Smith), for it often extends beyond these sulci in places and in others does not reach them. Thus the cuneus and the lingual gyrus participate in the striate area to variable extents, depending on the degree of infolding of the calcarine sulcus, that is to say on its depth; usually the latter is more involved, that is the striate area extends further ventrally from the calcarine sulcus than dorsally. Above the union of the parieto-occipital and calcarine sulci, and not infrequently even before, the dorsal striate area retreats entirely from the surface and disappears in the depths of the sulcus, while ventrally this usually happens further anteriorly. The anterior end of the area always lies in the depths of the calcarine sulcus and always in its ventral bank, but only exceptionally in such a way that the sulcus forms the rostral boundary (the *sulcus limitans anterior areae striatae* of Elliot Smith); the area scarcely reaches it until right at the most anterior end of the sulcus.

Elliot Smith has described in detail the individual variations of the striate area and the organisation of its sulci, while Bolton, in 1900, studied the characteristics of the area, his *visuo-sensory area*, in blind and anophthalmic patients. (I have dealt with the characteristics of this area in certain foreign races of man elsewhere; see my fifth communication for the Javanese brain).

Area 18 - the *occipital area* - is represented as a crown-shaped field, as in simians and prosimians, that surrounds the striate area laterally and medially as a sometimes wide, sometimes narrow ring-like formation. On the lateral surface it extends quite far anteriorly along the lateral (superior) occipital sulcus and covers a wide zone, while on the medial surface, especially in the most anterior parts of the calcarine sulcus, its area is drastically reduced and it forms a narrow fringe only detectable by examining serial sections.

Area 19 - the *preoccipital area* - surrounds the occipital area (18) like a ring, as the latter surrounds area 17; it is also much reduced in size on its medial aspect. It is quite difficult to demonstrate, especially in the depths of the calcarine sulcus, so that it often seems doubtful whether that part of the area lying dorsal to the sulcus really unites spatially with the ventral part. In contrast, on the lateral surface it covers a wide zone and extends anteriorly over the interoccipital and parieto-occipital sulci. Its exact boundaries are just as little related to sulci as those of area 18.

In the course of the intraparietal sulcus a narrow band of similar structure to area 19 extends forwards sagittally. Elliot Smith puts this band in direct contact with the postcentral area and calls this whole strip running along the intraparietal sulcus the "visuo-sensory band". Further, Elliot Smith differentiates, in addition to the striate area, an area parastriata and an area peristriata, which correspond approximately to my areas 18 and 19; lateral to the latter he separates two small poorly differentiated zones, an *area temporo-*

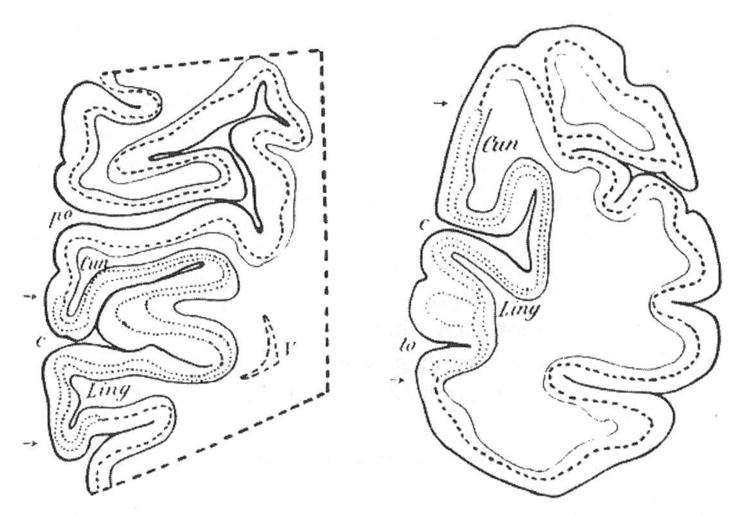

Fig. 87 and 88. Diagrams of coronal sections through the region of the calcarine sulcus. The double dotted line indicates the striate cortex; its extent is much wider within the sulcus than on the free surface.

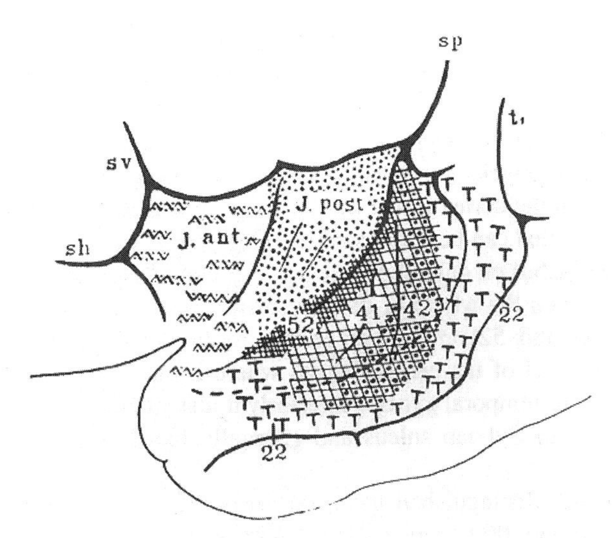

Fig. 89. Insular region and superior aspect of the superior temporal gyrus exposed. J. ant. = agranular anterior insular zone, J. post. = granular posterior insular zone, sp = posterior ramus of the Sylvian fissure, sv = vertical ramus of the Sylvian fissure, sh = horizontal ramus of the Sylvian fissure, t1 = superior temporal sulcus. On the superior aspect of the superior temporal gyrus are three areas: 52 = parainsular (*116) area, 41 = anterior or medial transverse temporal area, 42 = posterior or lateral transverse temporal area.

occipitalis and an *area parieto-occipitalis* ("often quite indistinguishable").

Campbell only distinguishes next to the striate area a large occipital structural zone that he calls the "visuo-psychic area" in contrast to his visuo-sensory area (this author's striate area).

Temporal region.

It also represents a morphologically well delimited and homogeneous region that, apart from its posterior border, is quite clearly circumscribed. It stretches from the posterior margin of the insula over the whole vertical extent of the temporal lobe to the rhinal sulcus or the temporal incisura (Retzius) and is thus the most voluminous region of the human brain after the frontal region. It is divisible into a large number of clearly different cytoarchitectonic areas of which certain, such as the transverse gyri and their surroundings, are extraordinarily characteristically structured and form a sort of subregion which, owing to their relationship to functional physiological centres, possess great practical importance. Those portions that directly abut the rhinencephalon (area 36) and the cortex of the temporal pole have a particularly variable structure.

We will undertake the description from medial to lateral on the brain map. (For area 35, see the Hippocampal region).

Area 36 - the *ectorhinal area* - lies, as its name implies, directly lateral to the rhinal sulcus and represents the first area of the neopallium adjacent to the archipallium, to which area 35 belongs. It possesses a markedly heterotypical architecture characterised by a distinct paucity of cells (and fibres) and also a massive development of the cells of layers V and VI, and forms a narrow band-like zone parallel to the outer edge of the rhinencephalon. Morphologically it represents the rostral extension of the lingual gyrus. I leave provisionally in abeyance whether the posterior portion of the zone delimited on our brain map as area 36 would be better differentiated as a "retrosubicular area", as in many animals.

Area 37 - *occipitotemporal area.* - Such is the concept of a rather wide, but poorly circumscribed transition zone between the adjacent occipital and temporal cortices, which lies on the most posterior part of the temporal lobe, partly laterally and partly mediobasally. It is sufficiently distinct from the preoccipital area 19 as well as from the temporal area 20 that it is justifiable to differentiate it as an entity. Elliot Smith also proposed a specific structure denominated "area paratemporalis" in a corresponding situation.

Area 38 - the *temporopolar area* - corresponds in its position, as its name suggests, grossly to the tip of the temporal lobe, without any clear external delimitation; the field fuses laterally with the adjacent caudally situated areas 20, 21 and 22, and medially with area 36, and is characterised by its great cross-sectional depth.

Area 20 - the *inferior temporal area* - corresponds essentially to the inferior temporal gyrus and blends rostrally and caudally with the neighbouring areas 37 and 38 without sharp borders.

Area 21 - the *middle temporal area* - is situated approximately in the middle temporal gyrus, although its borders do not precisely follow the sulci that demarcate the gyrus; it also blends gradually, especially anteriorly and posteriorly, with the neighbouring areas.

Area 22 - the *superior temporal area* - is differentiated from the aforementioned areas in its cyto- (and myelo-) architecture more than these two areas between themselves. Together with the cortex of the transverse gyri of Heschl it forms a homogeneous structural region that can be contrasted with the other temporal areas. The superior temporal area encroaches on only about the posterior two-thirds of the superior temporal gyrus, and not even the whole of its free surface that is partially occupied by the deep areas (41, 42 and 52) (*117), as shown in Figure 89. Anteriorly it reaches approximately the level of the central sulcus where it climbs partly onto the medial surface of the superior temporal gyrus; posteriorly it just attains the level of the vertical terminal branch of the Sylvian sulcus and gradually blends with the supramarginal area.

Elliot Smith has distinguished localised areas that correspond precisely to my areas 20, 21, 22, 37 and 38; Campbell on the other hand includes the whole temporal lobe (except T1) with the inferior parietal area as a single field, his "Temporal area".

It has long been known that the transverse temporal gyri (Heschl) possess a structure that is different from the rest of the temporal lobe. More exactly, Campbell (1905) first differentiated a special field approximately within the confines of this gyral formation, the "audito-sensory area", contrasting it with the other temporal gyri, or "audito-psychic area". Elliot Smith, in agreement with this, writes (1907): "The two transverse gyri of Heschl represent a sharply-defined anatomical area of this cortex" (*118), but gives no precise topographical description of the region. Rosenberg describes a specific structure in the anterior transverse gyrus and without hesitation considers it possible to regard it as corresponding to "auditory cortex", in agreement with Flechsig.

In my sixth communication (1907) I myself differentiated two specific cortical types in the region of the gyri of Heschl, one anteromedially and the other posterolaterally. Recently I was able to divide off another band-like zone, medial to the medial area just before the beginning of the insula itself and parallel to its posterior margin, that I had earlier simply accepted as a mixed or transitional region. However, I was able to convince myself that it is distinguishable as a homogeneous area, well characterised from both the insular and the remaining temporal cortex by its specific structure.

The superior surface of the superior temporal gyrus thus includes, apart from area 22, the following three separate areas one after the other from medial to lateral: 1. the parainsular area, 2. the medial (anterior) transverse temporal area, 3. the lateral (posterior) transverse temporal area. After comes 4. the superior temporal area. Their mutual relations and their approximate extent are visible in Figure 89.

Area 52 - the *parainsular area* - forms a narrow band-like zone on the superior bank of the superior temporal gyrus along the posterior margin of the insula and thus represents the transitional area between the temporal cortex and the actual insula. Anteriorly it extends almost to the limen of the insula, posteriorly it disappears gradually beneath its posterior margin near its posterior end.

Area 41 - the *medial (anterior) transverse temporal area* - corresponds approximately, but not precisely, to the anterior transverse gyrus and extends obliquely from anterolateral to posteromedial, descending gradually into the depths of the sulcus. It is bordered medially by the parainsular area from which it is sharply demarcated, while laterally area 42 forms an arc in contact with it. The transition to this last area is sometimes quite sharp, and in no way corresponds to the transverse sulcus, but lies in part in the middle of the apex of the anterior transverse gyrus. Both rostrally and laterally the area reaches beyond the anterior transverse gyrus.

Area 42 - the *lateral (posterior) transverse temporal area* - also extends obliquely from anterolateral to posteromedial over the superior bank of the superior temporal gyrus, but also lies on the free surface of the gyrus for a not inconsiderable distance. It forms a crescent along the lateral edge of area 41; caudally it extends deeply towards the posterior edge of the insula.

Insular region.

The insular cortex represents a well delimited, homogeneous region that is clearly differentiated from the surrounding regions thanks to an obviously recognisable specific laminar pattern (the claustrum). The region coincides approximately with the Sylvian fossa, but often extends over the margins of the circular sulcus of the insula and in particular may encroach partially on the under surface of the frontal and temporal opercula. The base of the insula also extends considerably beyond its anterior limiting sulcus, or at least the claustrum may be followed inwards as far as the orbital surface. One must therefore postulate a wider extent for the insula if one wishes to recognise the claustrum as an absolute criterion for identifying the insular cortex.

There are great difficulties in dividing it into individual fields, of which I described four in 1904. The most important aspect of differentiation within the insular cortex is to note that the insula divides basically into two separate halves along a line that is a prolongation of the central sulcus, one posterior and granular, the other anterior and agranular (Figures 36 and 37). Thus, like the central region, the insula is divisible according to the presence or absence of an inner granular layer into two totally different structural zones, the border of which lies in the prolongation of the central sulcus of Rolando but does not correspond exactly to the central sulcus of the insula. Figure 89 illustrates this relationship schematically. No other individual areas are illustrated. It should however be noted that on the edge of the anterior agranular half of the insula a cortical type of quite rudimentary structure can be distinguished, that I earlier called the olfactory portion of the insula. Also, the transitional zones against the orbital gyri again display a specific structural aspect. The precise localisation of these individual fields must await further investigation.

Campbell and Elliot Smith also differentiate between the anterior and posterior halves of the insula without, however, giving an exact description of their topographic relationships.

Cingulate region.

The crescent-shaped formation of the cingulate gyrus bordering the corpus callosum presents a strikingly rich cytoarchitectonic structure, although its physiological significance is the least well known of all the parts of the cortex. Comparative anatomy indicates that it represents a separate structural entity. As Figure 84 shows, we distinguish the most caudal part of the cingulate gyrus directly behind the splenium, the so-called "isthmus", as a special *retrosplenial region*. The remainder of the gyral formation is again divisible, like the insula, into two architectonically completely different portions, a *postcingulate subregion* and a *precingulate subregion*, of which the former exhibits the typical six layers with a distinct inner granular layer, while the latter (except area 32) does not possess an inner granular layer. Once again therefore there is here a sharp boundary between a granular and an agranular zone approximately at the level of the central sulcus. Thus the whole of the human cortical surface is divided by the central sulcus and its projections on the insula and cingulate gyrus into two structurally completely different halves, an anterior agranular and a posterior granular, a trend that is found similarly in lower mammals. I have recently become convinced that in the middle of the cingulate gyrus one can distinguish a special field, the *intermediate cingulate area*, that forms a transitional zone between the two halves, but it is not yet represented on the brain map.

Campbell and Elliot Smith describe sharply differing divisions of the cingulate gyrus, the former only distinguishing three fields within the retrospenial region and finding absolutely no division between anterior and posterior cingulate cortex.

Area 23 - the *ventral posterior cingulate area* - corresponds to the ventral part of the posterior half of the cingulate gyrus and lies directly above the corpus callosum insofar as it is not separated from it by the fields of the retrosplenial region. Caudally it forms an arc around the the splenium as far as the anterior bank of the parieto-occipital sulcus with which it gradually blends. Rostrally its border becomes variable and it fuses with the agranular formations of the precingulate subregion by means of a broad transitional zone that can be looked upon as a specific intermediate cingulate area.

Area 31 - the *dorsal posterior cingulate area* - includes the dorsal portion of the cingulate gyrus in its posterior half and forms an arc around area 23 as far as the parieto-occipital sulcus. There is no clear outer border with area 23 or with the parietal cortex (area 7); the subparietal sulcus lies partly within its extent and does not form its exact dorsal border.

Area 24 - the *ventral anterior cingulate area* - includes the ventral part of the anterior half of the cingulate gyrus that lies next to the corpus callosum, except the narrow band of rudimentary cortex in the depths of the callosal sulcus (heterogenetic cortex) belonging to area 33. Posteriorly it fuses gradually with the weakly granular transitional zone lying over the middle of the corpus callosum; anteriorly it extends as far as the region of the rostrum. Rostrodorsally it is often (but not always) limited by the medial branch of the callosomarginal sulcus. Its structure changes gradually from posterior to anterior.

Area 32 - the *dorsal anterior cingulate area* - is situated outside area 24 and, like it, describes a semicircle around the anterior end of the corpus callosum. Rostrally it extends partially over the anterior branch of the callosomarginal sulcus and reaches the vicinity of the margin of the frontal region. Ventrally it extends to near the superior rostral sulcus.

Area 33 - the *pregenual area* - is formed by the narrow strip of rudimentary cortex wholly hidden in the callosal sulcus that springs directly from the lateral longitudinal stria of the corpus callosum (*119). Anteroinferiorly it stretches round the end of the rostrum of the corpus callosum, while posterosuperiorly it extends quite far over the surface of the corpus callosum and gradually disappears in the depths of the callosal sulcus.

Area 25 - the *subgenual area* - lies in the narrow space between the precommissural area or the lamina terminalis on the one hand and the transverse rostral sulcus on the other; it would therefore also be appropriate to call this field the preterminal area. It stretches from the end of the rostrum to the inferomedial edge of the hemisphere, that is close to the olfactory trigone. Like the pregenual area it also possesses a rudimentarily developed (heterogenetic) laminar pattern.

Retrosplenial region.

This region forms a zone consisting of three crescent-shaped fields around the splenium of the corpus callosum, corresponding essentially to the isthmus of the cingulate gyrus. In the surface maps (Figures 84 and 86) the whole region is drawn relatively too wide in order to be able to represent the general situation of the fields. The cortex of the retrosplenial region belongs partly to the heterogenetic type; in particular, area 26 lying next to the corpus callosum is quite "defective" cortex in Meynert's sense with strikingly rudimentary lamination (especially layers II to V). In area 29 there is a unique development of the inner granular layer (IV) with a corresponding degeneration of layers II and III, and in area 30, on the contrary, the inner granular layer is degenerated while layers III and V are relatively well developed. The identification of these individual fields in man was very difficult as a result of the incomplete expression of their architectonic characters; I was only able to draw homologies in man after I had seen them in lower animals in which they are more clearly developed.

Area 26 - the *ectosplenial area* - is closely apposed to the posterior end of the corpus callosum, just as the pregenual area is to the anterior end, and like the latter remains completely hidden in the callosal sulcus. It only encroaches slightly on the dorsal surface of the body of the corpus callosum. Medially it forms a transition to the lateral longitudinal stria of the corpus callosum (*119), laterally it merges without a clear border with

Area 29 - the *granular retrolimbic area*. - It forms a quite narrow semicircular zone around the ectosplenial area and likewise lies to a great extent in the depths of the callosal sulcus; superiorly it does not reach far onto the corpus callosum, like the latter area.

Area 30 - the *agranular retrolimbic area* - essentially covers the edge of the surface of the isthmus of the cingulate gyrus, but also extends a short distance over the anterior bank of the calcarine sulcus. It forms a sort of arc around the aforementioned fields and in man is the most widespread of them.

Hippocampal region.

In addition to the head of the parahippocampal gyrus (*120) itself, that corresponds to the piriform lobe of macrosmatic animals, I also include in this region a narrow strip of cortex lying directly lateral to the subiculum of the hippocampus, partially within the hippocampal sulcus, and stretching dorsally to close to the splenium where the transition to the retrosplenial region takes place. Thus this region includes the whole of the cortical surface between the hippocampal sulcus on the one hand and the rhinal sulcus or the temporal incisura on the other, that is rather more than the so-called "rhinencephalon" of the literature [5]). Three to four different cortical types can be distinguished at most, all of which are of heterogenetic origin.

Area 27 - the *presubicular area* - follows on directly laterally from the actual subiculum from which it is separated by a sharp border (as seen in Figures 34 and 35 of Part I). The field forms an elongated, narrow zone stretching along the hippocampal sulcus from the uncus to the tail of the hippocampus just under the corpus callosum.

Area 28 - the *entorhinal area* - lies, as its name suggests, medial to the rhinal sulcus and includes the largest part of the head of the parahippocampal gyrus (*120) under its surface. Anteriorly it is limited by the temporal incisura, that may actually represent the remnant of the posterior rhinal sulcus of lower mammals. The cortex is especially richly laminated and easy to homologise throughout the whole mammalian class on account of its characteristic, atypical (heterogenetic) structure. It is possible to distinguish a modified cortical type in its ventrolateral part, the ventral entorhinal area, thus defining

Area 34 - the *dorsal entorhinal area*. It lies mainly medial to the inferior rhinencephalic sulcus (Retzius), so that this sulcus forms the approximate

---

[5]) Concerning nomenclature I agree with Gustaf Retzius "Das Menschengehirn" (1896) and "Zur äusseren Morphologie des Riechhirns der Säugertiere und des Menschen" (Biol. Unters. VIII. 1898) (*121).

border between the two types.

Area 35 - the *perirhinal area* - consists of a narrow, strip-like zone limited to the rhinal sulcus and its immediate surroundings, that follows this sulcus along its whole length, extending a little beyond it caudally. The cortical structure is characterised by a marked regression of the lamination in that the inner granular layer is missing. By its course and position this area forms the exact border between the archipallium and the neopallium, and it is difficult to decide whether it should be attributed to the one or the other; as it is obviously heterogenetic in structure, it can be looked upon as archicortex with more justification, although in many animals it lies partly lateral to the rhinal sulcus.

At the caudal end of the perirhinal area (35) and lateral to the presubicular area (27) one can also differentiate another special structural type in man, which I first recognised as such after it had struck me in lower animals, in which I described it as the retrosubicular area (48).

To the medial side of area 34 next to the uncus there are zones that have a yet different structure, such as the amygdala and the lunate gyrus (Retzius) that lies medial to the semiannular sulcus. More investigations must be envisaged later in order to determine the homologues of these structures in different animals. Even the olfactory tubercle, that is often wrongly identified with the olfactory trigone, still needs precise histological study, as does the anterior perforated substance. In lower orders I have grouped these structures together as the olfactory region.

II. Lower monkeys (guenon and marmoset).

I have drawn brain maps of two families of lower monkeys, the gyrencephalic cercopithecids (*122) and the lissencephalic callithricids (*123). On the one hand they share extensive basic similarities, and yet on the other marked individual differences, so that illustrate the principles of comparative surface localisation very well.

1. Cercopithicids (Figures 90 and 91).

A comparison with the human brain map reveals first of all a far smaller number of individual fields [6]), especially in the frontal and temporal lobes. In spite of this, the same regions as in man are found with similar locations and arrangements. For easier orientation the pre- and postcentral regions are represented in isolation in Figures 92 and 93; Figures 94 and 95 show the same regions of man in schematic outline.

---

[6]) One should again recall what was said in the introduction to Part II, that the same numbers and symbols on my brain maps often indicate only a relative and not an absolute homology of the relevant fields.

Postcentral region (Figures 92 and 93).

It concurs in its shape and the spatial relations of its sulci and gyri with the region of the same name in man. The homology is indisputable. Like the latter it lies between the central and the postcentral sulci, reaching medially as far as the callosomarginal sulcus and laterally as far as the Sylvian sulcus, thus encompassing the postcentral gyrus and a large caudal part of the paracentral lobule. It is similarly divisible into a number of extensively differentiated cortical types that can be distinguished as individual areas, as in man.

Area 1 - the *intermediate postcentral area* - corresponds to the apex of the postcentral gyrus.

Area 2 - the *caudal postcentral area* - on the posterior bank of the postcentral gyrus.

Area 3 - the *rostral postcentral area* - on the anterior bank of the postcentral gyrus. At the inferior end of the gyrus area 3 crosses the central sulcus rostrally to lie between it and the inferior precentral sulcus on the precentral gyrus (*125). Thus the central sulcus no longer forms the border between the pre- and postcentral regions here.

Originally I did not differentiate a specific **area 43** (*subcentral area*) in the guenon; I believed that I should interpret the zone at the inferior end of the central sulcus indicated by special hatching in Figure 90 as a mixed or transitional cytoarchitectonic form of areas 1 to 3. Thanks to new investigations of the myeloarchitecture of this region (T. Mauss) it has, however, been clearly determined that monkeys also possess a specific structural field on the Rolandic operculum corresponding to the subcentral area of man. Further study of my sections has led to my also recognising this area cytoarchitectonically.

Precentral region.

From its position and structure it corresponds entirely to the region of the same name in man; thus, as for the postcentral region, there is complete homology. In terms of surface area it occupies relatively much more of the total cortex than in man; it comprises almost the whole posterior half of the frontal lobe, whereas the corresponding region in man would make up scarcely one tenth of the surface of the frontal lobe. The whole length of its caudal border, apart from its most inferior portion, is formed by the central sulcus, but it is even more frequent in the lower monkeys than in man for the postcentral region to extend around the fundus of the sulcus partially onto the medial part of the postcentral gyrus. This is particularly true of the dorsal part of the sulcus (*126). At its inferior end, on the other hand, where it bends posteriorly in an arc, the border moves away rostrally from the sulcus, as we have already seen, and the whole zone comes to lie anterior to the inferior precentral sulcus. On the medial side its caudal border forms a line that is a direct extension of the central sulcus, separating the paracentral lobule into an anterior and a posterior half. The rostral boundary of the region coincides approximately, but

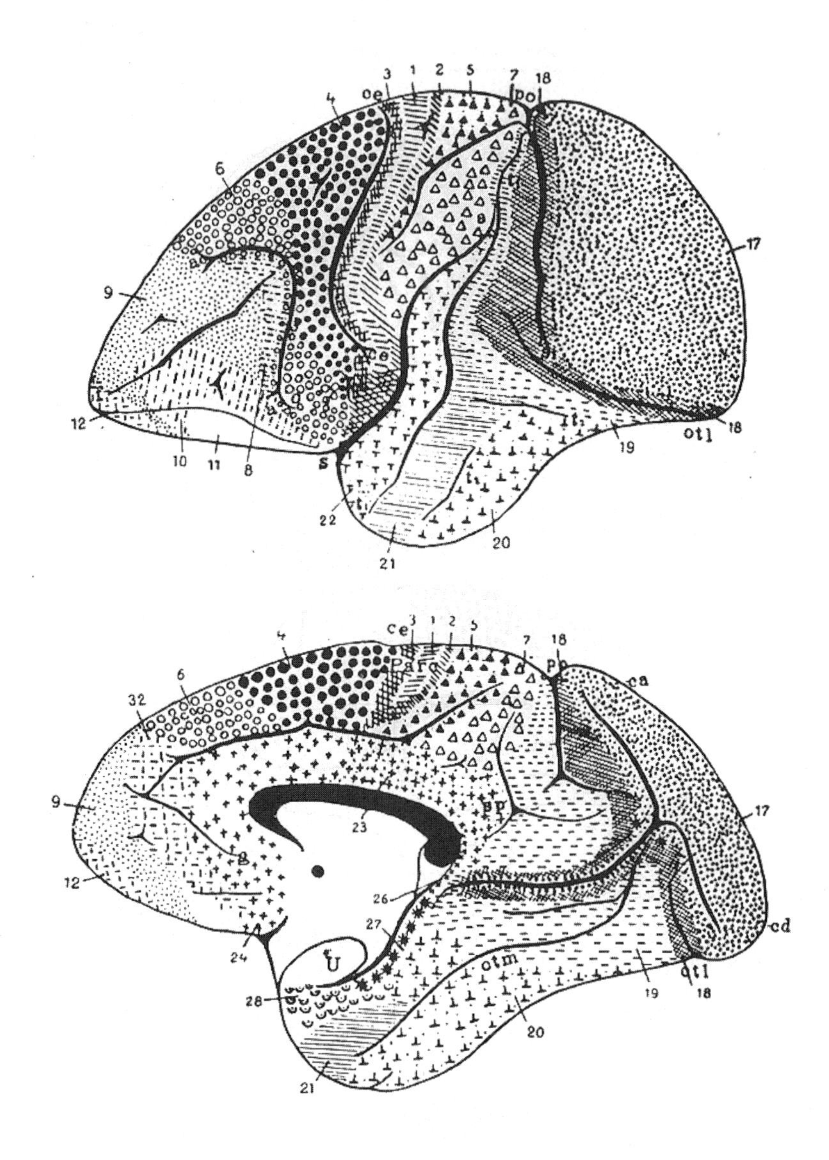

Fig. 90 and 91. Cytoarchitectonic cortical areas in the guenon. 1:1. (Reproduced unchanged from the third communication, 1904/1905.) (*124)

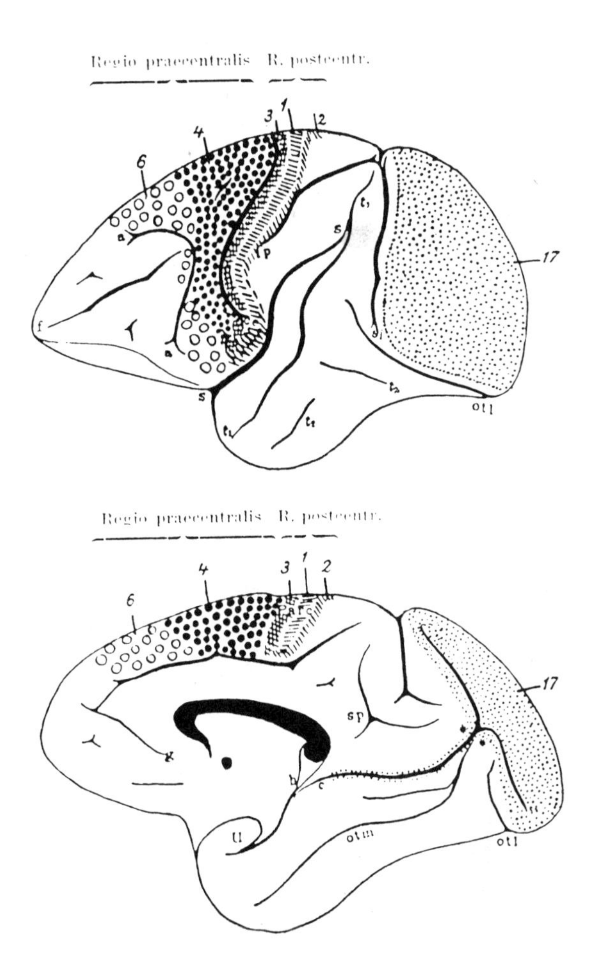

Fig. 92 and 93. The precentral and postcentral regions in the guenon. 17 = striate area. (cf. Figs. 94 and 95 of man.)

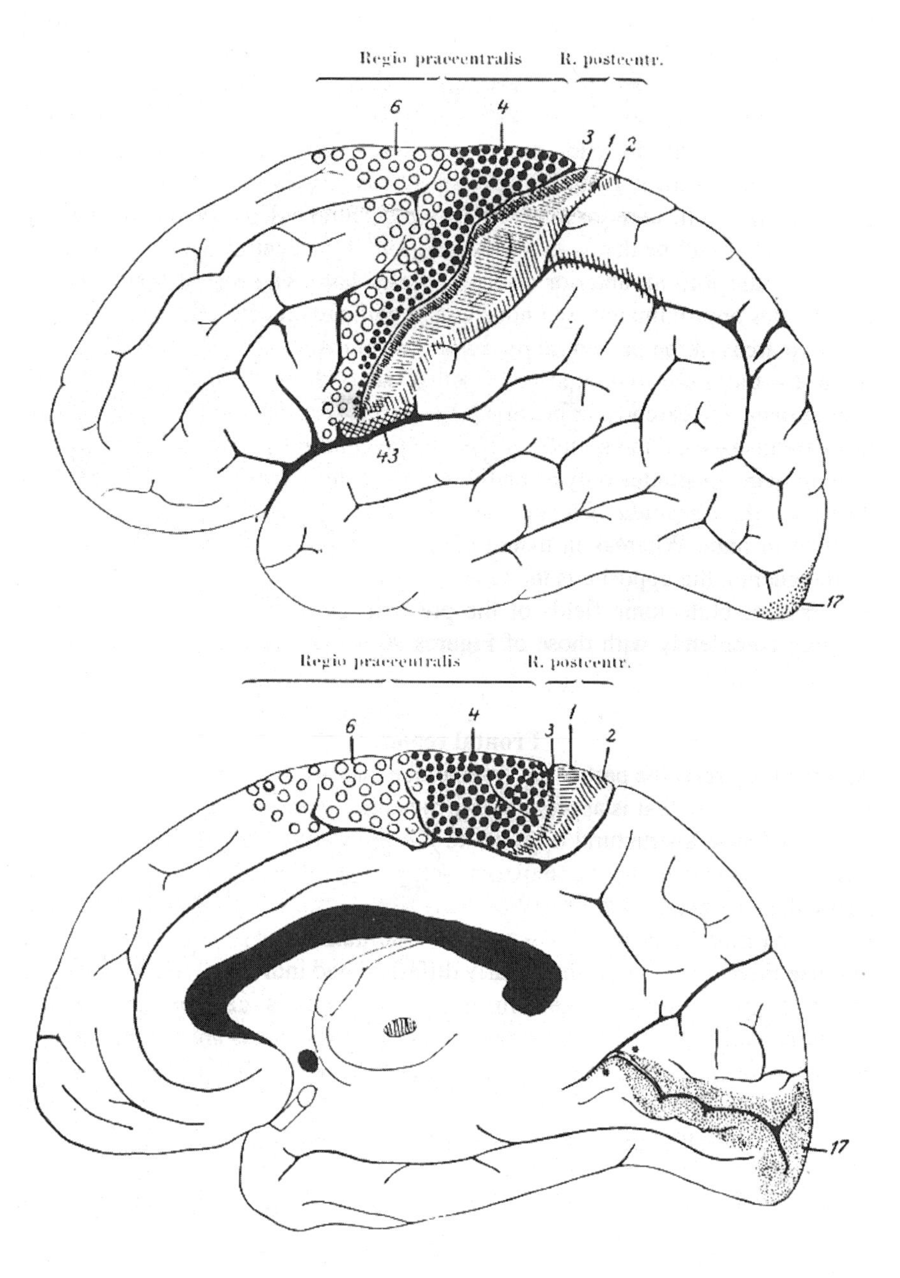

Fig. 94 and 95. The postcentral region (areas 1,2,3 and 43) and the precentral region (areas 4 and 6) of man (highly schematic). 17 = striate area.

not exactly, with the arcuate sulcus (*127).

The two areas of the precentral region are, as in man,

Area 4 - the *giant pyramidal area* - that has a relatively much greater surface area than in man, and extends far beyond the superior precentral sulcus (*128) rostrally and thus extends beyond the bounds of the true precentral gyrus, whereas the homologous area in man constitutes only a small fraction of this gyrus (Figures 44 and 59.)

Area 6 - the *agranular frontal area* - is, in comparison with area 4, relatively smaller than in man. Whereas in man area 6 considerably exceeds area 4 in surface area, in the guenon the opposite is the case.

The myeloarchitectonic fields of the pre- and postcentral regions described by Mauss agree completely with those of Figures 90 to 93 as far as position, shape and area are concerned.

Frontal region.

As such I interpret the part of the frontal lobe that is not occupied by the precentral region (areas 4 and 6), that is approximately the anterior half of the lobe in the guenon. As in man, it forms a structural opposite to the precentral region with all its cortical types characterised by an inner granular layer, while the latter is agranular. Its extent is much less than in man, encompassing scarcely half the frontal lobe in the guenon compared with more than four-fifths in man. The number of areas is limited to five, while in man there are eight or nine clearly differentiated individual fields (and according to O. Vogt a much larger number of subfields can be distinguished myeloarchitectonically). The detailed homologies of these areas are quite unclear; one can only say with certainty that in the monkey no structural equivalents of areas 44, 45 and 46 can be demonstrated. In general, then, the inferior frontal gyrus and the central part of the middle frontal gyrus are missing in the lower monkeys.

The individual areas of the frontal region are:

Area 8 - the *intermediate frontal area* - which forms a narrow band stretching along the anterior bank of the arcuate sulcus is therefore situated exclusively in the middle of the lateral surface of the hemisphere, thus occupying a quite different site from area 8 in man.

Mauss has described and delimited an analogous area under the same name on myeloarchitectonic grounds.

Area 9 - the *granular frontal area* - corresponds broadly from its position and structure to the granular frontal area and the frontopolar area (areas 9 and 10) of man, and encompasses a large part of the anterior half of the lateral and medial sides of the frontal lobe.

Area 10 - the *lateral orbital area* - forms approximately the lateral part of the orbital cortex and the adjacent lateral convexity.

Area 11 - the *medial orbital area* - occupies about the medial half of the orbital cortex.

Area 12 - the *frontopolar area* - includes the frontal pole; it corresponds rather to area 11 of the human cortex. Furthermore, the homology of this part of the frontal cortex between man and monkey is quite uncertain.

Insular region.

As in man, the fields that contain the insular type of cortex in the lower monkeys also lie almost entirely deep in the gyri of the insula, covered by its opercula. Therefore, for the sake of clarity, this region has not been drawn on the map. It is divided into four individual areas that in part can be sharply delimited from each other. The posterodorsal portions (area 13) possess a distinct inner granular layer, while the rostral and especially the ventral parts (areas 14 to 16) lack this and thus belong to the agranular cortex. The border between the granular and agranular parts of the insular cortex is formed by a line marking a prolongation of the central sulcus. This is of great importance, for the central sulcus also forms the dividing line between opposing granular and agranular structural regions. The delimitation of the region from neighbouring zones is also not sharp, for the claustrum often merges gradually with adjacent cortex.

Parietal region.

Compared with the massive parietal region of man, this region covers a relatively restricted area in the cercopithicids. It consists of only two individual fields, whereas in man four or five distinct areas can be distinguished. Nevertheless the whole region can be considered as a homologue of the parietal region of man, for its topographical relations to neighbouring zones are entirely similar.

Area 5 - the *preparietal area* - is certainly a homologous structural zone to area 5 of man. It is, however, relatively much more extensive and in the guenon encompasses almost the whole cortical surface dorsal to the intraparietal sulcus as far as the callosomarginal sulcus. Whereas in man it is limited to an anterosuperior wedge of the superior parietal lobe and just the caudal bank of the paracentral lobule, it here forms the whole zone defined morphologically as the superior parietal lobule and a not inconsiderable part of the paracentral lobule.

Area 7 - the *parietal area* - lies essentially ventral to the intraparietal sulcus, between this sulcus on the one hand and the Sylvian sulcus on the other. It thus includes the part of the surface morphologically defined as the inferior parietal lobule, with the limitation that a narrow dorsal zone in the depths of the intraparietal sulcus extends over the precuneus.

As the homology of area 5 leaves no room for doubt, the question arises as to which human field area 7 corresponds. On account of its position inferior to the intraparietal sulcus one might compare it with areas 39 and 40 of the inferior parietal lobule. However, on comparative anatomical grounds I believe that it corresponds to the whole parietal region of man and therefore represents a still undifferentiated precursor zone for all parietal areas (apart from area 5).

Temporal region.

The temporal region, made up of areas 20, 21 and 22, presents a morphologically better defined entity than in man, its only indistinct borders being with the occipital lobe. Although only structurally divisible into three individual fields, there is little doubt about its overall homology with the human region of the same name. The basic cytoarchitectonic characteristics of the temporal cortex are the same in both; the only difficulties concern the homologies of individual areas. Mauss has divided the region on the basis of myeloarchitecture in a similar way to me.

Area 20 - the *inferior temporal area* - corresponds essentially to the inferior temporal gyrus.

Area 21 - the *middle temporal area* - forms the middle temporal gyrus.

Area 22 - the *superior temporal area* - encompasses the superior temporal gyrus.

Thus in the monkey the human areas 36, 37, 38, 41, 42 and 52 are missing. In any case, it is certain that the two areas 41 and 42 belonging to the transverse temporal gyri (Heschl) do not have analogues in the monkey. This is demonstrated particularly well by the myeloarchitecture in cercopithecids which entirely fails to show the very fibre-rich structures of the human Heschl's transverse gyri. We shall again discuss the significance of this fact later in Part III. How much of the human areas 36, 37 and 38 is contained in undifferentiated form in the areas 20, 21 and 22 under consideration, cannot be determined.

Occipital region.

Compared with the relatively small occipital region of man, in all monkeys it extends very widely and constitutes a considerable part of the total cortical surface. It is the absolutely largest region of the brain of cercopithecids and its extent considerably exceeds all others, including the frontal region. There thus exists the opposite relationship to that in man, in whom the frontal cortex is strongly dominant. It is important that it includes the structurally most markedly differentiated and phylogenetically most constant mammalian neocortical type, the striate area or area 17 (Figures 69 and 78). This provides a powerful criterion for determining homologies in other, less obvious structural types.

Its individual areas 17, 18 and 19 are absolute homologues of the corresponding human areas. According to Mauss the myeloarchitectonic structure is exactly the same.

Area 17 - *striate area.* - It is the only area that exceeds that of man in surface area, being indeed not only relatively but absolutely larger (*129). As this is such an important area I shall repeat the detailed description given in my third and fifth communications on cortical histological localisation.

The striate area of the cercopithecids represents a sort of "calotte", a cap-like structure over the occipital pole. In contrast to man, it then spreads very widely over the lateral surface covering a broad territory limited by the simian sulcus (the sulcus lunatus of Elliott Smith) and the occipitotemporal sulcus, the so-called occipital operculum, only leaving a quite narrow strip free directly along these sulci. On the medial surface, adjacent to the occipital pole, it encompasses the gyri that flank the ascending and descending rami of the calcarine sulcus superficially but soon sinks into the depths of the sulcus at its main branch (at * * in Figure 93). It then extends forwards over the dorsal and ventral banks of the sulcus to terminate somewhat behind its anterior end, with individual variations, sometimes in the depths of the sulcus, sometimes slightly ventral and sometimes dorsal to it.

The topographic variation in different species and families of monkey are described in detail in my fifth communication, and at the same time the resulting inferences are discussed. The essential differences in the striate area of man can be summarised by the following statements:

1. The calcarine cortical type or striate area is, relative to the total size of the hemisphere, much more extensive in the monkey than in man.

2. Whereas in man it is wholly, or almost wholly, limited to the medial surface, it includes a large part of the lateral surface in the monkey.

3. On the medial surface in the cercopithecids the area only includes the superficial cortex of the two banks of the calcarine sulcus in its most caudal section near the occipital pole. Along the whole course of the main branch of this sulcus it involves exclusively the deep cortex leaving the lingual gyrus and cuneus completely free. In contrast, in man both these gyri contribute to a large extent to the formation of the striate area and the field only withdraws to the deep cortex at the level of the parieto-occipital sulcus.

Area 18 - the *occipital area* - forms, as does that of man, a closed ring-shaped coronal zone surrounding area 17 anteriorly like a belt of varying width on both its medial and lateral aspects.

Area 19 - the *preoccipital area* - extends rostrally as a coronal zone round the occipital area as the latter does round the striate area. At the anterior end of the calcarine sulcus it is usually so attenuated that the continuity of the dorsal and ventral parts seems interrupted.

Cingulate region.

In contrast to the richly organised cingulate region of man, this region here is essentially divisible into only three individual areas (23, 24 and 32). It is not unequivocally possible to decide to which human areas these are fully homologous, but it can be said that the granular area 23 must correspond broadly to the equally granular posterior half of the region in man, thus approximately to areas 23 and 31, whereas areas 24, 25 and 33 of the agranular anterior half of the human cingulate gyrus would be homologous to areas 24 and 32 of the guenon.

Area 23 - the *posterior (granular) cingulate area* - includes approximately the posterior half of the cingulate gyrus.

Area 24 - the *anterior (agranular) cingulate area* - encompasses the anterior half of this gyrus next to the corpus callosum. After I had distinguished a special structural area (25) anterior and inferior to the genu of the corpus callosum in other animals and in man, I found evidence of it in cercopithecids. This should, then, also be delineated from area 24.

Area 32 - the *prelimbic area* - lies in an arc round the rostral limit of area 24, similar to the area of the same name in man; nevertheless it is not homologous to the latter.

Retrosplenial region.

This is generally the smallest region of the monkey brain and, especially in cercopithecids, is topically and architectonically so poorly developed that it can easily be overlooked. Its homology can only be determined by comparison with those animals in which it is more clearly differentiated into individual fields and also has a greater area. The whole region is only represented by one area (26) in my brain map, whereas I was later able to distinguish several areas in related families (marmoset and lemur, Figures 96 and 98). Although it might be conceded that in the guenon there is a slight suggestion of differentiation within the region, as seen in these animals, I have nevertheless prefered not to undertake a topical separation into several areas to emphasise the contrast with the other brains.

Hippocampal region.

Like the retrosplenial region, this region of the lower monkeys lags behind that of other animals and of man in terms of extent and architectonic differentiation. It encompasses the anterior bank of the hippocampal sulcus and the head of the parahippocampal gyrus (*120) (the actual rhinencephalon) that is extremely reduced in all monkeys. The two areas that lie in this region belong to the heterotopic formations and their structure is so specific that their homology can be established with certainty in all animals.

Area 27 - the *presubicular area* - runs, as its name suggests, lateral to the subiculum of the hippocampus and lies almost completely in the depths of the hippocampal sulcus.

Area 28 - the *entorhinal area* - lies lateral to the extreme anterior end of the hippocampal sulcus. (In the map that was drawn some years ago, and that I intentionally use here unaltered, its position is imprecise; also missing is the weakly developed *perirhinal area* or area 35 [Figure 25].)

2. Marmosets - Callithricids (Figures 96 and 97).

In 1904 I pointed out in a publication that brains without sulci (lissencephalic) possessed great advantages over gyrencephalic ones for topical localisation, that is, the histological delimitation of cortical areas, for on the smooth surfaces of such hemispheres the shape, position and extent of the individual fields were more clearly visible than on folded surfaces, and that often it was only possible to understand the complex relationships of gyrencephalic hemispheres by reference to them. At that time I could only illustrate this by a few examples in lissencephalic marmosets. Meanwhile I have completed the topographic parcellation of the whole cortex in callithricids and it emerges, as evident from the brain maps of the lateral and medial sides, that the arrangement of structural zones is quite systematic and the principle of cortical areas is more clearly recognisable than in any other brain.

The surface of the hemisphere of the marmoset - especially laterally - is organised in a series of segment-like zones arranged one behind the other, usually in the form of stripes or bands, that either encircle the whole hemisphere both medially and laterally like a continuous belt, or encompass only part of the cortical surface like segments of a ring. Those that form *complete coronal segments* extending over the whole medial and lateral surfaces include areas 17, 18 and 19 in the occipital lobe, and areas 9, 10 and 12 in the frontal lobe. Areas 17 and 12 represent at the same time end-caps fitting over the occipital and frontal lobes. Areas 1, 4, 5, 6, 7, 8, 9, 20, 21 and 22 are *partial segments*, forming narrow or wide band-like zones extending over only a part of the lateral surface and sometimes encroaching partially onto the medial side. Exceptions to this arrangement are the cortical areas around the corpus callosum and those lying lateral to the hippocampal sulcus, that is to say the fields of the cingulate gyrus and the parahippocampal gyrus: areas 23 to 30, 35 and 48.

The regions distinguished in man and the gyrencephalic monkeys are also found in their entirety in the marmoset, but usually in simplified form and arrangement and sometimes with a reduced number of areas. Their precise topographic description is impossible as there is a lack of morphological landmarks for delimiting them in view of the absence of sulci. It must suffice to point out their general position and their mutual relationships.

It is of importance that, in spite of the lack of dividing sulci, there are often sharp borderlines between neighbouring structural regions. Thus, the (agranular) precentral region is quite sharply delimited from the (granular) postcentral region, just as is the latter from the parietal region (Figure 29). The knifesharp transition from the striate area (area 17, Figure 21) needs no further emphasis. Also, the cingulate region stands out clearly from the adjacent areas 1 to 8 and 19, although there is no callosomarginal sulcus to separate them.

Postcentral region.

The postcentral region of the marmoset, in contrast to all other monkeys and to man, is only represented by a single field, the *common postcentral area* (1-3). The differentiation into three or four individual fields typical of higher

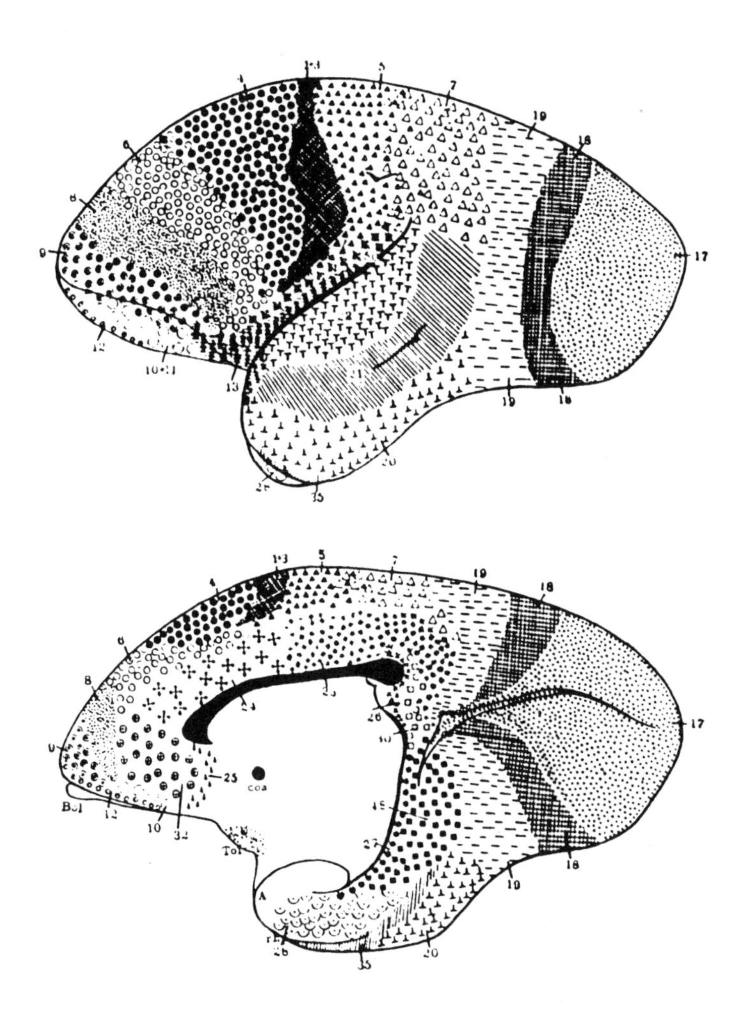

Fig. 96 and 97. The cortical areas in the marmoset (*Hapale*). 2:1.

primates has not yet developed in this lowest family of the monkeys. On the whole this area corresponds broadly in terms of its shape and position to the combined zone made up of these three juxtaposed areas. As in their case, it borders anteriorly on the unequivocally homologised area 4, and posteriorly on the equally easily identifiable area 5, so that there is no doubt about its homology. (In Figure 29 of Part I its spatial relationship to its two neighbouring areas is demonstrated by a microphotograph.) Its shape is not completely regular; it narrows superiorly and inferiorly, is wider in the middle, and forms a thin band-like stripe that begins on the medial surface - corresponding to the paracentral lobule - and crosses the superior edge of the hemisphere to extend over the whole lateral surface as far as the Sylvian sulcus.

The total area of this region is small and is perhaps somewhat magnified in the map.

The precentral region

as in higher primates consists of two individual fields, the *giant pyramidal area* (4) and the *agranular frontal area* (6), that resemble the areas of the same names in man and the guenon in all respects concerning their intrinsic cytoarchitectonics and also their external form and situation. The homology between these areas is complete. As far as the relative extent of the region is concerned it can be said that it forms an even greater proportion of the whole frontal lobe than in the cercopithecids. Its area occupies some three-fifths of the frontal cortex, as compared to about half in the latter. This zone has thus increased in size at the expense of the frontal region. Of the two fields, area 4 is far larger than area 6, in contrast to man where the opposite relationship exists. The shape of both areas is that of a wedge with the apex directed inferiorly. They are not as clearly demarcated from each other as in the higher monkeys, for their borders overlap somewhat, as indicated on the map.

The frontal region

of the marmoset only consists of four clearly individually distinguishable areas (8, 9, 10, 12) whose borders are fairly imprecise. The relative area of the region is much smaller than in the gyrencephalic monkeys and occupies scarcely two-fifths of the whole frontal cortex. The fields all have a typical segmental organisation as seen from the surface, and area 12 covers the orbitofrontal margin of the cortex like a cap.

The parietal region

is made up of two fields as in the cercopithecids, the *preparietal area* (5) and the *parietal area* (7), that fuse with each other across indistinct borders. Both posteriorly and laterally its structure gradually gives way to that of the temporal and occipital regions.

Area 5 - the *preparietal area* - is far more extensive than in all other monkeys and forms almost half the total parietal cortex. It stretches ventrally over the lateral surface to close to the superior bank of the Sylvian sulcus; on the medial side it covers only a narrow wedge near the upper margin of the cortex.

Area 7 - the *parietal area* - includes the rest of the parietal region and by far its largest part equally lies on the lateral surface where it stretches to the upper (posterior) end of the Sylvian sulcus.

The temporal region

stretches from the Sylvian sulcus dorsally to the rhinal sulcus ventroposteriorly, reaching the temporal pole anteriorly and merging with the occipital region posteriorly across ill-defined borders. It is organised simply as in the guenon and can be divided into *superior, middle* and *inferior temporal areas* (20, 21 and 22) of which the position and extent can be judged from the brain map.

Occipital region.

It is relatively even more extensive than in the pithecoid apes, making up approximately a fifth of the total cortical surface and thus exceeding in area all other regions of the marmoset cortex; it is also relatively larger than the same region of any other mammal. One can then make the general statement that in the marmoset there is a marked overdevelopment in the area of the occipital cortex.

Its organisation into individual segment-like fields is very clear in Figures 96 and 97. Three coronal structural zones surrounding the whole circumference of the hemisphere can be defined schematically more readily than in the same region of any other animal.

Area 17 - the *striate area* - is the most posterior of the coronally-placed segments and forms the caudal end-cap of the surface of the hemisphere. It is the largest field in the marmoset and covers about a tenth of the total cortical surface of this animal. (For its structure see Figure 70.)

Area 18 - the *occipital area* - surrounds the striate area anteriorly in a circle and also forms a closed belt-like zone, but throughout is very limited in width.

Area 19 - the *preoccipital area* - is on average somewhat wider than the last area and equally surrounds it anteriorly like a belt on both medial and lateral surfaces. However, it cannot be construed as a completely closed coronal structure, for on the medial side at the anterior end of the calcarine sulcus sure proof of continuity is lacking. In any case both areas 19 and 18 are severely narrowed here and in places they are scarcely recognisable.

The cingulate region

is rather differently organised from that of the cercopithecids. On the brain map, apart from areas 23 and 24 - the *posterior* and *anterior cingulate areas* -, another field, area 25, is delimited anterior to the genu of the corpus callosum and certainly represents the homologue of area 25 of lower mammals; it can be denominated the subgenual area. Also, area 32 can be differentiated as a special cytoarchitectonic zone in the anterior half of area 24, but its homology with area 32 of higher monkeys is doubtful.

Retrosplenial region.

It demonstrates a clearer differentiation into various structural zones, a feature that is scarcely discernible in the cercopithecids, so that the idea of a degree of topical specialisation seems justifiable. In the map, therefore, two separate areas (26 and 30) are indicated within the exceedingly narrow retrosplenial region, but in view of their restricted width their boundaries can only be established with difficulty. The intrinsic differentiation of three areas begins to emerge even more clearly than in man only in prosimians, where the region occupies a wider surface area.

In the hippocampal region

the *presubicular area* (area 27), the *perirhinal area* and the *entorhinal area* (area 28) again exhibit their typical features (Figure 24). In addition I was able to delimit an area 48 (retrosubicular area) the homology of which still proves difficult to determine.

III. The prosimians (lemurs). (Figures 98 and 99).

The surface topography of prosimians, especially the lemurs, shows many analogies with that of the lower monkeys. Basically it resembles a copy of the latter, and deviations only arise in details of size and position. There is a striking outward resemblance between the brain maps of these animals and that of the marmoset. The first surface map of the cortex of the lemur was presented in my sixth communication (1907). A precise topographic description of individual fields and their relations to sulci can be found in my seventh communication on histological localisation (1908). I shall follow it closely in the following description.

Since the publication of that first map, a paper by Mott and Kelley appeared, also dealing with details of surface localisation in the prosimian brain based on cortical structure. The authors emphasise that in general there are only minor differences between their results and mine, beyond the fact that my surface map illustrates a large number of "subdivisions" [7]). However, on closer critical comparison of the two organisational models, not inconsiderable

---

[7]) F.W. Mott and A.M. Kelley, Complete survey of the Cell Lamination of the Cerebral Cortex of the Lemur. - Proceed. of the Royal Society B, Vol. 80, 1908.

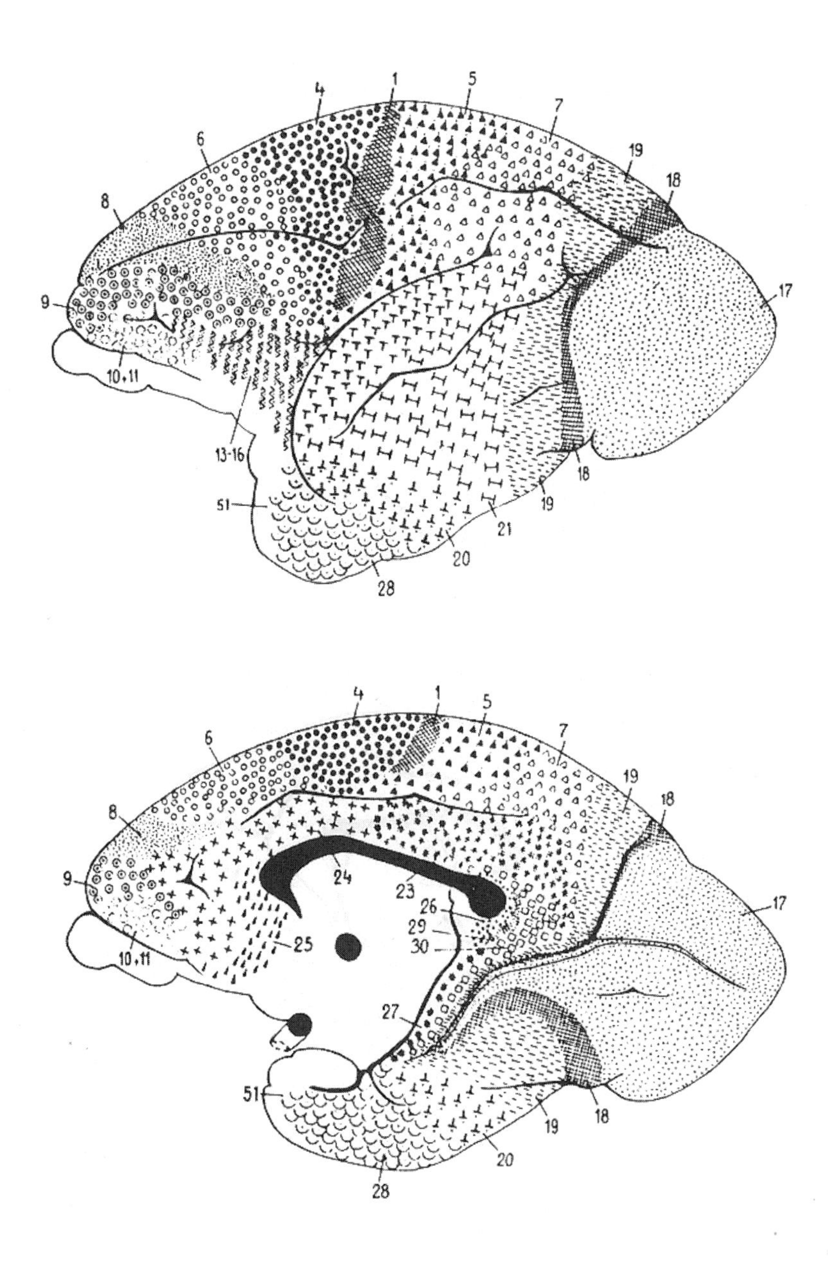

Fig. 98 and 99. The cortical areas of the lemur. 2:1. (Reproduced unchanged from the seventh communication; thus area 35, the perirhinal area, is absent.) (*130)

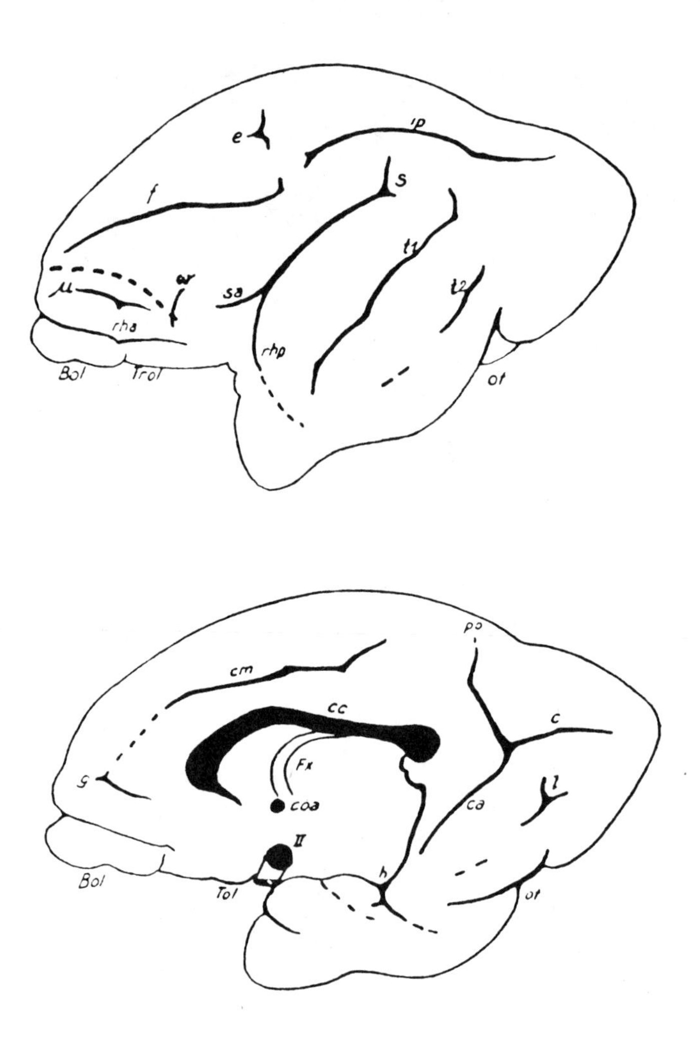

Fig. 100 and 101. The sulci of the lemur's brain.

differences appear. In particular Mott and Kelley have omitted to undertake a precise delimitation of the fields; all over their map they leave blank "intermediate areas" between the major areas for, as they maintain, the structure and extent of the border zones is too indistinct to permit a precise parcellation. This concept is only partly right, as I have emphasised repeatedly, for it is nevertheless possible to differentiate the distinct individual areas of the map from each other. I consider this essential in order to make possible a comparison with other animals, and especially to demonstrate the important relations of areas to sulci. Thus Mott and Kelley distinguish only eight or nine areas whereas my brain map illustrates 23 individual circumscribed areas.

The cytoarchitectonic regions of the lemur are the same as described above, the only variations being in their relative size and their mutual arrangement. The sulci are illustrated in Figures 100 and 101.

Postcentral region.

Like the lissencephalic monkeys (marmosets) the postcentral region of the lemur - and as far as I can tell the same is true of other prosimian families - consists of only a single field, the *common postcentral area* (1). This area represents an amalgamation of areas 1 to 3 found in the postcentral gyrus of cercopithecids, as is illustrated in detail in my seventh communication by means of microphotographs. Its position and outward shape correspond closely to those of the above three fields taken together. In both brains its anterior border is area 4, its posterior area 5, and it comprises a band-like zone stretching medially and laterally over the dorsal margin of the cortex of which by far the largest part lies on the lateral surface (*131). Its shape is not entirely regular; it forms overall a narrow strip running from superior to inferior and undergoes a sharp narrowing in the middle, where the intraparietal (ip) and frontal (f) sulci approach each other, broadening somewhat above and below this level. Its ventral end lies on the lateral surface near the Sylvian sulcus (s), but it extends only a little onto the medial side, not reaching the callosomarginal sulcus, and is sharply narrowed by areas 4 and 5. It is important to note that its rostral border does not correspond to the sulcus e, but lies considerably more posteriorly. The similarity to the homologous area 1 of the marmoset is very great (Figures 96 and 97).

What Mott and Kelley describe as "post-central area" does not correspond to our postcentral area, but rather includes in addition the preparietal area (5) and the parietal area (7) of my map.

Precentral region.

The similarity in position and shape with the region of the same name in man and monkey is very close, the only real divergence being in relative size, in that it has become bigger than in the monkey at the expense of the rest of the frontal lobe, that is to say of the frontal region. In the lemur it occupies considerably more than half the total frontal cortical surface, of which about two-thirds is on the lateral surface, whereas in monkey it includes only half to three-fifths of the total frontal cortex.

It consists of areas 4 and 6. (For area 4 see Figures 45 and 60.)

Area 4 - the *giant pyramidal area* - is very similar to the homologous area in the marmoset. It is widest on the margin of the cortex, narrowing rapidly inferiorly before again widening at the principal frontal sulcus as the cortex of this area spreads rostrally and caudally in the depths of this sulcus. Ventral to the frontal sulcus (f) it forms a strip that becomes narrower and narrower, running obliquely inferoposteriorly toward the Sylvian sulcus (s). Overall, then, it forms a wedge-shaped field of which the base lies superiorly and the apex inferiorly. On the medial surface the area narrows toward the callosomarginal sulcus, reaching the dorsal bank of this sulcus and merging with the most rostromedial wedge of area 5. A similar situation is found at the ventrolateral end of the area near the Sylvian sulcus (s). The sulcus e lies in large part within the area and does not form its caudal border. Thus we notice once again that the sulcus e of prosimians adopts a different relation to the giant pyramidal area than the central sulcus of primates; the latter is the posterior limiting sulcus of area 4, while the former lies within area 4. Thus there is no homology between the cortical segments that form the prosimian sulcus e and the central gyri that border the central sulcus of primates, for the corresponding zones are completely different in their morphological structure.

The posterior third of the frontal sulcus (f) is closely surrounded by area 4, only the short ascending end branch being an exception in that only its rostrodorsal bank is entirely within this area while the caudoventral bank falls partially in area 1. This ascending end branch of the frontal sulcus thus forms the border between area 4 and area 1 for a short distance and could therefore be more correctly considered as homologous to the central sulcus of the primate than the sulcus e (ε of Ziehen).

Area 6 - *agranular frontal area.* Like the previous area, area 6 lies for the most part on the lateral surface and less on the medial. It is widest at the upper edge of the hemisphere and extends over the lateral convexity as an inverted wedge-shaped zone across the middle third of the frontal sulcus (f), then ventrally almost to the Sylvian sulcus (s). Here it borders area 4 and 5 and is especially poorly demarcated from the former. On the medial surface it occupies a similar zone to area 4, reaching the callosomarginal sulcus; there is no really sharp border with area 4 on the superior bank of this sulcus.

The surface area of area 6 compared to area 4 is even smaller than in the monkey. Thus we observe that the relative size of areas 4 and 6 gradually shifts in favour of the former from man downwards. The giant pyramidal area (4) exceeds the agranular frontal area (6) in size more and more. Area 6 is also clearly structurally differentiated from area 4 in the lemur in that it does not possess Betz giant pyramids. The only common feature of both areas is the lack of the innner granular or fourth layer.

Mott and Kelley obviously overlooked this essential difference between the two neighbouring structural zones; they only recognise a "motor area", the homologue of my area 4, which they claim gradually gives way to their "frontal area" by way of an undifferentiated (!) transition zone. Thus, according to them, the whole frontal lobe is only divisible into two actual areas. On the other hand they separate the "motor area" into two subfields superoinferiorly, without sufficient evidence in my opinion, a "motor A" and a "motor B", of which the latter lies ventral to the frontal sulcus and is supposed to have a distinct granular layer. I consider this last statement erroneous. It is likewise doubtlessly due to an error when Mott and Kelley (*132) claim that the "motor area" (area 4) fuses caudally with the "post-central area" without obvious boundaries. As I believe I demonstrated convincingly in Part I the border between these two totally contrasting structural zones is formed in all animals by the sudden appearance of an inner granular layer, which leads to a sharp transition. The wider extent of Mott and Kelley's "motor area" is explicable by this lack of clarity in the topographic parcellation.

Frontal region.

In lemurs it occupies only about a third of the total frontal cortex, whereas in the cercopithecids it comprises nearly half. Three different zones can be distinguished with certainty, although their mutual delimitation is rather indistinct for all three. Their homologies with areas of the same number in the monkey is particularly uncertain.

Area 8 - the *granular frontal area* - has the shape of a horizontal wedge on the lateral surface with its base anterosuperiorly; the apex extends some distance caudally, almost to the Sylvian sulcus (s). A large part of the area lies within the frontal sulcus (f). The anterior end of this sulcus is entirely within it, while in monkeys it is the posterior end of the sulcus around which area 8 extends. Medially, this area occupies only a narrow zone just anterior to the callosomarginal sulcus.

Area 9 - the *prefrontal area* - includes the actual frontal pole and is much less extensive than in monkeys.

Areas 10 and 11 - *orbital area*. The easily distinguishable areas 10 and 11 of the guenon must be combined in a single type in prosimians as they cannot be divided into two separate structures with sufficient certainty, either architectonically or from their position. There is perhaps the greatest similarity with area 10 of cercopithecids. The orbital area (10) thus represents the same zone that is occupied in lower monkeys by areas 10 and 11, comprising essentially the medial and lateral orbital gyri. In our map it covers comparatively little surface area, mainly on the markedly shrunken orbital surface.

In prosimians it is not possible to demonstrate with certainty a cortical area corresponding to area 12 of the monkey.

Insular region.

Whereas in primates the actual insular cortex lies deep in the insular gyri completely covered by the opercula and is thus not represented in the surface map of the cercopithecids or of man, the insular areas of the lemurs extend over a wide extent of the free hemispheric surface. As already explained above, the different insular cortical structures are treated for the moment as a homogeneous region for the sake of clarity and because parcellation and homology still present difficulties. These areas 13 to 16 occupy a considerable expanse, as shown in Figure 98. They encompass the zone between the inferior end of the Sylvian sulcus (s) and the anterior rhinal sulcus (rha) and stretch anteriorly as far as the orbital sulcus, without any very clear boundary. Dorsally the field partly crosses the sulcus sa, and posteriorly it also crosses the depths of the Sylvian sulcus in places. The sulcus ω lies within the insular region.

The parietal region

encompasses a relatively large part of the free surface of the hemisphere, as in the guenon. Its demarcation from neighbouring regions presents great difficulties as its histological differentiation is indistinct. Also, areas 5 and 7 that make up the region cannot be exactly demarcated, particularly from temporal and occipital cortex.

Area 5 - the *preparietal area* - forms a wedge-shaped field with its base toward the margin of the hemisphere, of similar shape to area 4. On the lateral surface the apex of the area reaches rostroventrally to the dorsal bank of the Sylvian sulcus (s) and at its inferior end partly touches area 4 and partly area 6. Thus there is here a mixed zone in which the boundaries are frequently indistinct. It is similar on the medial surface where the area borders the callosomarginal sulcus and fuses spatially with areas 4 and 6. The wedge shape of the area on the lateral surface is somewhat altered by an anteroposterior expansion of its borders in the region of the intraparietal sulcus (ip). The map can only express this roughly.

It is worthy of note that there is an appreciable difference in the surface position of this area compared with the homologous area in lower monkeys. In monkeys (Figure 90) area 5 only reaches the ventral end of the intraparietal sulcus (ip), remaining well distant from the Sylvian sulcus (s); in the lemur, on the other hand (Figure 98), we see that our area 5 extends ventrally significantly beyond the rostral end of the intraparietal sulcus (ip), reaching close to the Sylvian sulcus (s). It is obvious from the surface maps reproduced from my seventh communication (Figures 11 to 15) that area 5 even encroaches quite widely on the deep cortex of this sulcus. This shows that the zone lying entirely dorsal to the intraparietal sulcus (ip), corresponding - at least in its anterior portion - to the superior parietal lobule, also extends in prosimians over that cortex lying ventral to the intraparietal sulcus and reaches the Sylvian sulcus. From this one can conclude that, in spite of their similar surface situation, the so-called intraparietal sulci of prosimians and primates are not homologous as they lie in morphologically quite different areas.

Area 7 - the *parietal area* - occupies a much wider zone on the free surface of the hemisphere than the homologous area of lower monkeys. One can appreciate from Figure 90 that area 7 of the monkey is sharply narrowed on the lateral surface near the edge of the hemisphere while area 5 reaches further posteriorly, and moreover that part of area 7 belongs to the deep cortex of the parieto-occipital incisura (po). In prosimians (Figure 98) this sulcus is absent and area 7 thus spreads further over the lateral surface. The area widens ventrally on both medial and lateral surfaces; it surrounds the dorsal end of the Sylvian sulcus (s) on the lateral convexity, and on the medial surface it crosses the posterior end of the callosomarginal sulcus and extends like a narrow wedge rather far postero-inferiorly to the isthmus of the cingulate gyrus.

Mott and Kelley include the whole parietal lobe in a single field, their "post-central area"; they did not undertake any further differentiation.

Occipital region.

As in monkeys, the size of the occipital region in lemurs is relatively large. It is structurally organised into three individual areas, 17, 18 and 19, whose borders are however not so strikingly contrasted as in the former. Nevertheless, once the differences have been seen in other brains it is not difficult to recognise them in lemurs (Figure 72).

Mott and Kelley's scheme only includes a single occipital field, their "visual area", the whole of the rest of the cortex being considered as an intermediate structural region without particular topical organisation, but they admit that in this region the homologues of Campbell's "*parietal*" and "*visuo-psychic area*" (my areas 7, 18 and 19) are probably included [8]). It has to be admitted that the transition of our fields to the temporal and parietal types is poorly marked and less obvious than in cercopithecids, but their separation is still unequivocal on histological grounds.

Area 17 - (calcarine cortex) *striate area*. - This area represents the caudal end-cap of the hemisphere as in the lower monkeys. However, in contrast to the lower monkeys (with the exception of the marmosets), more than half its extent lies on the medial surface. Whereas in primates (except the marmoset) the striate area at the occipital pole is restricted to the immediate vicinity of the calcarine sulcus, so much so that in many lower gyrencephalic monkeys, such as the guenon and the woolly monkey, it remains confined to the cortex deep in this sulcus and only covers a minute part of the cuneus and the lingual gyrus, in those prosimians that I have investigated (lemur, sifaka, potto, slow loris) this area encompasses a very wide expanse of the medial cortex including the whole surface of the cuneus and the lingual gyrus.

---

[8]) "The cortex of this region thus seems to be intermediate in structure - as it is in position - to the post-central, temporal and visual types" (p.495).

In its topical relationship to the calcarine sulcus the striate area of lemurs thus resembles that of the anthropoids and man rather than that of the cercopithecids (*133). This area shows the greatest similarities to that of the marmosets in which it similarly occupies a large part of the medial surface on both sides of the calcarine sulcus and only sinks deep into the sulcus quite far anteriorly. One should compare Figures 96 and 97 that show the extent of the striate area in the marmoset, and the slow loris, later (*134).

The area stretches far anteriorly along the main branch of the calcarine sulcus (c). However, after the union of this sulcus with the parieto-occipital sulcus (po) it no longer extends onto the dorsal bank of the sulcus but lies mainly ventral to it on the lingual gyrus. In this connection it is important to note that the calcarine sulcus (c) does not everywhere form the precise rostral boundary of the striate area. In places the area extends a little dorsally across the fundus of the sulcus, and elsewhere it does not quite reach it but lies only on the ventral bank of the calcarine sulcus [9]). In fibre preparations these characteristics stand out clearly macroscopically. The relationship of the striate area to the most rostral part of the calcarine sulcus is accordingly not absolutely regular and constant but quite variable, both between genera and species, and between individuals.

Area 18 - *occipital area.* - In its shape and position this area forms a quite similar cortical field to that described in the lower monkeys. It equally represents a coronal zone surrounding the hemisphere like a ring, forming the rostral border of the striate area over its whole extent, like a frame around it. Its caudal border is sharply defined everywhere by a sudden splitting of the inner granular layer at the transition to the calcarine cortex. However, its rostral boundary with the adjacent preoccipital area (area 19) is less clear.

This coronal area 18 is markedly narrower and more regular in shape than in lower monkeys. On the medial surface it is extraordinarily thin, and in both the parieto-occipital sulcus (po) and the anterior branch of the calcarine sulcus consists of such a narrow strip of cortex that the demonstration of this area in a section is often difficult. Nevertheless, by following serial sections it is possible to establish a spatial continuity even within the sulcus, and in the map a closed ring-shaped field is drawn, as in the monkey, surrounding the extreme rostral end of area 17 on the medial surface. On the lateral convexity the area is wider and forms a rather regular broad band.

Area 19 - the *preoccipital area* - is largely similar in shape and surface extent to the occipital area. It forms a coronal field that swings round the occipital area rostrally as the latter does round the striate area. It is wider on the lateral surface, and very narrow on the medial side, especially at the anterior end of the calcarine sulcus; it bulges at the dorsal and ventral margins of the hemisphere. Its total surface area is somewhat more than the occipital area, but

---

[9]) cf Figures 125-128 of my fifth communication (Journal f. Psycholog. u. Neurolog. vol. VI, p.332) and Figures 19-26 of my seventh communication (ibid. vol. X, p.311).

relatively less than the homologous cortical area of the lower monkeys.

The temporal region

is relatively considerably smaller than in monkeys or even in man. It is only clearly demarcated anteriorly and inferiorly, that is from the insular region (areas 13 to 16) and the hippocampal region (area 28). In contrast, it undergoes smooth transitions dorsally and caudally to the occipital and parietal cortex. I distinguish three areas. On the other hand Mott and Kelley only have one field in the temporal lobe, their "temporal area".

Area 20 - *inferior temporal area.* - As the middle temporal sulcus (t2) is less developed in prosimians than in monkeys and is rather variable, the surface boundaries of this area are less obvious, only being marked for a short distance by the posterior rhinal sulcus (rhp) anteriorly on the lateral surface. Thus it lies essentially lateral or superoposterior to the posterior rhinal sulcus and forms the field adjacent to the entorhinal (28) or perirhinal (35) areas dorsally as they lie respectively medial or inferior to this sulcus. Caudally it merges with area 21 or, on the medial surface, area 19. Medially it forms a narrow strip of cortex stretching to close to the hippocampal sulcus.

Area 21 - the *middle temporal area* - is only poorly localisable topically as its cytoarchitectonic boundaries are quite indistinct, especially dorsally and caudally. It merges with the parietal cortex (area 7) of the parietal lobe superiorly without obvious borders and fuses just as gradually with the preoccipital cortex (area 19) posteriorly. Rostrally it crosses the superior temporal sulcus and further extends over approximately the ventral third of the superior temporal gyrus.

Area 22 - *superior temporal area.* - Its position does not correspond exactly with the superior temporal gyrus, that is to say the cortical zone lying between the Sylvian sulcus and the superior temporal sulcus; rather it is essentially limited to the dorsal two-thirds of this gyrus while the ventral third, as explained above, belongs to the middle temporal area. This is in contrast to the situation in the cercopithecids where area 22 includes virtually the whole of the superior temporal gyrus.

The cingulate region

manifests more similarities with the region of the same name in callithricids than that of cercopithecids in terms of position and its division into individual fields. Dorsally it is more clearly demarcated from the areas of the lateral surface due to the presence of a callosomarginal sulcus. Three separate areas can be distinguished easily, of which the most caudal, area 23, is *granular*, that is it possesses an inner granular layer, whereas area 24 and 25 of the rostral half of the cingulate gyrus are *agranular*. The border between these two different structural types lies, as in monkeys, approximately along a line that can be considered an extension of the border between the giant pyramidal area

(4) and the postcentral area (1). It is relatively sharp, for the disappearance of the inner granular layer is quite sudden. Thus, in prosimians as in primates, the sharp border between an agranular anterior and a granular posterior division of the hemisphere, that coincides with the central sulcus in most monkeys, continues on the medial surface over the callosomarginal sulcus onto the cingulate gyrus and also divides this gyrus into an anterior and a posterior section. The appearance of such a boundary, common to both prosimians and primates and based on the disappearance of a whole cell layer, seems to me even more important in that it is independent of any sulcal development and stretches over the insula as far as the base of the brain thus encircling the entire circumference of the hemisphere.

Mott and Kelley integrate the whole cingulate gyrus, including the isthmus, in one field, their "limbic area". They do not even separate the retrospenial region from it.

Area 23 - *posterior cingulate area*. - This cortical area occupies essentially the caudal half of the cingulate gyrus with the exception of the part of the cortex of this gyrus that is situated in the most caudal portion of the callosal sulcus, which belongs to area 26 (the ectosplenial area). Its dorsal border is formed by the callosomarginal sulcus (cm) along the whole length of the area, except its posterior end; the ventral border is represented approximately by the dorsal bank of the cortex of the callosal sulcus. (In Figures 23/29 and Plate 7 of my seventh communication the transition from the posterior cingulate area to its ventral neighbour, the agranular retrolimbic area, is illustrated. It is sharply defined by the loss of the inner granular layer.)

This area extends beyond the posterior end of the callosomarginal sulcus caudally and approaches close to the parieto-occipital sulcus (po), only separated from it by a narrow fringe of cortex belonging to the occipital and preoccipital areas, and partially to the parietal area (areas 18, 19 and 7). The borders with these cortical areas are not very sharp. Especially at the level of the splenium of the corpus callosum, below which the posterior limbic area (*135) reaches like a pointed wedge, a mixed zone is created in which localisation is not always easy to unravel due to the various quite small areas that aggregate here. Anteriorly the area reaches about the middle of the cingulate gyrus and merges with the anterior cingulate area - area 24 - with the loss of the inner granular layer approximately where the pre- and postcentral regions are in contact.

Area 24 - *anterior cingulate area*. - Just as the posterior cingulate area occupies the posterior half of the cingulate gyrus, the anterior cingulate area covers approximately the anterior half. It is bordered dorsally, like the former, by the callosomarginal sulcus (cm) and ventrally by the callosal sulcus. Its posterior border coincides with the anterior border of area 23 as just discussed. Rostrally the area extends round the genu of the corpus callosum, crosses the genual sulcus rostrally somewhat, then stretches inferiorly in an arc as far as the inferior margin of the hemisphere. It is separated from the genu itself by the pregenual area that pushes in between them.

Area 25 - *pregenual area*. - I had already distinguished a specific area 25 in the anterior portion of the cingulate gyrus in the marmoset, separate from the anterior cingulate area (area 24). However, its differentiation is so weakly developed that a precisely localised parcellation was difficult to realise. In prosimians area 25 is more characteristically differentiated and the area is therefore more precisely discernable. I was able to ascertain, both in coronal and horizontal sections, that this area swings round the genu of the corpus callosum, only stretches a little anteriorly, and extends below the rostrum in an S-shaped arc. The area only climbs minimally over the dorsal surface of the corpus callosum.

Retrosplenial region.

In contrast to that of the monkey, this region has undergone a considerable development in prosimians; it covers a much wider area and is divisible into three clearly differentiated structural types that have well demarcated boundaries. Certain relationships emerge that are much more striking in the megachiropterans (Figures 102 and 103) and are even further developed in rodents.

Area 26 - the *ectosplenial area* - surrounds the splenium of the corpus callosum as a narrow arc-shaped band and runs for a short distance along the dorsal surface of the body of the corpus callosum itself. The ventral border of the area lies immediately beneath the splenium, where it merges with the presubicular area (27). Caudally it does not encroach on the cortex of the callosal sulcus but remains hidden in the depths of this sulcus. In the map it is indicated on the surface. Its posterior border coincides with the anterior border of the granular retrolimbic area and is not sharply demarcated. These two areas are so restricted in extent, and merge in the prosimians, that I only succeeded in separating them after I had demonstrated the corresponding areas in chiropterans, rodents and several ungulates, in which they are always characteristically differentiated and are more extensive. In monkeys, as we have already seen, both the architectonic differentiation and the surface area of this field are even more limited than in prosimians.

Area 29 - *granular retrolimbic area*. - It forms a very narrow fringe caudal to the ectosplenial area and extends in a crescent around it at the level of the splenium. The area hardly widens over the dorsal surface of the corpus callosum, and also only reaches a short distance below the splenium. Its maximum extent is thus only a few millimeters (Figure 41).

Area 30 - the *agranular retrolimbic area* - equally forms an arc-like cortical field of limited width surrounding the posterior end of the corpus callosum. At its point of maximum dorsoventral extent, this area nevertheless covers a wider zone than area 29. It stretches from the body of the corpus callosum, where it includes some of the ventral bank of the cingulate gyrus, over the caudal end of the cingulate gyrus down to the isthmus and runs along the course of the presubicular area as far as the rostroventral end of the calcarine sulcus.

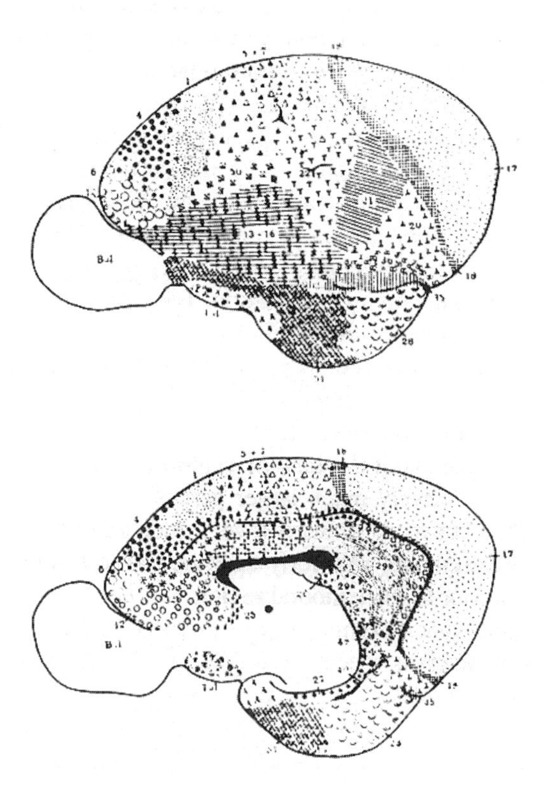

Fig. 102 and 103. The cortical areas of the flying fox (*Pteropus edwardsi*). 2:1.

The hippocampal region

is also better developed in prosimians than in monkeys in terms of surface area as well as histological differentiation. Both areas 27 and 28 are relatively and absolutely much larger than the corresponding areas in, for example, the guenon. It is possible to delimit a perirhinal area (35), a minute, narrow strip along the rhinal sulcus lateral to area 28 (see Figure 25), but it is not marked on the brain map.

Area 27 - the *presubicular area* - represents the ventral continuation of the ectosplenial area. On account of its position it forms mainly the posterior bank of the hippocampal sulcus but in places encroaches a little on the upper surface of the isthmus of the cingulate gyrus. (The latter is drawn intentionally wider in the map in order to be able to represent the crowded areas in this region).

Area 28 - the *entorhinal area* - includes a very considerable part of the cortical surface, involving the whole zone between the posterior rhinal sulcus and the hippocampal sulcus, that is the region that in many lower orders has become separated morphologically as a special piriform lobe. As the posterior rhinal sulcus lies on the lateral surface in lemurs, the entire anteroinferior part of the temporal lobe belongs to this area; posteriorly it crosses this sulcus significantly and is demarcated from the inferior temporal area (20) without an externally recognisable border.

(Mott and Kelley described a similar localisation for their "olfactory area".)

Olfactory region.

Anterior to area 28 in the maps in Figures 98 and 99 there is a white, unmarked zone; this corresponds broadly to the olfactory region of the other brain maps and encloses the architectonic zones that are shown separate in them: area 51 or prepiriform area, the olfactory tubercle, and the amygdala.

IV. Pteropus (flying fox). (Figures 102 and 103).

In the flying fox the macrosmatic character of the brain stands out not only in the structure of the piriform lobe and the strong development of the anterior olfactory lobe but also in its special cortical topography. Although the principle of field organisation is the same as in monkeys and prosimians, the localised differentiation of several regions shows definite divergences. As examples one may simply cite the cingulate, retrosplenial and hippocampal regions, as well as the intense development of the olfactory region that was not even drawn in the maps described so far on account of its rudimentary form. There is a much richer organisation into individual areas than in higher orders, including man, especially in the *cingulate region*, but also the *retrosplenial* and *hippocampal regions* demonstrate a massive expansion and cover a wide extent of the free cortical surface. In addition the *olfactory region* contributes an enormous area.

As otherwise the relationships that have been discussed in detail for the higher orders are repeated and, in particular, all the major regions are found again, I shall describe them only summarily. A detailed topographical description of individual areas is not justified, for as there is an almost total lack of sulci [10]) there are no landmarks to help with surface localisation; thus only a short account of the relative positions and sizes of the regions and areas is possible. In this context it should be noted that the borders of most areas are in reality not so sharp or linear as had to be indicated on the surface map, for the reasons that have been explained several times before.

The **precentral region**, as in primates and prosimians, is represented by the two agranular areas 4 and 6. It lies extraordinarily far forward, as can be ascertained from a comparison with the other brain maps; its total extent is much less, especially on the lateral surface. The two individual fields form indistinct borders with each other and their structures overlap at the transitional zones as indicated in the drawing. It is worthy of note that the giant pyramidal area (area 4) lies to a great extent on the medial surface and stretches far posteriorly on the dorsal bank of the splenial sulcus, a feature that I have not been able to demonstrate in any other animals. On the lateral surface both areas 4 and 6 form a sort of wedge-like zone that reaches to the superior margin of the insula, as we have become accustomed to see in the animals described above; area 6 extends as far as the frontal pole. A true frontal region anterior to the precentral region is absent.

The **postcentral region** continues caudally from the previous region after a fairly sharp border marked by the quite sudden reappearance of an inner granular layer in the sections. It is composed of a single structural area (1) that is homologous to the combined areas 1 to 3 in the primate, for reasons that were explained earlier. Posteriorly it gradually fuses with the parietal area.

The **parietal region** also consists strictly speaking of only one area (7). It is true that one finds a rather different cytoarchitecture in its anterior portion from that in its posterior part in that more, larger pyramidal cells are present in layer V, corresponding to the situation in the preparietal area of primates and prosimians. Nevertheless a separate area 5 cannot be distinguished, as the structural changes from anterior to posterior progress quite gradually and one can therefore never say where the border between the two areas might be. Moreover, the same situation exists with respect to the transition to the postcentral region. It is equally gradual so that the question arises whether it would not be more correct to regard the whole zone situated caudal to the agranular precentral region as a single entity. Its overall structure is very similar; in spite of this there are architectonic features in its anterior portions, that is to say just posterior to the agranular area 4, that are more related to areas 1 to 3 of primates and thus justify a separation from the parietal area.

---

[10]) On the medial side there appears only the *splenial sulcus*, and on the lateral side, apart from a short posterior rhinal sulcus, two small dimples, of which many authors consider the dorsal as a homologue of the *lateral sulcus* and the ventral as a fragment of the *Sylvian sulcus*. A true Sylvian sulcus is absent (Figures 102 and 103).

The **occipital region** consists mainly of the *striate area* - area 17. This once again represents a typical end-cap applied over the occipital pole, approximately equally distributed between the medial and lateral surfaces. Medially it reaches the splenial sulcus so that this sulcus forms the exact anterior border, while laterally there is no morphological border. Both laterally and medially a rather irregularly structured band-like strip follows the whole extent of its anterior boundary, a coronal field that spans the entire hemisphere just like the occipital or preoccipital area of primates. Based on the similarity of position it would be quite justifiable to consider this area as homologous to the occipital area - area 18 (and 19) - of primates. The striate area (17) is by far the largest cortical field of the flying fox.

The first thing to note about the **insular region** is that it lies entirely on the free surface and includes a relatively much larger part of the cortical surface than in the higher orders described above. It includes a very wide zone directly above the olfactory or piriform lobe, stretching anteriorly to close to the frontal pole. Posteriorly it pushes far caudally inferior to the temporal region.

The individual areas that make up the whole region are not shown separately in the brain map; however, it deserves special mention that even in the flying fox the insular region is divisible into two subregions, a caudal granular and a rostral agranular zone.

The **temporal region** (areas 20, 21, 22 and 35) is only weakly developed in the flying fox. Although it is clearly topically separated from the insular and hippocampal regions, its borders with the parietal and occipital cortex are not sharp. Also, only an approximate and indistinct parcellation of its individual areas is possible, with the exception of the heterogenetic area 35 that is situated ventrally just above the posterior rhinal sulcus. However, it is still doubtful whether it would be more accurate to include this with the structures of the piriform lobe, the more so since in other species it belongs spatially mainly or entirely to it. Apart from area 35 and perhaps area 20 their homology with the homogeneous (*136) fields of other animals cannot be established with certainty. It should be noted that area 22 is related partially to the shallow sulcus that is supposed to represent a rudiment of the Sylvian sulcus.

The **cingulate region** displays even more marked differences in field organisation from the brains described so far. Not only is the number of fields greater but also their intrinsic structure is sometimes so modified that it is impossible to identify given areas. Only the similarity of position provides a basis for homology. Thus it emerges that the whole caudal half of the cingulate gyrus is taken over by the massively developed retrosplenial area that climbs far over the dorsal surface of the corpus callosum. - A total of six areas are indicated on the map. Of these, area 23 is granular and should correspond to the posterior cingulate area (23) of other animals even if it is only a matter of relative homology. All the other areas are agranular; they are marked as 24, 25, 31a, 31b, 32a and 32b. Area 25 can be taken to be the certain homologue of area 25 of prosimians, but it is hardly possible to homologise the other structural zones in detail. It should be noted that a special cortical type is found in the depths of the splenial sulcus, and can be followed along the whole length of the horizontal branch of this sulcus, but which has a different structure in its caudal half from its rostral (indicated in the map by different signs and the numbers 30a/b and 31a/b) (*137).

In the organisation of the **retrosplenial region** essential fundamental differences from that in higher orders appear, both in the surface area of the whole region and in the number of separate individual areas. Whereas in primates and prosimians this region represents only a very small part of the cortex and is often only developed extremely rudimentarily, here it includes a very broad zone with five or six well differentiated areas (26, 29a, 29b, 30a and 30b) (*137). Obviously its massive expansion is related to the great width of the posterior portion of the cingulate gyrus. The same trend is found in the kinkajou and is even more pronounced in the rabbit and ground squirrel. What the physiological consequence of this singular development of the retrosplenial region could be can hardly even be the object of speculation for the moment.

As to the individual areas, it is not possible to suggest absolute and unequivocal homologies in all cases from the cell preparations. For this region myeloarchitectonics often provide the best decisive evidence. Area 26 (the ectosplenial area) is a small wedge-shaped field below and behind the splenium of the corpus callosum and is situated in a similar position in almost all animals investigated, even if its structure varies. In the place of area 29 (the granular retrosplenial area) of primates and prosimians, there are two separate structural types, indicated as areas 29a and 29b. Together they correspond to area 29 of higher animals and areas 29a-e of the rabbit (Figure 107). Area 30 (the agranular retrosplenial area) is, in contrast to that of the lemur and the kinkajou in which its width is relatively great, confined to a narrow strip on the anterior bank of the splenial sulcus, and can be further divided on structural grounds into two different sections, one posterior, 30a, that coincides with the vertical branch of the splenial sulcus, another anterior, 30b, that lies along the horizontal branch of this sulcus. Dorsal to area 30b, yet another narrow structural field can be distinguished lying in the deep cortex of the splenial sulcus (area 31), and whose association with either the retrosplenial region or the cingulate region must remain open; it is allotted to the latter above. (See Figure 39 for area 29, and Figure 64 for the whole retrosplenial region.)

The cortical region lying directly posterior to the splenium of the corpus callosum, that is only divisible into from one to three areas in primates and prosimians, can therefore be separated into six clearly different structural areas in the flying fox, each with more or less sharp borders. - In microchiropterans, incidentally, the relationships in this region are also much simpler.

The **hippocampal region** also displays richer differentiation than in higher orders. First, one finds a new area 48 ventral to areas 29a and 29b approximately corresponding to the isthmus and stretching over the hippocampal sulcus and containing a markedly reduced laminar organisation (heterogenetic cortex). - Area 27 (the *presubicular area*) lies in its usual position on the dorsal bank of the hippocampal sulcus, but does not reach as far caudally along this sulcus as in monkeys and prosimians. Parallel to this area I have defined a quite narrow zone, area 49 (the *parasubicular area*), which is otherwise only discernible in the rabbit (Figure 107), the hedgehog and the kinkajou (Figure 105). Area 28, the heterogenetic *entorhinal area*, is one of the largest areas of the whole cortical surface in the flying fox; it stretches quite far over the lateral surface reaching the posterior rhinal sulcus and then separates into two subfields (28a and 28b), indicated on the map by different symbols.

It is of importance that area 28 covers far less than the entire piriform lobe, but remains limited solely to its posterior half, while the anterior half, including a part of the anterior olfactory lobe, is occupied by a rudimentarily developed cortical formation - "defective cortex" in Meynert's sense. This region, that we name the **olfactory region**, is very extensive in the flying fox and we have indicated it as a special region in the map, but in the lemur it is only distinguishable at the extreme anterior part of the temporal pole (the unmarked white field in Figures 98 and 99). In the flying fox, in contrast, it covers most of the lateral surface of the piriform lobe and can be traced anteriorly to the vicinity of the olfactory bulb, always just ventral to the anterior rhinal sulcus.

Within this olfactory region one can distinguish three structural zones with obviously rudimentary (heterogenetic) laminar patterns, the *prepiriform area*, the *amygdaloid nucleus* (AA) and the *olfactory tubercle* (Tol) (*138).

V. The kinkajou (Cercoleptes caudivolvulus). (Figures 104 and 105).

I have only been able to complete a study of the organisation of the whole cortical surface so far in one animal belonging to the carnivores, the kinkajou. A more important and indispensible task for the future is the precise cortical localisation and parcellation of the brain of the dog, and perhaps also of the cat, that is those animals that physiologists have used mainly or exclusively as experimental subjects for localisational studies. Individual important areas of dog and cat have already been described briefly (my fourth and fifth communications). The complete study of localisation in these brains will be one of my next tasks. The brain of the kinkajou is particularly well suited for the production of a brain map in that on the one hand it is of moderate size and is therefore not too difficult to process technically, nor is it too primitive in its organisation, and on the other hand because its sulcal arrangement is simpler than in many of the larger carnivores [11]).

As a macrosmatic animal the kinkajou possesses, like the flying fox, an olfactory lobe of sizeable proportions and a strongly developed piriform lobe. Also, the cingulate gyrus, and especially its retrosplenial part, is strikingly wide

---

[11]) The sulci of the brain of the kinkajou are, medially: the *splenial sulcus (spl)*, which continues as the *cruciate sulcus (cr)* at the superior cortical margin, the *genual sulcus*, the posterior end of the *posterior rhinal sulcus (rhp)* and the *hippocampal sulcus (h)*; laterally: the *anterior and posterior rhinal sulci (rha, rhp)*, that unite with the main stem of the *Sylvian sulcus (s)*, the *presylvian sulcus (ps)*, the *cruciate sulcus (cr)*, the *lateral and postlateral sulcus (l, pl)*, the *coronal sulcus (co)* and the *suprasylvian sulcus (ss)*.

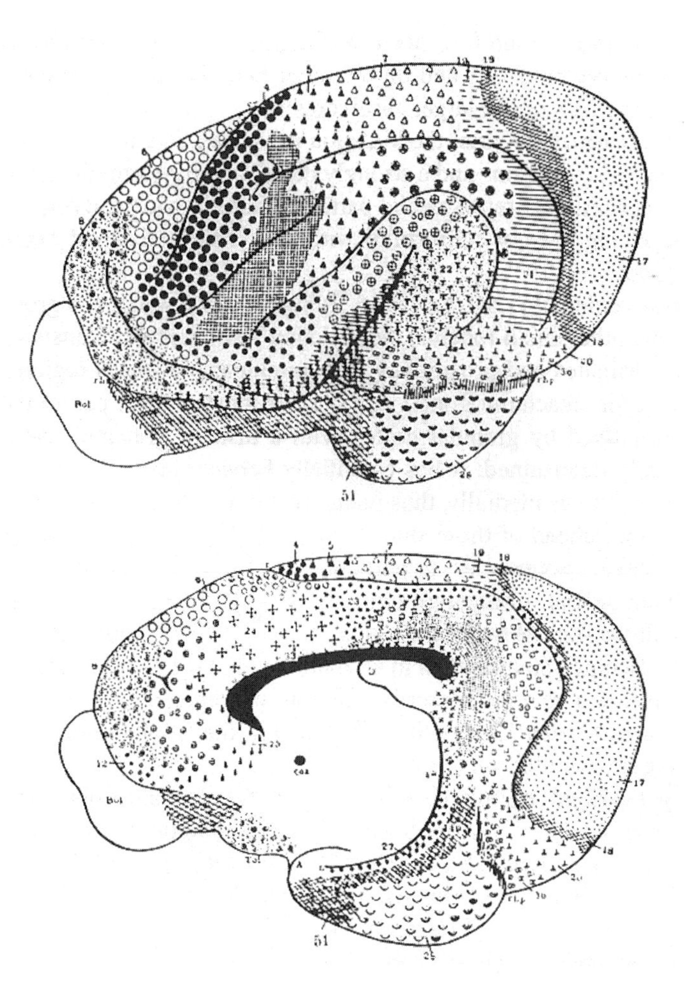

Fig. 104 and 105. The cortical areas of the kinkajou (*Cercoleptes caudivolvulus*). 1:1.

and relatively even bigger than in the flying fox. In conformity with this is the rich areal organisation of this cortical region.

Frontal region. In contrast to the flying fox, the rabbit, the ground squirrel, the hedgehog and other small rodents, the kinkajou possesses an extensive granular frontal region, like primates and prosimians. The agranular precentral region (areas 4 and 6) does not therefore reach the frontal pole, but the most anterior part of the frontal lobe is again encompassed by granular cortex with a distinct granular layer. Its position is fairly precisely determined: it lies essentially between the presylvian sulcus laterally and the genual sulcus medially, thus including the whole cortical surface on the medial and lateral sides ahead of these sulci as far as the frontal pole. However, the sulci do not form the exact boundaries; in particular the region extends partially caudally over the presylvian sulcus. Compared with the massive frontal region of man, this region represents only a minute fraction of the surface of the hemisphere and is also relatively and absolutely much smaller than in all monkeys and the bigger lemurs; nevertheless it can be demarcated as a circumscribed region of homogeneous structure. Whether, in addition to area 8, area 12 should be included in the frontal region or is better in the cingulate region, I prefer not to decide.

The **precentral region** directly adjoins the previous one caudally and thus extends immediately posterior to the presylvian sulcus. In the kinkajou, as in all carnivores, it has undergone a degree of architectonic specialisation that is scarcely found elsewhere among mammals, including the primates. The whole region is distinguished in striking fashion from the entire remainder of the cortex by the total absence of the inner granular layer, by its sparse cellularity, by the domination of large cells not organised in laminae, and by the considerable depth of the cortex, and it can be identified at a glance by the expert. It consists of the two characteristically organised agranular areas 4 (Figures 46 and 61) and 6 (the *giant pyramidal area* and the *agranular frontal area*) that we have found with similar characteristics in all orders so far.

Area 4 lies mainly posterior to the cruciate sulcus, between it and the coronal sulcus, that is essentially on the posterior sigmoid gyrus, but the agranular frontal area, area 6, lies anterior to the cruciate sulcus, mainly on the anterior sigmoid gyrus. At the lower end of the coronal sulcus area 4 bends at an acute angle and continues as a narrow strip around the suprasylvian sulcus onto the anterior sylvian gyrus dorsal to the insular cortex. Whereas area 4 in primates and prosimians is generally vertically oriented in a wedge, tapering from superior to inferior, in the kinkajou its lower third bends sharply inferiorly, a feature that I described earlier in the dog, cat and stone marten. The cytoarchitecture of the suprainsular portion of the giant pyramidal area (4a) is rather different from the dorsal part of the area (4). Area 6 is also partly involved in the angular bend.

The borders of the precentral region are relatively sharp. Anteriorly it extends to the presylvian sulcus but leaves the dorsocaudal bank free. Posteriorly the morphological border is less distinct: medially it only encroaches slightly on the marginal gyrus, while on the lateral surface it extends upwards to near the anterior end of the lateral sulcus but does not stretch caudally as far as the coronal sulcus, rather reaching this sulcus only at its middle before crossing it to run inferiorly. The region extends posteriorly as a narrow zone over the presylvian sulcus and widens on the anterior Sylvian gyrus to form a broad area (4a) without very clear boundaries. In relation to the overall situation of this region it can be said that it lies to a great extent around a principal sulcus, with one half anterior to and the other half posterior to the cruciate sulcus. Thus it displays a contrary organisation to that of primates, in which the precentral region is placed essentially anterior to a principal sulcus, the central sulcus. It is particularly important to note that none of the sulci in question represent the caudal border of area 4 (that is, of the precentral region) and that therefore neither the coronal sulcus nor the cruciate sulcus can be considered as the homolgues of the cruciate sulcus in primates (*139).

The **postcentral region** extends caudally from the precentral region in an irregularly shaped strip (area 1) and is characterised by the reappearance of an inner granular layer and a markedly reduced sectional depth, with histologically sharp borders, whereas posteriorly it merges gradually with the parietal region. Its surface relations are not strictly determined by particular sulci; it extends from superior to inferior across the lateral sulcus and the coronal sulcus and stretches ventrally to the suprasylvian sulcus. Here the postcentral region is wedged in the angle formed by the bending round of the giant pyramidal area, thus lying partially within area 4 surrounded by it on two sides, whereas in man, monkeys and prosimians it is situated directly caudal to it. A noteworthy feature that should be emphasised is that the region does not reach quite to the dorsal edge of the cortex, not extending at all onto the medial surface, whereas this is the case in all other brain maps.

The **parietal region** is composed of three areas, 5, 7 and 52, and is inserted, without distinct boundaries, between the postcentral region on one side and the temporal and occipital regions on the other. It includes about the middle third of the suprasylvian and marginal gyri. Only area 5 is sufficiently distinctly differentiated that is can be easily homologised; from its cellular structure it corresponds to the preparietal area of higher mammals. The homology of area 7 can only be ascertained by the coincidence of its position, while there is no equivalent of area 51 (*140) in other brains. The border of the region on the medial side corresponds to the splenial sulcus, and on the lateral side to the suprasylvian sulcus ventrally with no morphological boundaries rostrally and caudally. Area 5 is sharply demarcated from the giant pyramidal area at the superior margin of the cortex, similar to the situation on the paracentral lobule in man and many monkeys; it merges gradually with area 7 (Figure 17).

In the **occipital region** three areas can also be distinguished, among which is the certainly homologisable *striate area* (area 17). The region forms a typical end-cap and extends between the postlateral sulcus and the splenial sulcus. The majority of the surface is covered by area 17 of which by far the largest part lies on the medial side, as in man; on the lateral surface it does not quite reach the lateral sulcus. Area 18 surrounds the striate area in a circular fashion on its anterior aspect, like the occipital area in primates and prosimians. The narrow zone indicated as area 19 directly over the lateral sulcus might well correspond to a fragment of the preoccipital area, but the homology is not certain. (For the cytoarchitecture see Figures 73 and 55.)

In the **temporal region** one encounters the greatest difficulties with structural homologies. In the accompanying map (Figure 104) I include areas 20, 21, 22, 50 and 36 in this region. Of these, however, only the ventral area 36 has a certain homology in that it corresponds to the *ectorhinal area* of other brains. On the other hand it is not unequivocally sure whether area 20 represents the inferior temporal area; it is not at all possible to decide which of the three areas 21, 22 and 50 correspond to the middle and superior temporal gyri, and thus the middle and superior temporal areas, of primates. The cytoarchitectonic specialisation of the temporal cortex has here reached such a degree that morphological relationships can no longer be recognised from the histological structure alone. Only the position and relations to neighbouring zones can serve to provide landmarks. One can, however, deduce that, apart from the posterior limb of the suprasylvian gyrus, almost the whole Sylvian gyrus, including even most of the part lying anterior to the Sylvian sulcus, belongs to the temporal region. Assuming that this proposition is correct, the important conclusion for experimental physiology emerges that the auditory cortex of carnivores, specifically the kinkajou, has hooked around the upper end of the Sylvian sulcus and has thus come to lie partly in front of and above this sulcus. Support for this interpretation is provided by certain myeloarchitectonic findings that will be further discussed in Part III.

The **insular region** displays a simpler organisation in terms of localisation and allows more definite parcellation thanks to its heterotypical structure (claustrum). I have only evaluated the region as a whole and provisionally avoided the separation of the insula into individual areas; it is however easy to distinguish a dorsal granular portion from the rostroventral agranular main part. The zone indicated as area 13 lies mainly in the deep cortex of the Sylvian sulcus, but does in fact climb a little out of the sulcus onto the free upper surface of the anterior and posterior Sylvian gyri both anteriorly and posteriorly. Thus the inferior corner of the morphological temporal lobe must be counted as part of the insula, as also indicated in our map. Rostrally the insular region extends more obviously beyond the anterior rhinal sulcus and continues anteriorly as a narrow strip.

In the **cingulate region** a granular posterior section is easily differentiated from the agranular anterior one. The transition between the two is about in the middle of the body of the corpus callosum and is rather indistinct. The smaller granular subregion consists only of area 23, while the larger agranular zone is composed of areas 24, 25, 32 and 33. Its position needs no particular explanation, being evident from the map; its boundaries are not sharp. The homology with the areas of the same name in other mammals is only partial.

The **retrosplenial region** is represented by the very extensive cortical surface lying above the entire posterior half of the corpus callosum and whose homologue in primates, as we have seen, forms merely an extremely rudimentary zone next to the splenium. It is composed of the three characteristically organised and unequivocally identifiable areas 26, 29 and 30. The kinkajou occupies an intermediate position between the lemurs on the one hand and the rodents on the other in terms of size and histological differentiation of these areas. On the whole this region closely resembles that of the flying fox, except that in the latter the agranular retrosplenial area (area 30) is smaller and yet divisible into several subareas.

The **hippocampal region** also reveals a high degree of histotopical development that is manifested on the one hand by a marked expansion in area and on the other by a rich architectonic differentiation. It involves the absolutely largest surface area of all the brains studied so far and forms the major part of the very large *piriform lobe* that is clearly demarcated by the posterior rhinal sulcus. One can without difficulty distinguish the *presubicular area* (27), the *retrosubicular area* (48), the *parasubicular area* (49), the *perirhinal area* (35) and the *entorhinal area* (28). As in the rabbit and the flying fox, the last of these can be separated into two clearly different subareas, the medial and lateral entorhinal areas, as is also indicated on the map. Area 49 inserts itself between areas 27 and 28 to form a narrow dividing zone; in my material it has only also been demonstrable in the flying fox, the rabbit and the hedgehog, but in these remains inferior in size to that of the kinkajou.

The **olfactory region** forms approximately the anterior third of the piriform lobe and the largest part of the anterior olfactory lobe; it thus occupies a very large surface area, as in the flying fox, the rabbit and the ground squirrel, and is only exceeded in size by that of the hedgehog. Dorsally its border coincides with the rhinal sulcus, caudally it fuses gradually with area 28, and it reaches as far rostrally as the olfactory bulb. Within the olfactory region the specific rudimentary cortical formations of the *prepiriform area* (51), the *amygdaloid nucleus* (A) and the *olfactory tubercle* (Tol) can be distinguished.

VI. Rodents (rabbit and ground squirrel). (Figures 106-109). (*98)

I have completed the study of the cerebral cortical field organisation of two rodents, the rabbit and the ground squirrel and summarised the results in maps. Isolated blocks from other families have also been examined. The description of localisation will be restricted to the rabbit brain, and only essential new findings will be considered. Any differences in the ground squirrel will be mentioned.

Little more need be said about the majority of the regions beyond what has already been discussed in relation to the flying fox. The main difference in surface topography in both rabbit and ground squirrel compared with higher mammals resides in the extraordinary expansion of the surface area and the rich differentiation of the retrosplenial region, the hippocampal region and the olfactory region that exceed even those of the flying fox. The retrosplenial

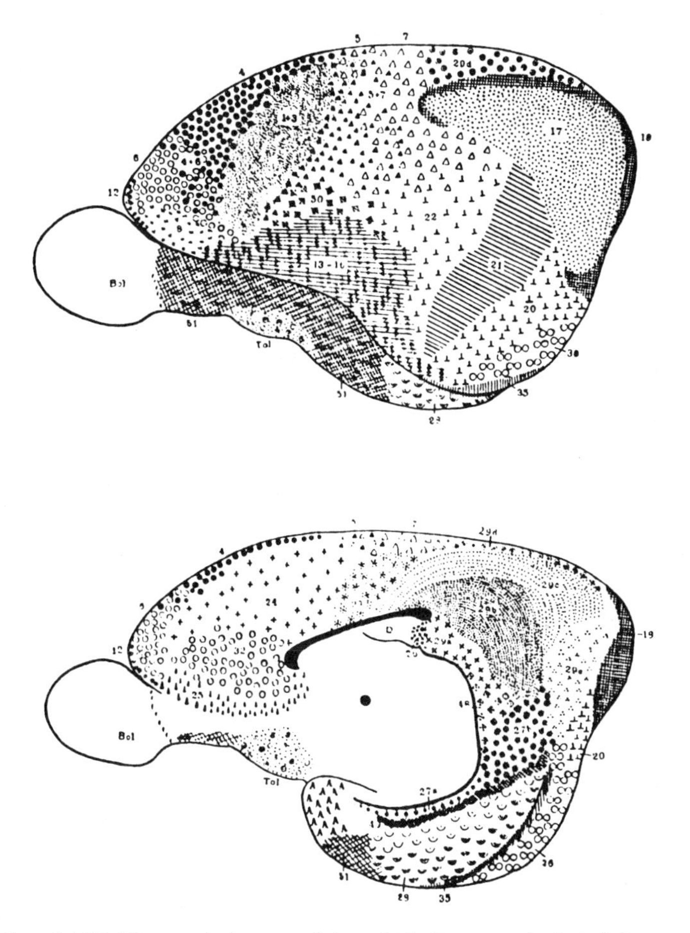

Fig. 106 and 107. The cortical areas of the rabbit (*Lepus cuniculus*). 2:1.

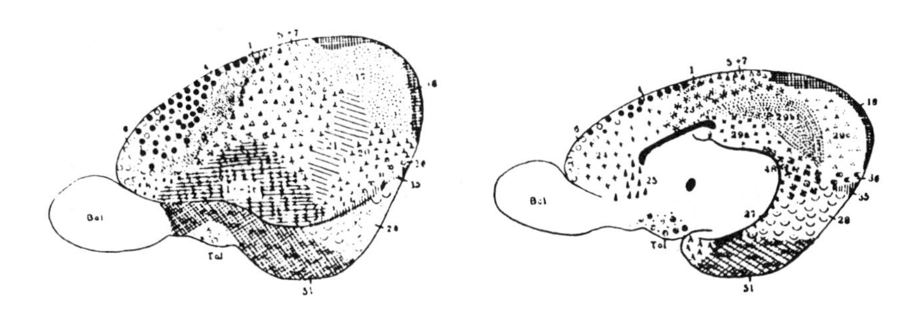

Fig. 108 and 109. The cortical areas of the ground squirrel (*Spermophilus citillus*). 2:1.

region alone in the rabbit [12]) can be divided into six different areas.

The **precentral region** lies very close to the frontal pole and stretches mainly horizontally along the superior margin of the hemisphere. It scarcely encroaches on the medial side as the cingulate region occupies most of the space above the corpus callosum. The laminar pattern of both areas 4 and 6 that characterises the region in primates is also suggested here, but actual separation of the areas can only be accomplished with difficulty. Therefore these areas are drawn as broadly superimposed in the maps, especially in the ground squirrel (Figures 108 and 109). At the frontal pole, yet another small field with a specific structure and shaped like an end-cap, area 12, but I cannot detect it in the ground squirrel. Whether this is the homologue of area 12 in the marmoset and the kinkajou I dare not decide in spite of its similar position.

Without doubt the whole agranular frontal cortex of both rodents (except perhaps area 12) belongs to the precentral region. There is no granular frontal region.

Concerning the **postcentral region** and the **parietal region** reference can be made to the description of the flying fox. They represent a large combined zone that is somewhat different in structure in its posterior portion from its anterior and, although certain features of the clearly separate individual areas of higher animals are revealed, they do not justify a precise spatial segregation into areas. In the brain maps the equivalents of areas 1, 5 and 7 are drawn with considerable overlap. (For areas 1 and 4 see Figure 63.)

The **occipital region** lies almost entirely on the lateral surface; it has obviously been forced from the medial side over the occipital margin of the hemisphere onto the lateral side by the overwhelming development of the retrosplenial region.

Once again the striate area (17) is the major field, as we saw in Part I, although in a considerably modified and simplified form (Figure 76). Caudally a crescent-shaped field borders area 17, that we have indicated as area 18, without wishing to insist upon its homology with the occipital area of higher mammals. (For area 17 see Figure 76.)

The **insular region** has undergone an even greater expansion than in the flying fox and the kinkajou. Due to the absence of any sulci it is entirely on the free surface which allows an easy estimation of its total area. It occupies at least a third of the vertical height of the hemisphere and its rostrocaudal length amounts to more than half that of the hemisphere. Ventrally it is sharply separated from the olfactory lobe by the rhinal sulcus; its other borders are indistinct, for its main architectonic feature, the claustrum, gradually merges with neighbouring zones. Even here a caudal granular and a rostral agranular subregion can be distinguished, each with two individual areas. Whether area 50, lying at the upper border of area 13, should also be counted in the

---

[12]) Zunino has been able to demonstrate a myeloarchitectonic zone corresponding to each of the cytoarchitectonic areas that I have distinguished. - Journal f. Psychol. u. Neurol., XIV, 1909. (*141)

insular region I cannot decide for the present; I am equally doubtful about the allocation of area 8, that also demonstrates a quite specific and individual cellular structure, to any of the regions under consideration.

The **temporal region** is composed of four individual areas (20, 21, 22 and 36) and is characterised, as in the flying fox, by quite indistinct boundaries with the parietal and occipital regions. Areas 20, 21 and 22 are not very specifically differentiated and there are gradual transitions between them, but on the contrary the ventral areas 35 and 36, that lie across the rhinal sulcus, are differentiated very characteristically and it can be accepted with certainty that area 35 corresponds to the perirhinal area of man and area 36 to the ectorhinal area. Area 35, lying partially lateral to the rhinal sulcus and therefore in the temporal lobe, is better considered with the hippocampal region judging from its structure.

In the **cingulate region** of the rabbit, areas 23, 24, 25 and 32 of the flying fox are again found, but in the ground squirrel only 23, 24 and 25 are sure. As to the small stripe-like zones that surround the splenial sulcus in the flying fox (areas 30a, 30b, 31a and 31b of Figure 103), demonstrable homologues are absent in both the rabbit and the ground squirrel. There is little of significance to say about the position of the individual areas, except that area 25 extends very far rostrally, as far as the level of the frontal pole, unlike the situation in the previously mentioned animals.

The **retrosplenial region** is very differently constructed in the two closely related animals, the ground squirrel and the rabbit. Whereas in the former only four areas (26, 29a, 29b and 30) are distinguishable with certainty, one can clearly demarcate six structurally different areas (26 and 29a-e) in the equivalent region of the rabbit. Of these, area 26 corresponds to the ectosplenial area, while areas 29a-e must be considered as produced by further differentiation of the granular retrosplenial area of other species on account of their related structure. Thus area 29 has here differentiated into five subareas with specific structural features, although all (except area 29e) show marked evidence of being interrelated (Figures 65 and 66). Consequently the extent of this combined zone is unusually great. The retrosplenial region of the rabbit includes the whole medial surface of the occipital lobe, that is to say those extensive zones that in higher mammals belong to quite different cortical formations, notably the occipital region with the striate area, and which, looked at purely externally, are represented in man by the cuneus, the lingual gyrus and, partially, the precuneus. Its area in the rabbit represents at least about a tenth of the total cortical surface, whereas the homologous region in man cannot amount to any more than one three-hundreth of the surface. A cortical type corresponding to the agranular retrosplenial area (area 30) is absent in rodents. Area 29c takes on a peculiar aspect, especially due to its marked poverty in fibres (Zunino). Whether it corresponds to area 30 of prosimians is very questionable, but in any case it contrasts sharply from it cytologically (*142).

The **hippocampal region** also reveals an extraordinarily rich differentiation. In the ground squirrel its predominant feature is its relative size, corresponding to the marked development of the piriform lobe; on the other hand the number of areas is greater in the rabbit. We distinguish in the latter one more area than in the flying fox, in that another special cortical type of related and yet quite specific structure is recognisable at the caudal end of the presubicular area (27), for which we propose the name *ectosubicular area* (27b). This area encroaches further along the hippocampal sulcus between areas 27 and 48, thus forming a sort of transition zone between them.

The other areas, 27, 28, 35, 48 and 49, are arranged in similar fashion to those of the flying fox and nothing essentially new can be said about them, except that it is worthy of note that area 28 demonstrates two clear architectonic variations so that it is justifiable to distinguish a *lateral entorhinal area* (28a) and a *medial entorhinal area* (28b) that are spatially sharply separated. In the ground squirrel no sure homology can be established for either area 49 or 27b, and even a similar division of area 28 cannot be accomplished easily.

The **olfactory region** of the rabbit, and even more so of the ground squirrel, is relatively more extensive than in the flying fox, but is likewise composed of three individual fields: 51, the amygdaloid nucleus (AA) and the olfactory tubercle (Tol). Area 51, or the *prepiriform area*, is of unusually large size, especially in the ground squirrel. The amygdaloid nucleus and the olfactory tubercle emerge very characteristically as circumscribed cortical fields in the rabbit and ground squirrel thanks to their atypical (heterogenetic), rudimentary cellular structure.

VII. The hedgehog (Erinaceus europaeus). (Figures 110 and 111).

The brain of the hedgehog occupies a special place in my research material in terms of its cortical architectonics as well as in the organisation of the cortical surface. The arrangement of the cellular laminae and the field distribution are so completely altered that great difficulties are encountered in trying to relate them to those of higher species. Only the study of intermediate situations and comparison with numerous other groups clarifies the organisational plan. This is an example of the importance and superiority of comparative anatomical methods. Anyone wishing to investigate the brain of the hedgehog or other insectivores in isolation without knowledge of the cortical structure of other orders, and understand their localisational organisation, would fail at the task. The only way to recognise the general principles of mammalian cortical structure is by means of ample comparative anatomical material comprehending the whole class. Therefore I consider the histological cortical parcellation of the insectivores proposed by Watson as generally erroneous, in spite of many correct isolated findings.

A characteristic of the hedgehog brain is the unusually large archipallium. The piriform lobe and the anterior olfactory lobe together account for between half and two-thirds of the total cortical surface. Corresponding to the massive expansion of this part of the brain one also finds very extensive heterogenetic

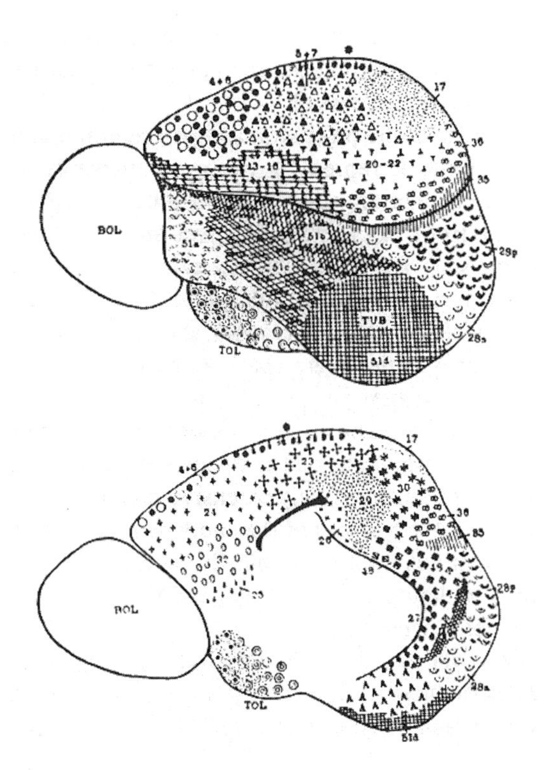

Fig. 110 and 111. The cortical areas of the hedgehog (*Erinaceus europaeus*). 2:1.

and rudimentary cortical zones, whereas the neocortex forms only a small percentage of the total cortical surface. The olfactory, hippocampal, retrosplenial and cingulate regions together form over three-quarters of the total cortex, while their area in man accounts for only scarcely a fiftieth of the cortex.

Even separating the regions within the neopallium from each other presents considerable difficulties. The borders indicated in the map are therefore only relative, so gradually do the different structural zones merge with each other. Even the differentiation of the precentral region from the postcentral or parietal regions is not as conspicuous in insectivores, specifically the hedgehog, as in other species, including marsupials and monotremes, for on the one hand the inner granular layer is only weakly developed and on the other hand even the homologues of the Betz giant pyramids are scarcely evident. The structural features of these two regions that are so clearly expressed in higher species are thus quite rudimentary in the hedgehog. It is similar in other regions; only the insula is sharply defined as a special region in the neopallium. The trend is different with the regions of the archipallium and also partially those of the cingulate gyrus. These are not only easy to distinguish from each other, but also their individual areas possess a reasonable degree of spatial differentiation.

The agranular **precentral region** is drawn on the map as a uniform area (4+6). It fills approximately the anterior third of the lateral surface, a little more along the superior margin of the hemisphere, and reaches the frontal pole. A topical separation into a giant pyramidal area (4) and an agranular frontal area (6) is not possible; the two spatially completely separate areas of higher species have not yet differentiated here, but form a common zone.

A similar situation holds for the **postcentral and parietal regions**. Whereas both these exhibit a certain topical specialisation in rodents, notably the rabbit and ground squirrel as we have seen, their separation is totally lacking in the hedgehog. Rather, in our brain map caudal to the precentral region there follows a homogeneously structured zone (5+7) that cannot be further divided into individual areas and must wholly represent these two regions; thus in it lie the combined areas 1 to 3, 5 and 7 of other mammals, undifferentiated or regressed.

At the inferior border of the precentral and postcentral regions the **insular region** forms a rather large zone. It represents the most markedly differentiated and, on account of the claustral formation, easily recognisable neopallial cortical region of the hedgehog brain, apart from the ventral temporal areas (35 and 36). Anteriorly, the insular region stretches as far as the frontal pole and here encroaches somewhat medially over the orbital surface of the frontal lobe. It runs posteriorly along the rhinal sulcus on the dorsal bank of which it reaches about the level of the junction of the middle and posterior thirds of the long axis of the hemisphere.

Caudally the insular region is bordered by a structural zone that likewise lies lateral to the rhinal sulcus over about its posterior third, representing the whole of the **temporal region**. Areas 35 and 36, that is to say the perirhinal and ectorhinal areas, can be homologised within this relatively extensive region on the basis of their cellular structure; they contain all the features of the equivalent cortical types of higher orders. However, dorsal to them lies a rather atypically built area 20-22 that merges with its neighbouring regions without distinct borders and that may be the equivalent of the upper temporal fields of other species.

Posterosuperior to the temporal region and caudal to the parietal region lies a small, indistinctly demarcated area 17, whose cellular structure bears certain features of occipital cortex, although considerably modified. It is situated mainly on the lateral surface and scarcely extends to the margin of the occipital lobe. Everything points to its being the rudimentary homologue of the striate area, that is what remains of the **occipital region**. A sure decision about this homology, as about so many other questions, will only be made possible by a myeloarchitectonic study, or even more so by a systematic study of fibres (*143).

The **cingulate region** is easier to evaluate. Its position next to the corpus callosum leaves no doubt as to its homology, even if its cytoarchitecture is hardly striking. It includes the whole extent of the medial surface of the hemispheres superior and anterior to the corpus callosum and thus covers a relatively large expanse. One can distinguish four different structural types in it, that are indicated in the brain map as areas 23, 24, 32 and 25. The homologies of the individual areas are not without doubt, only area 25 being certainly identical to the area of the same name of other brains. The precentral area stretches over the cingulate region in the rostral half of the superior margin of the hemisphere. On the other hand the caudal half is covered by a cortical type which I could not attribute to any of the neighbouring regions. I do not consider it totally impossible that the narrow cortical strip in question (* in the map) represents an architectonic variant continuing the precentral region caudally, such that it would reach almost to the occipital pole; but it is more probable that this area belongs to the cingulate region and represents the agranular homologue of one of these areas (perhaps 30). This problem cannot be decided purely cytoarchitectonically.

The **retrosplenial region** is not cytoarchitectonically as clearly developed as in most other macrosmatic animals (flying fox, rabbit etc.), but nevertheless covers a relatively large surface area. It consists of three individual areas 26, 29 and 30, of which 26 and 29 are probably homologous to the ectosplenial and granular retrosplenial areas of other mammals respectively, while area 30 possesses a specific non-homologous structure.

In the **hippocampal region**, that covers a relatively large expanse, areas 27, 28a, 28p, 48, 49 and perhaps 35 can be clearly distinguished. Of these the first three possess such a characteristic structure that they can immediately be homologised with the presubicular area (27) and the entorhinal area (28). Area 28 is again divisible into two different fields, a posterior (28p) and an anterior entorhinal area (28a) that are more sharply demarcated from each other than in most other species. Area 27 is separated from area 28 by a narrow strip of cortex of heterogenetic structure, that justifies distinguishing as a specific area

49 (parasubicular area). Area 48, or the postsubicular area, has a particular structure and is sharply differentiated; it forms the dorsocaudal prolongation of the subicular area. Area 35 certainly corresponds to the perirhinal area of other animals and forms the boundary zone between the archipallium and the neopallium in the rhinal sulcus and its adjacent cortex.

The **olfactory region** has undergone a massive expansion in the hedgehog. It includes nearly a third of the total cortical surface of the hedgehog brain and manifests, apart from the amygdaloid nucleus (AA) and the olfactory tubercle (Tol), four distinctively differentiated, well circumscribed fields that are indicated as areas 51a, 51b, 51c and 51d, and that together represent the homologue of the prepiriform area of other brains. Their position can be seen in the map. Area 51d corresponds to the piriform tubercle (Tub). The olfactory tubercle is itself composed of three different parts, the anterior, middle and posterior nuclei, also indicated in the brain map, so that in the olfactory region of the hedgehog at least eight individual fields can be distinguished, whereas in primates it can hardly even be identified as a region, and certainly not as containing individual parts.

If we summarise the findings of the above descriptions of individual brains in answer to the question asked at the beginning, we come to the conclusion that in principle there is a broad agreement regarding topographical cortical localisation among all the animals investigated, but that, in spite of these similarities in the basic features, considerable variations emerge in numerous details even between closely related species. Such great differences are seen, for example, in the primates between the cercopithicids and the callithricids (Figures 90 to 91 and 96 to 97), and in the rodents between the rabbit and the ground squirrel (Figures 106 to 107 and 108 to 109).

Overall, then, one has to recognise the similarities and differences, or constant and inconstant features, in the cortical cytoarchitectonic topography of different mammals.

Constancy of features is the expression of a similar developmental direction, whether in a phylogenetic sense or in the sense of a convergence; differences, on the other hand, reflect morphological, and related functional, specialisation of individual brains.

We wish to discuss specifically the common features and variations in cortical field patterns in the following chapters.

Chapter V.

Common features in cortical architectonics.

These are essentially expressed in the following three ways:
1. in similarity of overall position,
2. in correspondence between structural regions,
3. in persistence of individual areas throughout the whole mammalian class.

1. Similarity of position.

If the preceding brain maps are compared objectively, it must be admitted that there exist overwhelming similarities in the overall patterns of topical parcellation of the hemispheric surface as a whole in all these different animals. Equally, whether we are dealing with a brain with complex sulcal development, like that of man, or one with smooth surfaces, like that of the marmoset or rabbit or ground squirrel, the same fundamental structural subdivisions are always found. In all the brain maps segmental zones recur in the form of complete segments, partial segments, coronal fields and end-caps. Brains, especially of closely related animals, are often extraordinarily similar in terms of the mutual relationships and the sequence of the areas that, as we have seen, are especially evident in the horizontal plane. Only the shape, size, specific position and, even more so, the number of individual segments or fragments of segments differ considerably. Partial segments are commonest and complete segments rare.

We have only identified complete belt-like segments, encircling the whole hemisphere, in the occipital lobe of the higher orders, including areas 17, 18 and, in part, 19, and in areas 9 and 10 of the frontal cortex of several animals (*144). In the partial segments one can include areas 1 to 9, areas 20 to 22 of the temporal lobe and areas 5 and 7 of the parietal lobe. In the brain map they stretch across a more or less extensive part of the hemispheric surface in the form of band-like zones, mostly astride the superior margin of the cortex. There are also great similarities with regard to the parcellation of the cingulate and hippocampal gyri. The more individual brains one compares, the more obvious are the similarities in the overall layout of cortical surface topography.

Thus we can state:

The essentials of cerebral cortical areal parcellation are the same in all mammalian orders examined so far; it is influenced by a *principle of segmentation*, that is more or less clearly expressed. Its basic premise, stated briefly, is that the cortical surface is divided into a large number of circumscribed structural zones, that are arranged broadly rostrocaudally behind each other and take the form of segments or fragments of a segment [1]). In many brains, especially the simpler structured ones of certain lissencephalic animals, this principle emerges clearly and systematically, while in others it is complicated, and thus weakened, either by marked folding of the surface or by further differentiation of certain regions and the formation of subareas, and in yet others it is only imperfectly expressed owing to more primitive development or as a result of regression.

2. Constancy of regions

As was explained above, we understand as regions those large, clearly defined histological cortical zones that contain several individual areas, that can be distinguished from each other by striking structural features, and that recur in their essentials in all or most mammalian classes, even if modified in detail and often quite vestigial. Figures 83 and 84 show diagrammatically the cortical regions of man. If one compares them with the surface maps of the other brains, it is clear that most of these regions are surprisingly constant throughout the whole mammalian class, although their arrangement, the number and shape of their individual areas and, most of all, their size and position, may vary markedly. The special features that distinguish different animals are described in the foregoing chapters; here we shall summarise briefly the common characteristics of the individual regions as a whole.

a) *The precentral region.*

It is chiefly characterised by two architectonic criteria: on the one hand the

[1]) Of course this does not suggest a relationship to the metameric segments of the spinal cord, but is merely a superficial analogy.

loss of the inner granular layer and the subsequent lack of a clear laminar organisation, and on the other hand the considerable depth of the cortex. It is composed in almost all brains of two more or less sharply demarcated individual areas, the giant pyramidal area (4) and the agranular frontal area (6). Only in a few primitively structured, lower species with very small brains (small rodents, insectivores, microchiropterans) can a real division of the precentral region into two structural areas not be demonstrated, on account of their architectonic features having completely regressed or, alternatively, not having differentiated sufficiently. The extent of the precentral region varies remarkably between different animals. In general, though, it includes a well-defined zone of the pallium that is bounded laterally by the Sylvian sulcus or, where this sulcus is missing, by the superior margin of the insula, and extends medially over the superior margin of the hemisphere as far as the callosomarginal sulcus or the cingulate region. The caudal border is usually sharp, marked by the sudden appearance of the inner granular layer in the neighbouring postcentral region. In primates it coincides fairly precisely with the central sulcus (Figures 92 and 94), while in lissencephalic brains there are no externally recognisable boundaries (see Figures 96 to 103 and 106 to 111 of marmoset, lemur, flying fox, rabbit, ground squirrel and hedgehog). Anteriorly its border is rather unclear. In many species, such as the flying fox, the ground squirrel and the hedgehog, it reaches the frontal pole, while in others, including man, there are no external boundary features apart from, in some, its approximation to a sulcus, for example the arcuate sulcus in the guenon (Figure 90) and the presylvian sulcus in the kinkajou (Figure 104).

The great differences in the cortical surface areas have been stressed in the description of the individual brain maps. The absolutely largest precentral region of the mammals that I have investigated belongs to man; on the contrary, its relative size - compared to the total cortical area - is least in man. In broad terms the relative size of the precentral region should increase rather than decrease with decreasing brain size as one descends the mammalian class; no strict rule can be formulated, however, as there are undoubtedly many exceptions. Reliable data for comparative quantification of size can only be obtained by the systematic measurement of cortical surface area.

As to the relative size of areas 4 and 6 that make up this region, we were able to determine that in man area 6 dominates, while the opposite is true in other mammals, and that even in monkeys area 4 has the larger surface area. These relative sizes cannot be determined exactly in lower animals as the two areas either have gradual transitions so that their borders overlap considerably, or in many cases actually largely fuse, as in the ground squirrel and the hedgehog (Figures 108 and 110).

b) *The granular frontal region.*

Its occurence in mammals is not as constant as that of the precentral region; a corresponding structural region is not demonstrable at all in a series of lower animals such as insectivores, microchiropterans and many rodents. In these the precentral agranular region includes the whole frontal lobe and stretches anteriorly as far as the frontal pole. However, in the majority of the species studied it represents a quite characteristic and regionally easily definable cortical zone that is clearly demarcated from the adjacent regions.

Its main structural feature is the reappearance of a definite inner granular layer anterior to the agranular precentral region (in addition to the ill-defined lamination, the greater average cortical thickness, the lesser cell density and the gradual transition to the white matter in the latter).

As an entity, this region is one of most variable of the whole cerebral cortex in terms of its size and position, and its particular composition of individual areas. In man it makes up a considerable portion of the whole pallium, having by far the greatest surface area, at least about three-quarters of the whole frontal cortex, and is composed of eight or nine clearly demarcated cytoarchitectonic fields, areas 8 to 11 and 44 to 47, of which most can be divided into several smaller myeloarchitectonic subareas [2]). In the lower monkeys it is already much smaller; it no longer exceeds the precentral region in size and is composed of only four or five individual areas (Figures 90 and 91). In the lemurs the pattern is even less impressive. Here it lags well behind the precentral agranular region in extent and includes the three still separated areas 8 to 11. In the kinkajou it consists of only the single area 8 which covers an even smaller surface than in the lemur (Figure 104). The ungulates and the pinnipeds show the same trend as the carnivores. In other, mainly lower, mammals (except many marsupials and, as far as I can determine, the echidna) a granular region is no longer formed at all at the anterior end of the frontal lobe.

Thus, concerning the frontal cortex, we observe very variable relationships. In broad terms one can say that the granular frontal region becomes smaller from man downwards and that hand in hand with this goes a simplification of its anatomical differentiation, which is manifested as a reduction in the number of cytoarchitectonic areas.

Whereas in man this region, at a rough estimate, makes up some three-quarters of the total frontal cortex anterior to the central sulcus, and the precentral region only one quarter, it is about of equal size in the lower monkeys. In lemurs it is actually smaller than the latter and in the lowest mammals it amounts to merely a minute fraction of the volume of the frontal cortex, and even disappears entirely in some species [3]).

---

[2]) O. Vogt differentiates about 50 myeloarchitectonic areas in the frontal cortex, of which several are included in one of my cytoarchitectonic areas. Therefore, this a case of further differentiation of the cell areas into smaller architectonic units.

[3]) One should note explicitly that precise data for a comparative quantitative study of regions or individual areas can only be obtained by systematic measurement of cortical surface area. For preliminary orientation the proportions can be judged by eye. Professor R. Henneberg is at the moment occupied with such studies in the Neurobiological Institute and will soon be able to conclude them (*145).

[4]) cf. the electrical cortical stimulation by Sherrington and Grünbaum (*146) and C. and O. Vogt in great apes, and by F. Krause in man.

Thus the agranular precentral region comprising areas 4 and 6 together which, according to our present knowledge [4]), is intimately concerned with motor functions, forms a much smaller portion of the frontal cortex, and indeed of the whole cortical surface, in the highest mammals, and especially in man, than even in the next lowest primates. Thus the quantitative importance of these "centres" compared with other functional areas decreases sharply in higher species (man, monkeys and prosimians), while in lower species the opposite is the case and the motor centres dominate progressively, an observation that correlates well with our general physiological and clinical concepts.

c) *The postcentral and parietal regions.*

As in many lower mammals there is an extensive fusion or superimposition of these two regions, it is convenient to discuss them together.

We have discussed in detail above that in all gyrencephalic primates the postcentral region can be divided into three separate, well differentiated areas, there even being four individual areas in man, the great apes and many guenons, and that on the other hand the whole region consists of only one area in mammals from the lissencephalic marmoset down. We have also seen that in man, monkeys, prosimians and the kinkajou, the parietal region is sharply demarcated from the postcentral region and itself presents several different areas. If one takes as a comparison a lower group, such as small rodents, insectivores or bats, it emerges that the two regions are almost completely fused and, so to speak, superimposed. In the brain maps of the rabbit, the ground squirrel and the hedgehog this trend is expressed by the diacritic symbols being intermingled (Figures 106 to 111). The structural features of not only the parietal areas 5 and 7 but also of the two regions are mixed and fuse in such a way that spatial separation and demarcation become impossible.

Seen developmentally, there thus exists a primitive state in these animals, in that the topical specialisation of this cortical zone, that has led to the separation of two heterogeneous regions in the more highly organised animals, and especially in primates, has not yet taken place, such that there is only a single homogeneous, poorly differentiated zone as an equivalent or homologue of the two regions in man or monkeys.

Of great importance and significance for our interpretation is the fact that such a large, homogeneous zone exists posterior to the agranular precentral region, without exceptions in the whole mammalian class, and that this postcentral (parietal) region contrasts structurally with the former in a similar way in all animals. Whereas the precentral region is agranular, for the inner granular layer is missing in all mammals, a distinct, thick inner granular layer always appears in the postcentral region, with a quite abrupt transition. Thus we have a further constant feature of regional cortical structure in mammals, in spite of individual discrepancies and certain rather far-reaching specialisations within this zone.

d) *The insular region.*

The most constant and, as an entity, most striking large, homogeneous structural zone in mammals is the insula or "insular region" (areas 13 to 16). As we have seen, the insular cortex is recognisable structurally because of the formation of the claustrum as a specific cellular sublayer derived from the innermost cortical lamina, the multiform layer (VI) (*64). Morphologically the whole insula is characterised by its essential homogeneity brought about by the existence of this special cell layer, and there can be no doubt about its homologies. As a result of this feature it is easily demarcated from neighbouring regions and always easily demonstrable.

I have never noted its absence in any brain. Its borders and extent, however, are subject to large variations in different animal groups. In man and monkeys it is completely hidden in the depths of the insula, and thus not indicated on the relevant brain maps, except in the marmoset where it extends somewhat onto the orbital surface near the inferior end of the Sylvian sulcus. In the lemur it is on the free surface for most of its extent but, as in the flying fox, it also includes the deep cortex of the Sylvian sulcus and part of the superior bank of the posterior and anterior rhinal sulci. Its extent is greatest in rodents and insectivores where it spreads over a large part of the free cortical surface.

An ever constant feature that should again be particularly emphasised is the separation of the insula into an anterior agranular and a posterior granular half, as already mentioned several times. The further differentiation of the insular region into individual cytoarchitectonic areas will be the object of future studies.

e) *The occipital region.*

What was said for the postcentral region is also applicable to this region *mutatis mutandis*. That is, it is always demonstrable as a regional entity of clearly recognisable structure, but is subject to major modifications in detail and relative to the number of individual areas. Its identification in most major mammalian groups is simple on account of its containing the extremely easily recognisable and absolutely constant striate area, or area 17 - the histological "visual cortex" of the literature. The main feature is the extraordinarily massive development of the inner granular layer, together with the thinness of the cortex, the high cell density and the generally small cellular elements. Its position is always at the occipital pole and in its vicinity, sometimes more on the medial and sometimes more on the lateral surface of the hemisphere. It is vestigial in the hedgehog.

f) *The temporal region.*

It can also only be homologised with certainty throughout the mammalian class as a general structural region, but not in terms of its individual areas. As a region it is always consistent, and is recognisable in all orders from its position alone, although its cellular structure possesses few characteristic features, except in man, and varies widely in most animals. Compared with man and the primates, the relative extent of its surface area is markedly decreased in lower species, as seen by comparing areas 20, 21 and 22 in the brain maps of the various animals.

g) *The cingulate region.*

The cingulate region possesses an even greater constancy of position and general structure than the last region. Its basic cytoarchitecture distinguishes it from the rest of the cortex, in spite of numerous subareas that are present in many animals. In addition it is situated so consistently directly around the corpus callosum, that there are never insuperable difficulties in demonstrating it, even in the acallosal marsupials and monotremes. On the basis of its laminar organisation, it is always divisible, like the insular region, into two completely different major sections, an agranular precingulate subregion and a granular postcingulate subregion, each of which is further differentiated into a more or less large number of different structural types in individual species.

h) *The retrosplenial region*

is demarcated from the most posterior section of the cingulate region as it forms a zone characterised by the regression of several basic layers and a widely variable size and architectonic development. On the basis of its position, this structural region corresponds broadly to the isthmus of the cingulate gyrus. In many macrosmatic species it extends extraordinarily widely, and is subdivided into numerous areas, seven or eight in the rabbit for example (Figure 107); in others, especially in primates, its development is quite rudimentary and it consists of only one, or up to a maximum of three areas (Figures 86, 91 and 97). In keeping with the very different spatial extent of the region, its surface position naturally also varies markedly, as demonstrated by comparing the brain maps.

i) *The hippocampal region*

is, with the precentral and insular regions, the most absolutely constant of the major, homogeneous structural regions in mammals. It belongs to the heterogenetic formations and is immediately recognisable in all mammalian brains by its atypical lamination. It includes the cortex medial to the posterior rhinal sulcus as far as the hippocampal sulcus, that is to say the head of the parahippocampal gyrus (*120) or part of the piriform lobule or its homologues, and also the deep cortex of the hippocampal sulcus itself; it consists of areas 27, 28, 35, 48 and 49 in our maps, as well as their subdivisions.

k) *The olfactory region*

is formed of rudimentary cortical structures, of which the laminar pattern is quite atypical and cannot be traced back to the primitive cortical type. Sometimes a sort of layer I and VI can be distinguished, but all true layers are usually absent (olfactory tubercle and amygdaloid nucleus). It thus follows the trend of the hippocampal cortex, except that its layers are even less differentiated. Its incidence, and even more so its size, is very inconstant in mammals. It is particularly massively developed in many otherwise simply organised macrosmatic animals, such as the hedgehog, in which it comprises about a third of the total cortical surface area and covers almost the whole piriform lobe and anterior olfactory lobe. Even in small rodents, it is well developed and includes half or more (ground squirrel) of the piriform lobe. In contrast, it is quite vestigial in microsmatic species, and in lemurs, for instance, is limited to the extreme anterior pole of the temporal lobe, and in primates I was quite unable to even identify a sure homologue.

3. Persistence of individual areas.

Like the regions, many individual histological areas also prove very constant throughout the animal kingdom, either by persisting through the whole mammalian class or by being present in most orders, although absent in others. In order for the principle of cortical differentiation to have meaning, it is important that cortical areas in which the lamination has undergone the quite specific modifications of the basic pattern that were described earlier for heterotypical and heterogenetic formations, are among the most constant ones. In my fifth communication on cortical histological localisation I already pointed out two homologous areas in most mammalian orders, the striate area and the giant pyramidal area, and have been able to determine their detailed topical localisation for a large number of animals.

The following areas can be designated as certainly homologous in the brains examined so far:

area 1 (or 1 to 3) = *postcentral area*: corresponding to the postcentral gyrus in primates;

area 4 = *giant pyramidal area*: including all or part of the precentral gyrus of primates;

area 6 = *agranular frontal area*: lying in the part of the frontal lobe immediately rostral to area 4;

area 5 = *preparietal area*: in the anteriormost section of the parietal lobe; only certainly present in a proportion of the brains and often fusing with area 7 or area 1;

areas 13 to 16 = *insular area*: the aggregate areas of the insular region;

area 17 = *striate area*: in the calcarine cortex of primates and its morphological homologues;

area 18 = *occipital area*: anterior to area 17 in the occipital lobe; only certainly demonstrable in higher mammals;

area 26 = *retrosplenial area*: at the posterior margin of the splenium;

area 27 = *presubicular area*: lateral to the subiculum, absolutely constant throughout the mammalian class;

area 28 = *entorhinal area*: medial to the posterior rhinal sulcus, thus in the piriform lobe and its homologues, equally consistently present in all animals studied;

area 35 = *perirhinal area*: transitional cortex between archipallium and neopallium;

area 51 = *prepiriform area*: sometimes massively developed in macrosmatic animals.

The homology of these fields, as should once again be expressly stated, does not only rely on the similarity of the cellular lamination, but in many cases more on the coincidence of their position and even more on their spatial relationships to homologous neighbouring areas. One should not forget that cortical structural types usually undergo considerable modifications throughout a developmental series, and that because of this the particular cytoarchitecture of an area in an individual of a given species or family may be modified so profoundly that there is scarcely any obvious similarity with the corresponding area in closely related groups, to say nothing of it being possible to identify the area merely on the strength of its histological structure. In such cases demonstration of the homology of a cortical area depends less on similarity of histological structure than on coincidence of position and relations to neighbouring areas, and perhaps on developmental criteria. Areal homologies in different species is unequivocal if adjacent areas are the same in all animals, even in the face of considerable differences in intrinsic structure. As an example of this we have already mentioned area 1 of the marmoset and the lemur (pages 136 and 143).

The situation is similar in many other areas, and even with homologies between larger regional zones, such as the temporal region, for example. Here again, in many cases homologies cannot be determined unequivocally from the sectional cytoarchitecture alone without reference to the global topography. But after all this, there still remains a not inconsiderable number of areas for which homologies with related animals are not deducible with certainty, or even not at all (for example area 8 in rabbit and ground squirrel). There is still very much painstaking comparative anatomical work to be performed.

Chapter VI.

Variations in cortical architectonics.

Variations in areal parcellation of the cerebral cortex are, from the very nature of the subject, more numerous than the similarities between different animals. They concern both the larger structural entities, such as regions, and also the smaller areas, as is evident from the descriptions of the individual brains. Factors involved include differences in shape, position or size of these entities, that could be termed extrinsic features. Otherwise, there may be specific architectonic differences between cortical zones, such that in some species new areas are formed while, on the contrary, there may be regression, atrophy or fusion of areas, groups of areas or whole regions.

It is therefore justifiable from the outset to separate essential and non-essential differences in cortical parcellation. To the former belong variations in size, position and shape of individual areas, to the latter the appearance of new areas or regions in certain groups, whether due to new developments on the one hand or to fusion or regression on the other. Of course these different factors, that determine the particular pattern of cortical topography in a given animal, cannot be sharply segregated, as they usually compete in individual cases.

1. Non-essential variations. (*147)

The differences in external shape and position of the individual areas described so far are so numerous and so varied, on account of the form of the hemispheres and the specific sulcal patterns, that it is impossible to enumerate all cases in detail. The important features emerge from the foregoing descriptions of the brain maps and their mutual comparison; several of the more important discrepancies between the regions are also described in Chapter V. It is thus sufficient to single out a few examples to illustrate the principles of comparative areal topography. Obviously, for this only those areas are suitable that are absolutely consistent on the one hand and whose homology leaves no doubt on the other. Such suitable areas include areas 4, 6, 17, 27, 28, 35 and the retrosplenial region.

We prefer to limit ourselves essentially to the physiologically important areas 4 and 17, and to examine area 28 in addition, but more cursorily. The variability of individual regions will be dealt with when discussing the essential differences in localisation.

Area 4, or the *giant pyramidal area*, forms a major part of the motor centre of the surface of the hemisphere, of which the details are known thanks to recent research, without however coinciding exactly with it (Sherrington and Grünbaum (*146), C. and O. Vogt, Mott and Halliburton). As a result of this relationship to a well localised functional centre, our area is also of considerable physiological interest; indeed, it has often been referred to as the "motor cortex", without, however, necessarily reflecting that it only forms part of the latter, especially on its anterior aspect. We have now seen that, with regard to the extrinsic position and overall configuration of this area, there exists a certain consistency throughout the mammalian class in that it always occupies the middle or anterior third of the hemisphere, more or less close to the frontal pole, and also extends mainly over the lateral convexity of the hemisphere. Nevertheless, specific differences in location are so important that one could define an individual localisational type for each major group. We shall see in more detail later that different relationships to the sulci also always play an important role.

There is a great similarity in the external shape of the area among the primates, including man and monkeys, and the prosimians. Approximately the same surface area is demarcated in all these, in spite of certain detailed differences that have already been partially discussed earlier. That is to say it forms an almost vertically disposed, wedge-shaped field with its apex directed inferiorly, although slanting slightly posteriorly, that sits astride the dorsal margin of the cortex and whose wider base stretches more or less onto the medial surface (Figures 112 and 113; see also Figure 130 of the marmoset). The pattern is similar in the kangaroo (Figures 114 and 115), and even in carnivores I observe an essentially wedge-shaped form to the area, except that here its lower tip is bent backwards at an angle (cf. Figures 116 and 117 of the stone marten). In contrast, this area in ungulates is situated horizontally along the

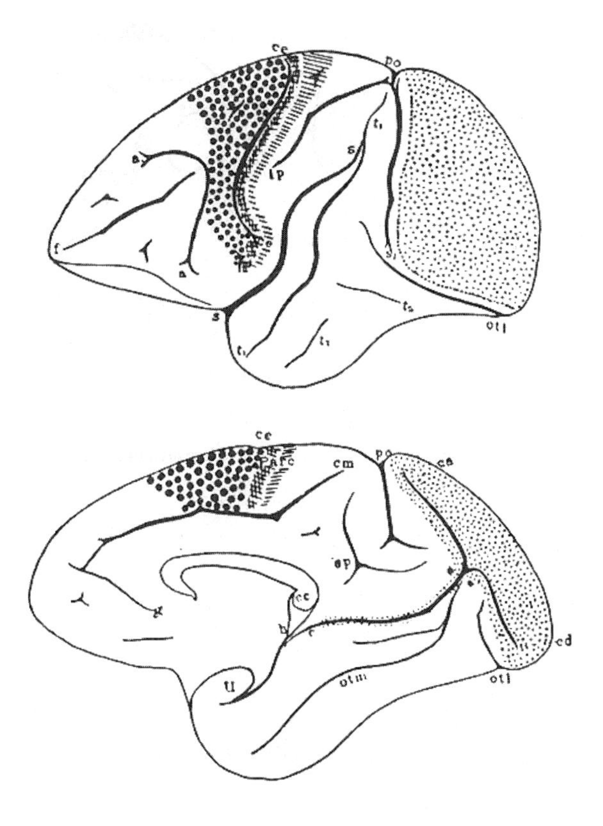

Fig. 112 and 113. Location of the giant pyramidal area (4), the striate area (17) and the postcentral region of the cercopithicids. The symbols for the individual areas are same as in the previous brain maps.

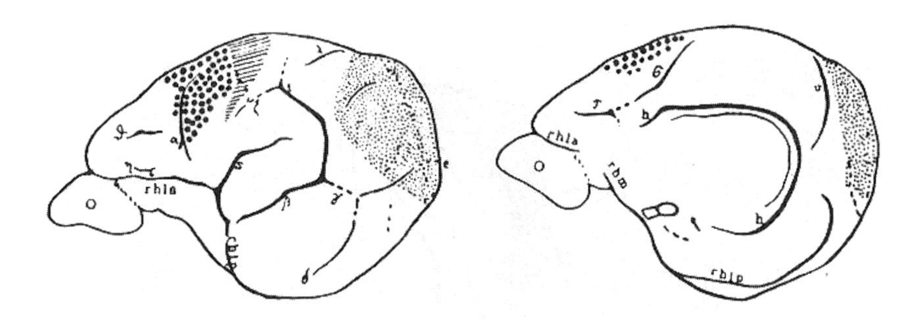

Fig. 114 and 115. Location of the giant pyramidal area and the striate area in the wallaby (*Macropus pennicillatus*). See my fifth communication, 1906.

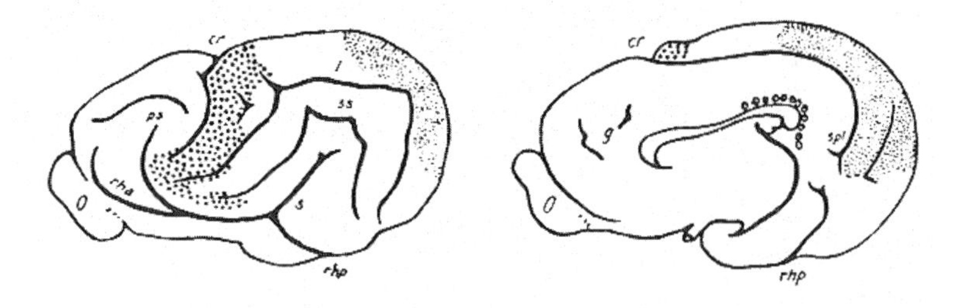

Fig. 116 and 117. The same as the previous figures for the stone martin (*Mustela foina*).

superior margin of the cortex and mainly medial to the coronal sulcus (goat and pig, Figures 118 and 119). In insectivores and rodents (hedgehog, rabbit, ground squirrel) it is quite horizontal and extends mainly along and around the dorsal margin of the hemisphere, and finally in the flying fox it adopts an intermediate position, but lies to a considerable extent on the medial surface.

In terms of relative position of area 4, man stands out in that his area 4 is displaced furthest caudally of all mammals, as a result of the massive development of the frontal lobe. In man it is situated in the middle third of the anteroposterior length of the hemisphere, whereas in monkeys and prosimians it lies at about the junction of the anterior and middle thirds, and in chiropterans, insectivores and rodents it is exclusively in the anterior third, directly behind the frontal pole. The carnivores, the ungulates and, of the marsupials, the kangaroos, occupy an intermediate position between the extremes mentioned above. It should also be noted that the proportion on the medial side is very variable. In all primates, and especially in man, the giant pyramidal area extends considerably medial to the dorsal margin of the cortex, on the paracentral lobule and its homologues. In the flying fox this is relatively even more marked, but less so in lemurs, even less in the hedgehog, rabbit and ground squirrel, and insignificantly in many carnivores such as the dog, stone marten, kinkajou, although more again in the ungulates (goat, pig).

Comparison of different mammalian species also reveals interesting general relationships with respect to relative and absolute size. Without doubt, the total surface area of area 4 is absolutely greatest in man (of the animals in which I have studied cortical localisation), but the relative size - compared with the total cortical surface area of a hemisphere - must be about the smallest in man [1]). Whereas the giant pyramidal area of lower monkeys (cercopithecids, marmosets) and lemurs forms on average approximately a tenth to a twentieth of the total cortex [2]), in man it involves scarcely a hundreth of it, that is to say comparatively much less of the total cortical volume. The area is relatively extensive in carnivores, such as the kinkajou, but strikingly small in the flying fox and in the rabbit, ground squirrel and hedgehog.

The implications of these observations will be further developed later in Part III. Also, the relations of the area to certain sulci will best be considered there in a logical sequence.

Area 17 - *the striate area* - also enjoys great physiological importance as it corresponds to the cortex that on clinical and pathological grounds is intimately related to the so-called "visual area" of the literature (Henschen). The variations in its spatial localisation are even more extensive than for area 4. They result from its being associated with a very pronounced sulcus, the calcarine sulcus, in such animals as primates and prosimians, whereas in the

[1]) Whether these relationships are different in some very large mammals such as the horse, cow, elephant, and certain carnivores, and especially whether the absolute size of this area exceeds that of man, must be decided by later research.

[2]) Precise absolute figures are being established at the moment in the Neurobiological Laboratory by systematic measurement.

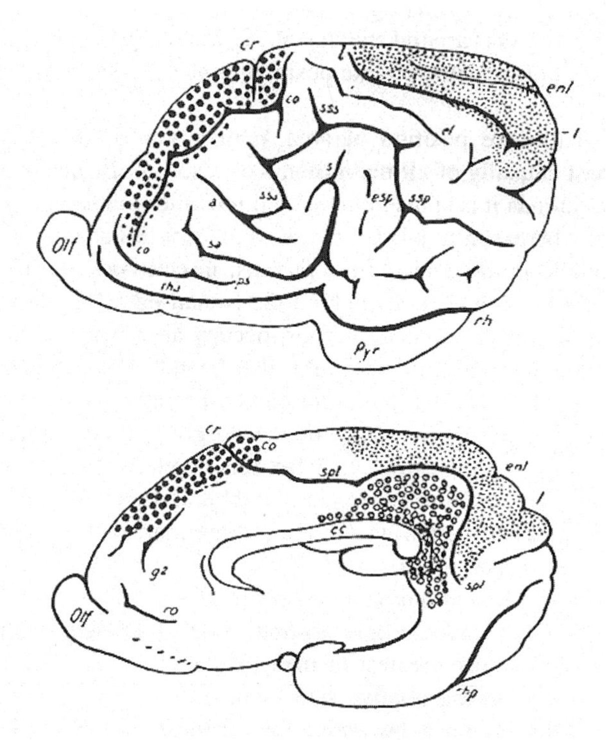

Fig. 118 and 119. The same for the goat (*Capra hircus*).

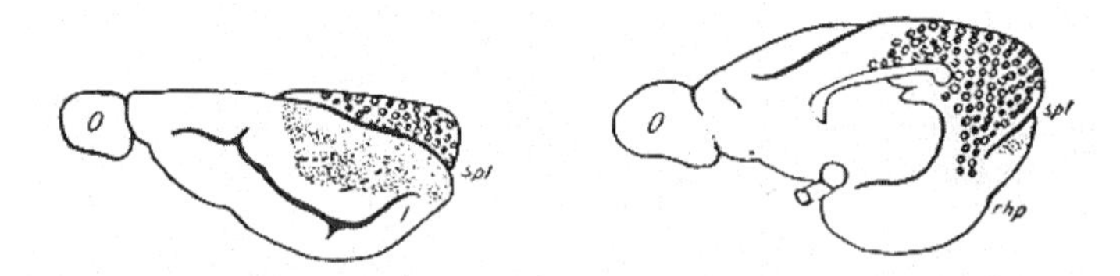

Fig. 120 and 121. Striate area of the dwarf musk-deer (*Tragulus minima*). The zone indicated with circles posterior to the corpus callosum corresponds to the retrosplenial region. It is very large and climbs perceptibly onto the lateral surface.

majority of orders this sulcus is totally absent.

With respect to position, there is a certain general correspondence in all mammals in that the area is always localised at the occipital pole and in its vicinity [3]). On the other hand there are certain radical differences in individual animals, and even not infrequently between closely related species, particularly concerning the distribution between the medial and lateral surfaces of the hemispheres. Here, we must restrict ourselves to illustrating the essentials. Later we shall come back to racial differences in man.

Judged from its external form, and apart from a few exceptions, the striate area represents a field stuck on the occipital pole like a cap, a sort of end-cap, that extends medially and laterally to very differing degrees and thus occupies a very changing position. Whereas in man (Europeans, Figures 94 and 95) it is almost entirely limited to the medial side, and especially to the cortex of the calcarine sulcus, in monkeys (except the marmoset), and particularly in the great apes (Figure 122), by far its largest extent is on the lateral convexity, and it is divided approximately equally betwen medial and lateral surfaces in marmosets, lemurs and macrochiropterans. In carnivores (Figures 116 and 117) the medial portion of the area is usually larger than the lateral (to a notable degree in the dog), as in man, and this is also the case in many ungulates (Figures 118 and 119). In contrast, in many of the rodents that I have studied, such as the rabbit and ground squirrel (Figures 106 to 109), and in addition in the chevrotain (Figures 120 and 121), the area is moved entirely to the convexity. Marsupials, represented by the kangaroo (Figures 114 and 115), have a similar pattern to the monkeys. It should be specially noted that in the last-mentioned group the striate area is at the same time pushed frankly up to the dorsal margin of the hemisphere, while it lies more ventrally in man. This is obviously related to the stronger development of the retrolimbic areas, as well as the piriform lobe, in these animals, the overwhelming growth of which, in my opinion, forces the adjacent cortical areas upwards.

Just how much the position of areas, especially in relation to particular sulci, can vary in different families, or even in different members of the same family, can be appreciated by comparing different species of monkeys.

In Figures 122 and 123 of the orang-utan the whole cuneus on the medial side, that is the gyral surface lying between the calcarine sulcus (c) and the parieto-occipital sulcus (po), is covered by area 17, as is the major part of the so-called occipital operculum that stretches like a tongue caudal to the simian sulcus (si) laterally. Figures 124 and 125 of man should be compared.

In Figures 126 and 127 of the langur the whole operculum on the lateral convexity is covered, but on the medial side only the major part of the cuneus. In the guenon and the macaque (Figures 112 and 113) this trend appears more markedly in that the portion of the cuneus opposite the massively developed lateral opercular surface regresses even more, and in the woolly monkey (Figure

---

[3]) The differing views of Köppen and Löwenstein in ungulates and carnivores, as of Watson in insectivores, are based on mistaken homologies (cf. my fifth communication).

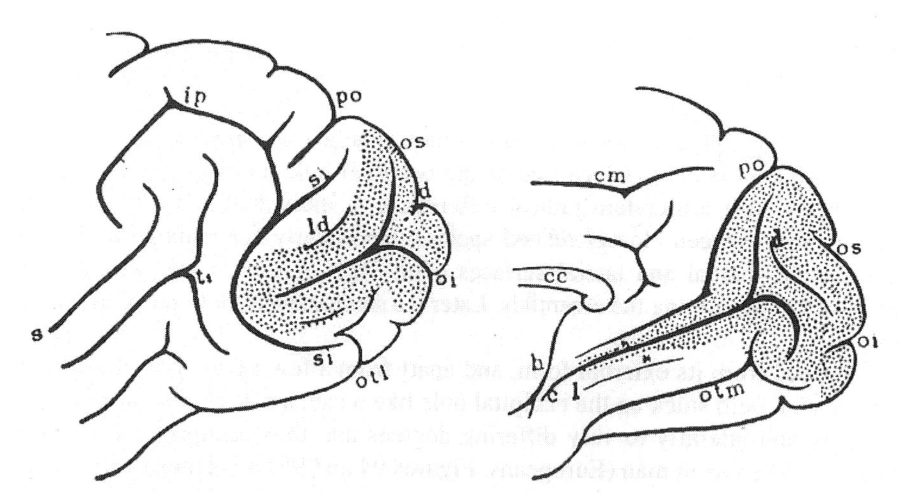

Fig. 122 and 123. Striate area of the orang-utan (*Simia satyrus*). At ** the boundaries of the area sink completely into the depths of the sulcus. cc = corpus callosum, c = calcarine sulcus, cm = callosomarginal sulcus, po = parieto-occipital sulcus, si = simian sulcus, os and oi = superior and inferior opercular sulci, d = descending sulcus, otm = medial occipitotemporal sulcus, otl = lateral occipitotemporal sulcus, ld = dorsal limiting sulcus, l = lingual sulcus, s = Sylvian sulcus, t1 = superior temporal sulcus. The labels are also valid for the following figures.

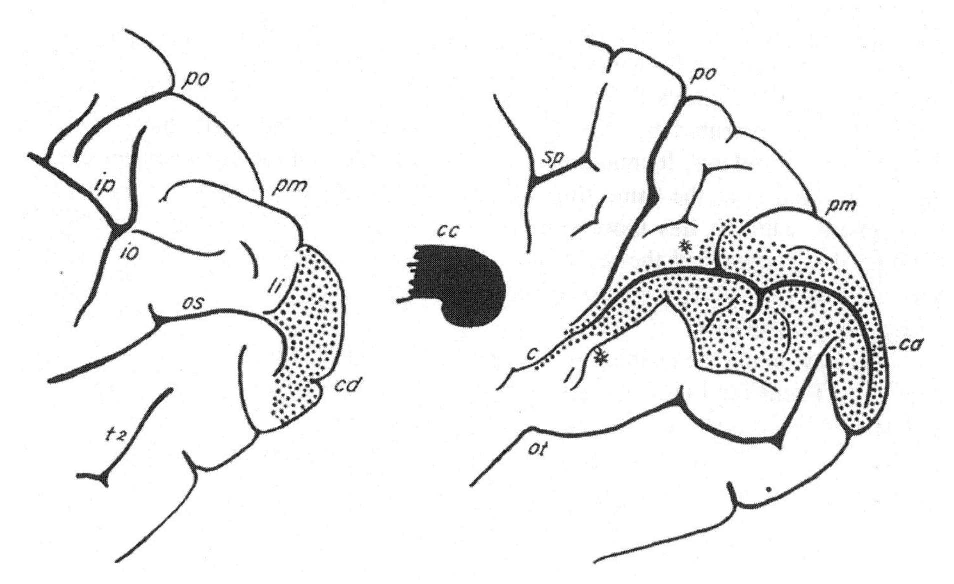

Fig. 124 and 125. Striate area in man (European). Left: lateral view; right: medial view. The area lies almost exclusively on the medial side and hardly encroaches beyond the occipital pole onto the lateral surface. The lateral extent in Fig. 124 is even drawn too big in terms of perspective. At the sites marked by stars the area disappears into the depths of the sulcus c. See also Figs. 94 and 95.

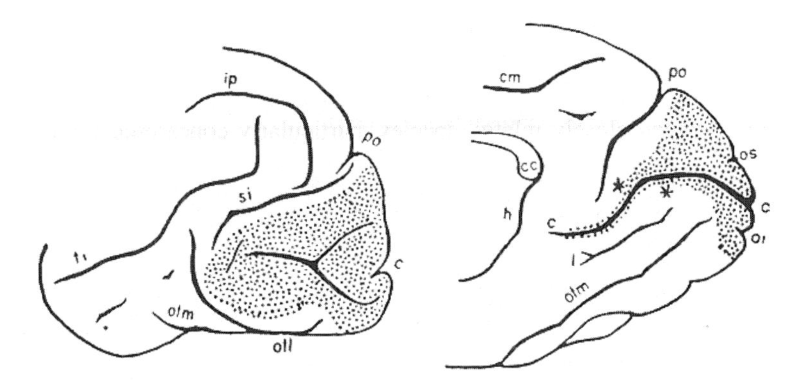

Fig. 126 and 127. Striate area of *Semnopithecus leucoprymnus*.

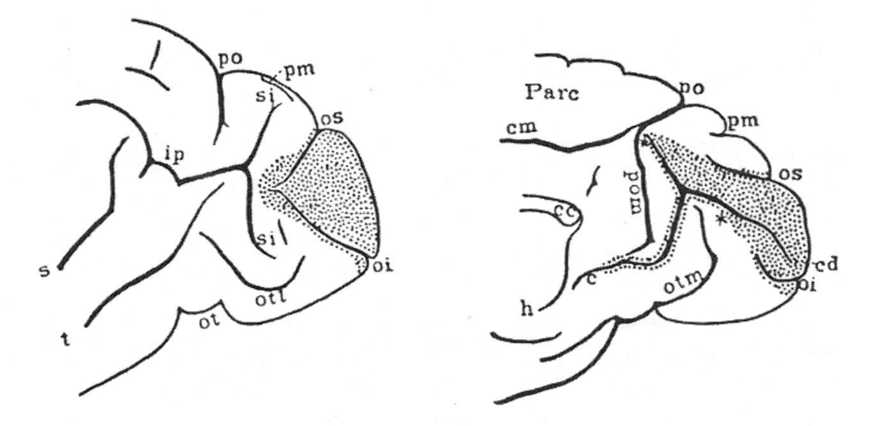

Fig. 128 and 129. Striate area of the woolly monkey (*Lagothrix lagothrica*).

128 and 129) neither the operculum nor the cuneus are fully covered by area 17. The marmosets *Callithrix jacchus* and *pennicillata* (Figures 130 and 131) follow closely the trend in the orang-utan and thus resemble the prosimian lemur (Figure 132 and 133) and slow loris (Figures 134 and 135). In the latter the similarity with the orang-utan is greater in that a simian sulcus (si) appears on the lateral surface of the occipital lobe so that a sort of occipital operculum is formed that is entirely included in our area 17 in both cases.

These localisational peculiarities, especially the relationship to sulci, are even better demonstrated on cross-sectional diagrams. Figures 136 to 139 represent frontal sections through the occipital lobe in the region of the greatest extent of the striate area in three families of monkey, and Figures 140 to 143 represent four different orders (*149). In the orang-utan, the langur and the macaque, area 17 covers at least as extensive an area on the lateral side as on the medial; in the orang-utan the lateral part of the area is indeed much larger than the medial on account of the formation of a deep opercular fossa (o) in which a great part of the cortical surface is buried. In contrast to this, in the marmoset and lemur (Figures 140 and 141), and especially in the cat, the medial surface representation is much more extensive. In the flying fox and the cat (Figures 142 and 143) the area lies extremely dorsally (*148).

The differences in size or surface extent of area 17 in individual mammalian groups are no less, as even a superficial examination of the above brain maps clearly reveals. Unequivocal information can however only be obtained by systematic measurement. As this is not yet available, we must be satisfied provisionally with naked-eye estimations, which nevertheless provide considerable reliable data concerning obvious differences in size. Of the animals I have studied, the great apes possess the absolutely largest striate area, especially the orang-utan. In its case, through the considerable development of the occipital operculum and the appearance of a true opercular fossa in it (Figure 136), a massive lateral cortical area has grown up next to the medial area and easily exceeds in size that of the human calcarine cortex (Figure 87). Just as the total cortical surface area is smaller in smaller animals, the absolute size of the striate area also diminishes in smaller brains; this is true for most cercopithecids and even more so for lemurs. In even smaller brains the cortical surface area decreases correspondingly more. Whether in really large mammals, for instance in certain very big carnivores, ungulates and cetaceans, the absolute figures are higher than in the monkeys described here or in man, only later research can decide.

In my opinion the relative surface area of a cortical area is more important, and it emerges - as far as my material allows conclusions to be drawn - that on the one hand the relative size of the striate area is never in any way proportional to the size of the brain, in terms of volume or cortical surface area, and that on the other hand it is equally not possible to determine any constant relationship with the level of organisation of the corresponding species. Man indisputably possesses a relatively very small striate area; its surface area must amount to about a fiftieth, or 2%, of the total cortical area. In many monkeys

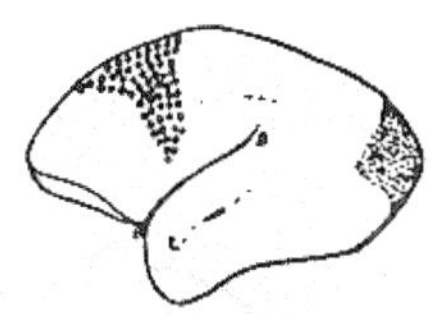

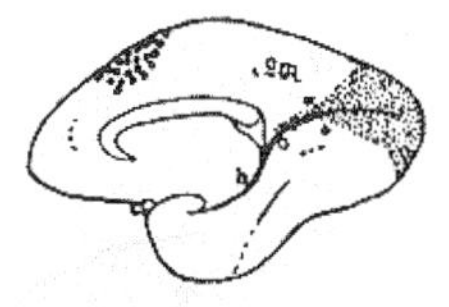

Fig. 130 and 131. Striate area of the marmoset (*Hapale jacchus*).

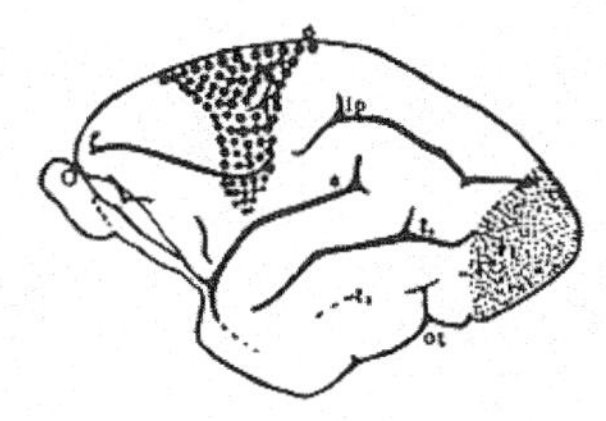

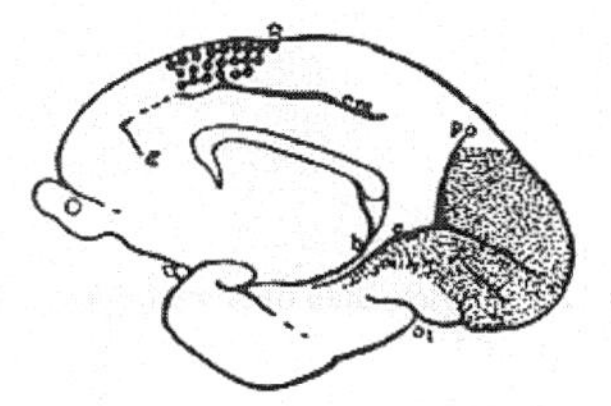

Fig. 132 and 133. Striate area of the black lemur (*Lemur macaco*).

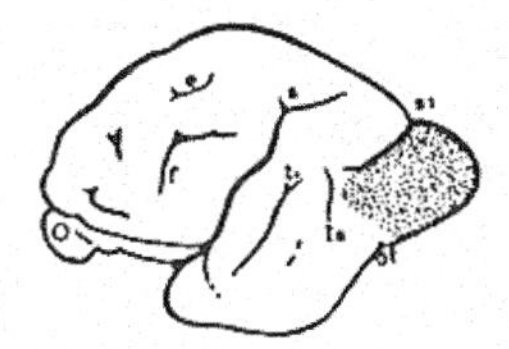

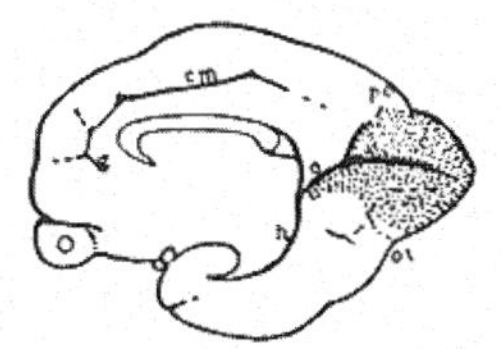

Fig. 134 and 135. Striate area of the slow loris (*Nycticebus tardigradus*).

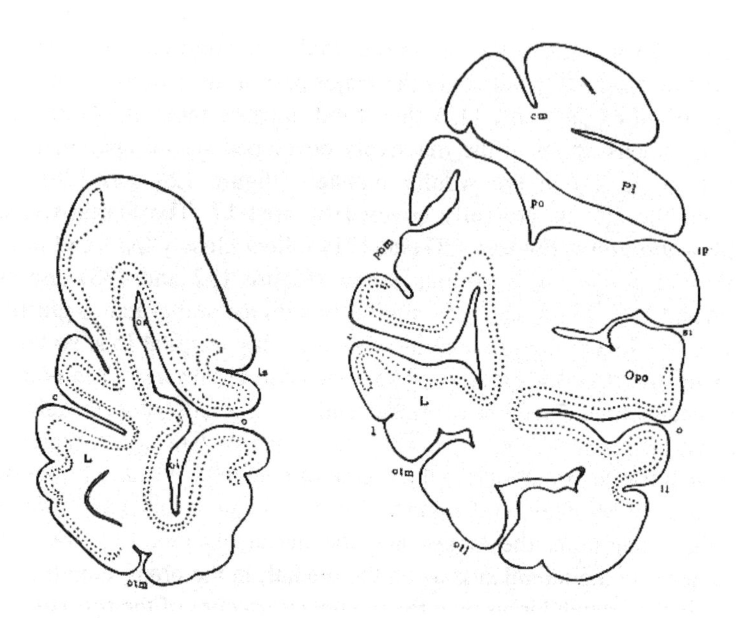

Fig. 136 and 137. Two coronal sections through the occipital region of the orang-utan brain at different distances from the pole. - Extent of the striate area (calcarine cortex) indicated by double dotted lines. The left side of the figures corresponds to the medial side of the hemisphere. (cf. the corresponding sections of man, Fig. 87, page 119.)

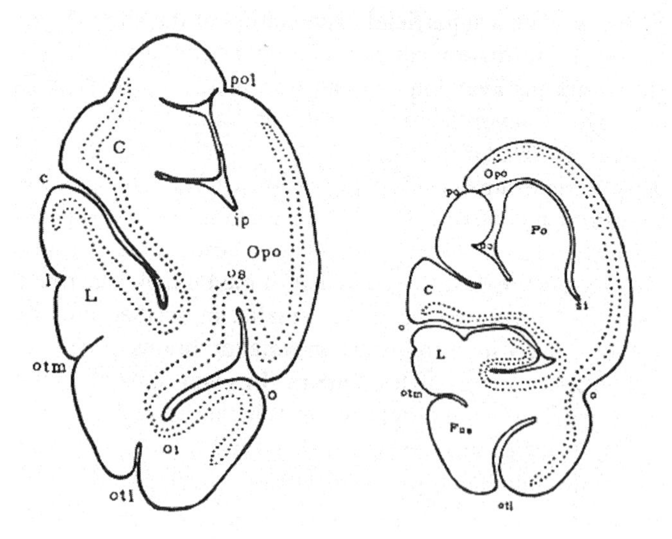

Fig. 138 and 139. The same as the previous figures for two lower monkeys (*Semnopithecus* and *Macacus*). For the abbreviations see Fig. 122. L = lingual gyrus, C = cuneus.

this proportion is entirely different, being at least 10% of the total cortex, that is ten times more (*150). In lemurs it is similar or even higher; one would not be far wrong to assume that in their case the striate area includes close to 15% of the hemispheric surface area. Its relative area is also large in the flying fox, but undoubtedly less in the kinkajou, as indeed throughout the carnivores and the ungulates, and even smaller in insectivores and rodents due to the great dominance of the archipallium.

The possible physiological significance of the localisational differences described here, especially the considerable variations in size of the striate area, cannot be determined in detail for the moment. It is however worthy of note that in monkeys and prosimians, that can be considered to be animals with good vision, the field in question has undergone a relatively great expansion in surface area (to 15% of the total cortex) and at the same time a very characteristic differentiation of its intrinsic laminar structure (as also in man), while on the other hand the rodents, and even more so the hedgehog [4]), possess a striate area that is only poorly developed in surface area and is little differentiated cytoarchitectonically, corresponding to their inferior visual capacity.

On the contrary, we find that animals with a highly developed sense of smell, such as rodents and insectivores, whose brain organisation is in addition rather simple, have developed an extraordinarily extensive and richly differentiated olfactory cortex that occupies about half the total cortical surface, whereas animals with poor olfaction, such as primates, only possess an extremely small olfactory region.

Area 28 - the *entorhinal area* - represents, with the presubicular area, the most constant structural zone of the cerebral cortex throughout the mammalian class, as we have already seen; it is expressed in a typical way in all orders, unless it is absent in the cetaceans that I have not studied [5]). In spite of this absolute constancy, it varies enormously in individual species, and there is no doubt that its development in macrosmatic animals is substantially greater than in microsmatic ones. Accordingly, the area is relatively smallest in primates, its development being particularly poor in the lower monkeys. In man its surface area represents no more than approximately a one- or two-hundredth (1% to 1/2%) of the total cortex; in lower monkeys the proportion is relatively more on account of the smaller brain volume, reaching about a fiftieth, that is 2%. The prosimians, which have even developed a sort of piriform lobe, possess an entorhinal area that stretches quite far over the lateral convexity and occupies, for instance in the lemur, at least a twentieth of the surface area of the hemisphere. In the macrochiropterans there is an entorhinal area that

---

[4]) In the mole, in which the visual system has degenerated to a great extent, I could not find a cytoarchitectonic homologue of the striate area with certainty. Perhaps myeloarchitectonics will solve many problems on this subject later.

[5]) In V. Bianchi's work "Il mantello cerebrale del Delfino" (*151) (Naples, 1905) there appears an illustration of the hippocampal cortex. The figure, however, does not allow one to determine whether this represents the architectonic homologue of my entorhinal area, and a precise localisation of the illustrated cortex is lacking.

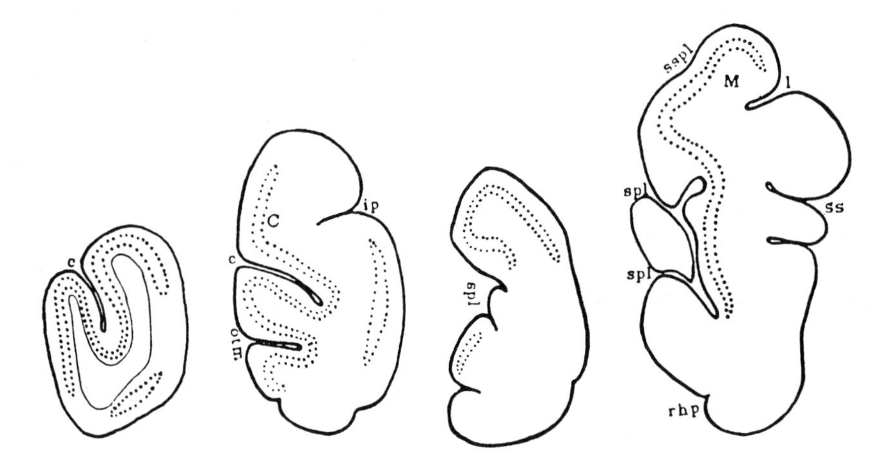

Fig. 140, 141, 142 and 143. The same as the previous figures for the marmoset, lemur, flying fox and cat. Note especially in the last the extensive spread on the medial surface (*148).

includes over half the massively expanded piriform lobe and equally covers about a twentieth of the cortex. The situation is similar in many carnivores, represented by the kinkajou. Rodents display an even more marked relative development of the area; in the rabbit and the ground squirrel it forms the absolutely largest area of the whole brain and about a fifteenth of the total cortical surface. Finally in insectivores (hedgehog) I estimate its surface area as about a tenth of the cortex.

The position of the area on the hemispheric surface is equally subject to very great variability, as will now be outlined, depending on the differing structures of the archipallium and especially of the piriform lobe. In man and all the monkeys, it lies entirely on the medial surface of the temporal lobe and represents an exceedingly small zone at the anterior end of the hippocampal sulcus, corresponding to the head of the parahippocampal gyrus (Figures 86 and 91). In lemurs it is equally divided between the medial and lateral sides and includes the whole pole of the temporal lobe (Figures 98 and 99). In the flying fox it forms approximately the posterior half (Figures 102 and 103), and in the kinkajou the posterior two-thirds, of the piriform lobe on the lateral and medial surfaces (Figures 104 and 105). In rodents (Figures 106 to 109), and even more so in the hedgehog (Figures 110 and 111), it has undergone a considerable displacement backwards and upwards because of the dominance of the olfactory region and now lies quite close to the occipital pole, particularly in the latter, partly medially and partly laterally.

From this comparison it can be seen that the brain of these macrosmatic species has undergone a rotation of its inferior margin in a dorsocaudal direction such that the position of the entorhinal area is pushed more and more upwards and backwards, corresponding to an increasingly caudalward migration of the whole archipallium. It is worthy of note that in many marsupials (kangaroo, phalanger) this displacement has hardly occurred. In spite of the relatively large surface covered by the area in these, it thus retains its position well anteriorly and, as in man, entirely on the medial surface. A similar pattern seems to occur in the monotremes (echidna).

2. Essential variations of areal parcellation. (*152)

As essential we include those differences in areal localisation that involve the emergence of structural regions in a species or a family that are absent in other species and families or that are only represented rudimentarily, and on the other hand the disappearance or regression of areas that are present in other groups. Some of these differences have already been touched upon above in the discussion of cortical regions, when it was mentioned that a different number of individual areas can constitute such regions.

I shall summarise here partly the formation of new anatomical cortical areas and partly their regression or involution in the context of cortical genesis. Which of these two processes dominates, or has played the dominant role, in individual cases in which differences in areal parcellation occur in two closely related groups, cannot always be determined with certainty. Doubtless, both processes are at work throughout, such that progressive and regressive differentiation compete during the development of cortical areas.

If, as a result of specific extrinsic factors, an animal is forced to engage in a particular developmental direction, for example as a predominantly olfactory, visual or auditory animal, the nervous mechanisms that correspond to the functional constraints imposed by the new circumstances, including the relevant cortical centres, must undergo a degree of specific expansion and hypertrophy. In contrast, other centres that do not progress as rapidly, as is the case for most, they remain relatively retarded in growth and regress simply because of developmental constraints.

On the other hand, when a functional mechanism, such as a sensory system, degenerates in a family or species because of diminished activity, the regression in function will necessarily lead to an anatomical regression, and in extreme cases to the complete disappearance of the corresponding central and even cortical apparatus. As a consequence, this in turn leads to the relative dominance of other centres and systems.

From this we see that quite complicated conditions govern definitive structural localisation in the cerebral cortex of an animal, and their interaction and coordination explain the large differences in surface parcellation even between closely related species.

In any case, as long as the basic physiology of the mammalian cortex is not better known and, further, as long as we do not know more details of the sensory abilities of individual animals than today, it will be impossible in most cases to say what functional influences have caused the emergence of the specific localisational organisation that we discover in a brain. We can simply record the facts and try to interpret the differences between the different species in a comparative anatomical way as either progressive development or, on the contrary, as regressive reorganisation in the sense discussed above.

Progressive development is certainly present in those cases in which, in addition to an increase in surface area of a zone, such as a cortical architectonic region, there is an increase in the number of individually differentiated areas within the zone in a given species.

The extreme macrosmatic animals provide convincing examples of this type. For instance, in the hedgehog the marked development of olfactory function is associated with a particular organisation of certain parts of the archipallium, especially our olfactory region. This is manifested not only by the extraordinarily great increase in surface area of the whole region, that constitutes about a third of the whole hemispheric surface, but even more by the rich variety of local differences in intrinsic structure that has led to the spatial separation of new areas within this region, that are absent in other species (Figures 110 and 111, areas 51a-d). A suggestion of this parcellation is also found in many closely related macrosmatic species, but nowhere is it as pronounced in my material than in the hedgehog; in most mammals the region consists, apart from the amygdala and the olfactory tubercle, only of a single area that is not further differentiated. Indeed, important areas, such as the olfactory tubercle of primates, are frequently so rudimentarily organised that their homologies cannot be established with certainty, whereas the tubercle of the hedgehog represents an organ of considerable size that can be further divided into three different zones.

Thus this region of the hedgehog undoubtedly demonstrates an expansion of specifically differentiated cortical matter, recognisable on the one hand by the increase in surface area and on the other hand by the larger number of differentiated fields; in other words, there has been a new acquisition of cortex that has led to the establishment of four secondary areas in the place of a homogeneously structured prepiriform area.

A similar situation exists in the retrosplenial area of many large rodents. We shall examine the rabbit as an example (*98) (Figures 106 and 107). In this animal the region can be divided into six or seven clearly isolated and spatially sharply demarcated individual areas [6]), while in all other mammals the number of differentiable areas is much smaller: in the closely related ground squirrel only three or four are present, in the hedgehog three, in the flying fox four, in prosimians and primates three, and in many families even only one or two. The great variations in the region have already been discussed above.

The cingulate region also manifests similar particular developments in many species. Thus, in the flying fox seven or eight structural areas can be distinguished in this region, whereas most other mammals only demonstrate at most three to five individual areas.

In the hippocampal region of the flying fox, the rabbit, the hedgehog and the kinkajou one can find a structurally defined parasubicular area (49), that cannot be recognised in most of the other animals that I have examined. Similarly, in addition to area 27 the rabbit possesses an area 27a, that I cannot detect in other animals. And again the brain maps of the flying fox, the kinkajou, the hedgehog, the rabbit and the ground squirrel contain an area 48, for which no obvious homologue could be demonstrated in higher orders (except perhaps the marmoset).

Still more important than the increase of individual areas within a region is the emergence of a whole new region in certain species. We have seen that the (agranular) precentral region occupies the whole frontal cortex in many of the species examined, stretching as far as the frontal pole. In contrast, in carnivores, ungulates, pinnipeds, prosimians and primates there appears ahead of this region a zone of varying extent that is granular and either, as in the first three of these orders (*153), consists merely of one extensive area or, as in prosimians, monkeys and man, has further differentiated into a larger number of areas.

In the same category also belongs the splitting off of three or four structurally differentiated areas within the postcentral region of primates, whereas this region in all other lower orders consists of only one area, and in

---

[6]) The myeloarchitectonic differences between the individual areas are even more striking than the cytoarchitectonic ones. S. Zunino, 1909.

many species is even extensively fused with the parietal region.

Among the most important are naturally the regional modifications of the cerebral cortex in man. As a glance at the brain map will show, here it is not simply a matter of a greater volume of cortical substance, but an increase in specifically differentiated cortical areas. The new additions to the human cerebrum affect exclusively the neopallium, and predominantly the parietal lobe, the temporal lobe and a large part of the frontal lobe. Here structural zones appear that do not have homologues in any other mammals. We shall return to the clinical and psychological importance of this fact in more detail in Part III; thus here it will suffice to briefly reiterate just the essential features.

In the frontal lobe it is mainly the inferior frontal gyrus that has differentiated into a whole series of new areas, which cannot even be demonstrated in monkeys. These are areas 44, 45, 46 and 47. The parietal lobe is divisible into the four areas 5, 7, 39 and 40 that correspond to only two (5 and 7) in most other brains, or often to only one (7). Finally the temporal lobe is characterised mainly by the three areas 41, 42 and 52 on the superior aspect of the superior temporal gyrus that are not comparable to any formations in other mammals, even in the closely related higher monkeys. One should also refer again briefly to the rich areal differentiation of the temporal lobe in general, as well as the cingulate gyrus, compared to those of monkeys.

In all these cases there undoubtedly emerges an increase in the specifically differentiated cortical mass that is manifested on the one hand by a greater surface area and on the other hand by a larger number of differentiated areas; in other words, there are *new additions* to the cortex. That such localisational transformations of particular cortical zones should also take place in lower, less well organised species, is a proof that we are dealing here with real progressive differentiation, that is to say a specific modification of anatomical organisation dependent on physiological adaptation, manifested as a corresponding functional (sensory, motor, intellectual) specialisation of the mammalian group in question.

It will prove more difficult in most cases to demonstrate unequivocally regressive modifications of cortical regions or a single cortical area than to recognise true progressive differentiation. When their origin is considered, such vestigial parts of the body can be regarded as "*cataplastic* or *involuted*" organs (Haeckel [7]); in extreme cases there are physiologically "degenerate" parts, like tools that are "out of service".

Naturally, the simplest way of proving a degenerative development would be to argue physiologically, if it were possible to determine that a part of the body, such as a particular section of cortex, did not exert its respective role and thus, as it were, no longer existed physiologically. For the central nervous system, and especially for the cerebral cortex, this is only possible in a few exceptional cases, namely when there is absence or extreme involution of a whole peripheral organ and its afferent cortical projection. Such an occurence

---

[7]) E. Haeckel, Generelle Morphologie der Organismen. Vol. II. Allgemeine Entwicklungsgeschichte der Organismen. Berlin 1866. See especially pages 124ff. and 266ff. (*154)

is however, as stated, extremely rare; one is thus mainly obliged to resort to morphological reasoning and possibly to conclusions drawn from analogies.

The morphological proof of cataplastic development in a cortical zone relies on three factors: firstly the empirically determined process of individual embryological development, when a vestigial part will display a better organisation in embryonic life than in its mature form; second, the comparative anatomical study of the same part in related groups, and third the establishment of the degree of morphological differentiation of a given section of cortex in relation to the whole organ in a particular animal.

The last criterion is not always unequivocal and therefore, in general, hardly a deciding factor. On the one hand far-reaching morphological divergence occurs between different cortical zones, and on the other hand there are intrinsic similarities of these divisions among themselves. This usually does not allow one to decide with certainty whether, in a given case, a variation in architectonic differentiation, or even an in itself very considerable reduction in the surface extent of a cortical area, is to be seen necessarily as the expression of a real regression, that is to say as an "atrophic" tendency, rather than as a differently directed formative process. In any case, caution in interpreting the significance of such findings is called for.

Embryological and comparative anatomical data promise to be more conclusive, but for the moment there is an almost total lack of suitable material in just these domains.

Nevertheless, in a very few cases we can justifiably speak of a true regression of cortical areas, and in other cases such a process is at least very probable.

We can consider as certainly regressive or rudimentary those cortical areas that can be demonstrated to be directly connected to an obviously atrophied peripheral organ. Comparative anatomy recognises an abundance of such rudimentary or degenerate organs, in the morphological sense, throughout the biological world. For our purposes those cases are best suited in which the atrophy involves an organ with a quite specific function, such as a sensory organ.

A prime example is the atrophy of the eyes in certain underground [and parasitic] animals including, in mammals, moles (common mole, golden mole) and blind rodents (mole-rat, tucu tucu [8])). As might be expected, in the mole there is no trace in the occipital region of any cortical architectonic type that could be compared with the calcarine cortex of other mammals. The architectonic differentiation of the occipital cortex is in no way like that of the well defined structural type found elsewhere, and no corresponding field can be distinguished in the whole region. Therefore one may unhesitatingly speak of the loss of part of the cortex corresponding to the disappearance

---

[8]) In the tucu tucu, a rodent that lives entirely underground, the severely atrophied eye is still undergoing an active process of regression, according to Darwin on the basis on his own observations. (Cited by Haeckel, Generelle Morphologie, Vol. II, p.275.) This feature makes this animal a very valuable model for research into our problem, for its striate cortex may possibly represent an architectonic transitional stage.

of a peripheral organ. There is even a certain regression of this area in the hedgehog, that is equally known to have poor vision, and the quite meagre differentiation of the striate cortex in many rodents, ungulates and carnivores (such as the dog) can be seen as an indication of the same regressive process in these animals that, compared with primates, prosimians and certain marsupials (kangaroo), possess visual capacities that, as far as we know, are to say the least not very fully developed.

The pinnipeds (and cetaceans) offer an analagous situation in relation to another sensory system, olfaction. It has long been known to morphologists that those sections of the hemispheres that show a massive development even in their external appearance in macrosmatic animals, are highly rudimentary in these anosmatic animals [9]). In agreement with this, I find a quite meagre differentiation of the relevant areas in microsmatic animals, especially the primates, compared to the hypertrophic development in macrosmatic mammals. One only has to recall the rudimentary structure of certain sections of the rhinencephalon in the former, especially the olfactory tubercle, the prepiriform area and the amygdaloid nucleus, as has been pointed out several times during the earlier descriptions of the brain maps. Whether still other parts of the rhinencephalon, such as areas 27, 28 and 35, and perhaps even the hippocampus itself, should be regarded as partially regressive, I dare not decide.

However, one can say with confidence that the retrosplenial region is atrophied in primates; in man and monkeys not only is its surface extent reduced, but its histological structure is poorly differentiated compared with most other mammals, including both lower orders and closely related groups. This regression seems most marked in the cercopithicids. In many mammals those parts of the cortex directly related to the corpus callosum are equally clearly atrophied and certainly partly functionless, parts that Meynert distinguished as "defective cortex". They include the *lateral and medial longitudinal striae of the corpus callosum* (*156), the *septum pellucidum* and the *preterminal area* (*157).

Numerous other examples of rudimentary or regressive parts of the cortex will undoubtedly emerge if suitable material becomes available and research expands to include a larger number of animals that are closely related, but whose organisation diverges markedly in particular respects. For the present purpose these few examples must suffice.

We see from this that progressive and regressive developmental processes interact in a complex way in one and the same brain and that together they bring about most important modifications concerned with the specific cortical development of an animal.

The degeneration of an organ is related directly to its differentiation, as

---

[9]) See particularly: E. Zuckerkandl, Über das Riechhirn. Eine vergleichend-anatomische Studie. Stuttgart 1887. - G. Retzius, Zur äusseren Morphologie des Riechhirns der Säugetiere und des Menschen. Biologische Untersuchungen. Vol. VIII, p. 23ff., 1898. (*155)

Haeckel stated in his "Generelle Morphologie", and is often inseparable from it. The struggle for existence through natural selection not only produces progressive changes, but also regressive ones, if the latter are more useful to the organism than the former. Thus, the perfection of a composite organ is frequently directly related to a regression of its individual constituent parts, and the degeneration of a particular part is often a decisive and overwhelming advantage for the progressive development of the other parts and therefore the whole.

Part III.

Synthesis:
Hypothesis of the cortex as a morphological, physiological and pathological organ.

In the foregoing chapters an attempt has been made to give a coherent description of the basic features of the histological structure of the mammalian cerebral cortex based on the distribution of its neurons, that is to say the topographic, architectonic pattern of the cortex with the exclusion of its fibrous components in the neuropil. This work was intended mainly to explain the general principles of cortical structure in mammals, and form a basis for a comparative approach to localisation, the details of which still remain to be clarified, so there has necessarily been an accumulation of minute data which detracts from this general aim. Thus it was necessary to select, from the endless mass of morphological structures in the cortex of the different mammalian groups, those patterns that could be useful in answering certain questions of decisive importance for the progress of the study.

This is why, in the above descriptions, the important stands next to the seemingly unimportant and why multiple histological details, that may appear incidental to the morphological problem, are treated comprehensively. Such detailed data were indispensible to define organisational rules and to sketch the broad outlines of cortical structure. Thus the richly varied patterns, their relation to the general priciples of differentiation, and the uniformity of the organisational plan only emerge after minute analysis of detail. However, anyone wishing to study the problem of histological localisation more profoundly or tackle localisational questions independently will also have to master apparently trivial histological and topographical details [1]).

Whereas Parts I and II have thus had to be mainly devoted to a descriptive and analytical collection of data, we wish to address some more general questions in the last Part; in particular we shall deal briefly with the importance of our findings for the problem of defining an "organ" as well as their relationship to physiology and histopathology.

---

[1]) Just how much comparative studies lacking in the necessary circumspection and critique lead to error in our young science, and how many momentous mistakes can arise from faulty observation, especially from a too cursory consideration of the domain of comparative anatomy or incomplete research material, can be determined from numerous examples in the literature. One may once again recall the appalling errors in the study of lamination in which no serious attempts at all were made to thoroughly homologise the individual layers. Even more important are the numerous demonstrable mistakes in the homologies of cortical structural areas that have already been made and have given rise to far-reaching functional hypotheses.

Chapter VII.

Histological cortical localisation in relation to morphology.

As we have seen repeatedly, until recently the concept of the morphological homogeneity of the cerebral cortex was generally popular and, most importantly, was recognised universally by anatomists and physiologists. Even neuropathology could not rid itself of this notion, in spite of ever increasing contradictory clinical observations, and to this day one finds the view from time to time, even in new textbooks, that, apart from trivial local structural modifications, the cerebral cortex is a completely homogeneously built organ containing no sharp divisions in its histological structure.

In view of this it seems justifiable to present the essential points of our observations once again, even if they are not all original and new, and despite previous necessary repetitions, and discuss briefly the validity of the principles of localisation that we have discovered for the whole mammalian class.

1. The basic principles of localisation.

a) *The principle of regional differentiation.*

If one examines sections of the cerebral hemisphere of any animal one finds, even on macroscopic observation, a number of distinct regions within the cortex, each of which demonstrates a quite unique specific structure that allows one to distinguish it from adjacent cortex and from other regions [2]).

When examining a series of sections of the cortex of man or other mammals under the microscope, one can recognise that each small portion of the cortical surface, whether its external shape is that of a lobule or a gyrus, whether it only forms part of a gyral formation, or whether it represents a flat section of a lissencephalic hemisphere, is distinguished by a characteristic cytoarchitectonic (and myeloarchitectonic) structure.

As we have seen, the structural differences between different locations depend on very varied architectonic features. They may be quantitative and consist of differences of cortical thickness or in variations of the individual basic layers which may be more densely or more loosely structured, or wider or narrower. However, the variations may also be qualitative, with the appearance of new cell forms, the loss of basic layers or, on the contrary, the organisation of the cells into quite new laminar aggregates.

Thus, in opposition to the old concept, we arrive at a principle of regional differentiation of the cerebral cortex. The same principle applies to the whole mammalian class and is just as well expressed in the lowliest aplacentals as in the highest placentals. In summary it implies that the cerebral cortex of all mammals displays a far-reaching regional variability in its cellular (and fibre) structure, expressed as local structural differences in number, size, shape, grouping and mutual arrangement of the neurons, as well as in the relative size of the constituent layers.

As a result of this variability of the layers an animal's cerebral cortex manifests an unimaginable regional diversity in its intrinsic organisation, and one may say without exaggeration that histological specificity of individual parts is perhaps more highly developed in the cerebral cortex than in any other organ or organ system, and scarcely anywhere else are the different elements structurally so sharply demarcated from each other. Because of this it becomes possible to distinguish a considerable number of structural zones within the cerebral cortex of all mammals and, in other words, to produce a plan of topical localisation on the basis of morphological features. The height of regional differentiation and anatomical complication is achieved by the human cortex.

b) *The principle of similar differentiation* of sections of the cortex in different mammals (homology of regions and areas).

A second finding that we have established is that, in all mammals, consistent laminar patterns with characteristic structures are found regularly in constant locations on the hemispheric surface. This leads to the establishment

---

[2]) Elliot Smith pointed out this fact years ago and more recently has published many findings concerning localisation based on purely macroscopic observations, most of which agree with my own research. See: Notes upon the natural subdivision of the cerebral hemisphere. Journ. of Anat. and Physiol. 35, 1901, p.431. Also: New studies on the folding of the visual cortex and the significance of the occipital sulci in the human brain. ibid 41, 1907, p.198. Also: A new topographical survey of the human cerebral cortex. ibid 41, 1907, p.237.

and demarcation of anatomically similar (homologous) zones throughout the mammalian class, and thus a concept of homologies of hemispheric surface structures.

As a first step one must distinguish a number of major homologous architectonic zones from the basic features of their cellular lamination. These can then be divided into structural sub-groups or individual areas that have undergone very different development and degrees of differentiation in different species. From the ontogenetic point of view, two major forms of cortical structure can be distinguished, described above as homogenetic and heterogenetic cortex (Figures 144 to 146).

1. We include in **homogenetic cortex** all those cytoarchitectonic forms whose cellular structure displays a common laminar pattern, the basic six-layered tectogenetic type, or that can be traced back to such a pattern, or that will lead to it. Homogenetic cortex can further be divided into two major forms, homotypical and heterotypical.

a) *Homotypical cortical areas* are those which keep the original six-layered pattern throughout life in a more or less distinct form and display only minor architectonic deviations during ontogeny and between individual examples [3]).

b) *Heterotypical formations* on the other hand are those that do not display a six-layered structure in the mature cortex, although it was laid down initially. This may be either because during ontogeny there has been a dissolution or fusing of the basic layers originally present (loss of layers), or because new layers have emerged secondarily by division or splitting of primitive embryonic layers (increase in layers).

2. We include in the **heterogenetic cortical formations** all those cortical zones that, in contrast to homogenetic ones, display a structure other than a six-layered one from their earliest Anlage, that is from early foetal stages when the ultimate architectonics are beginning to be established. They too can be subdivided into several subgroups according to the degree of advancement of the cytoarchitecture, and three major types can be distinguished: a primitive cortex, a rudimentary cortex, and a (heterogenetic) striate cortex (*161).

a) We term *primitive cortex* those cortical zones that never display a cellular lamination comparable to the rest of the cortex, either during ontogenetic or phylogenetic development, but rather display from the very beginning an unordered, extremely primitive structure consisting of more or less irregular agglomerations of neurons. Usually these formations lack even a clearly demarcated subcortical white matter, and that they belong to the cortex at all can only be determined on developmental grounds, especially when, as in

[3]) I know very well that Bronn (Morphologische Studien über die Gestaltungsgesetze der Naturkörper überhaupt und der organischen im besonderen, 1858) (*158) has used the expression "*homotypical*" in a different sense, to denote the major morphological units of an organism that lie side by side and that Haeckel calls "*antimeres*" (*159), in contrast to "*homonymous*" organs, the parts of the body that lie longitudinally, Haeckel's metameres (*160). As Bronn's expression has not generally displaced Haeckel's it is justified to introduce it again here with a new meaning. Furthermore, Gegenbaur has retained the concept of homotypical to describe an aspect of his "*general homology*".

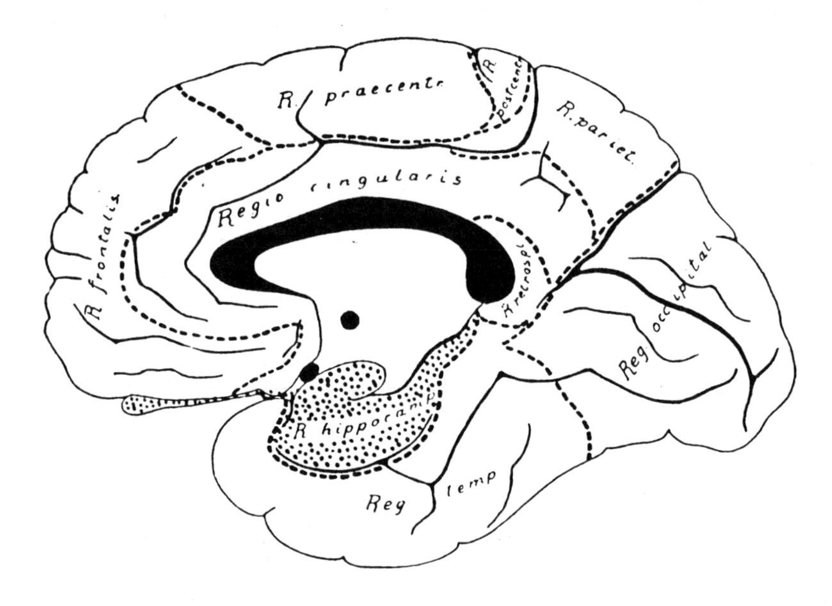

Fig. 144. Medial aspect of a human hemisphere. The dotted surfaces indicate the approximate extent of heterogenetic cortex in man. The indusium griseum, the septum pellucidem and the preterminal area are not indicated.

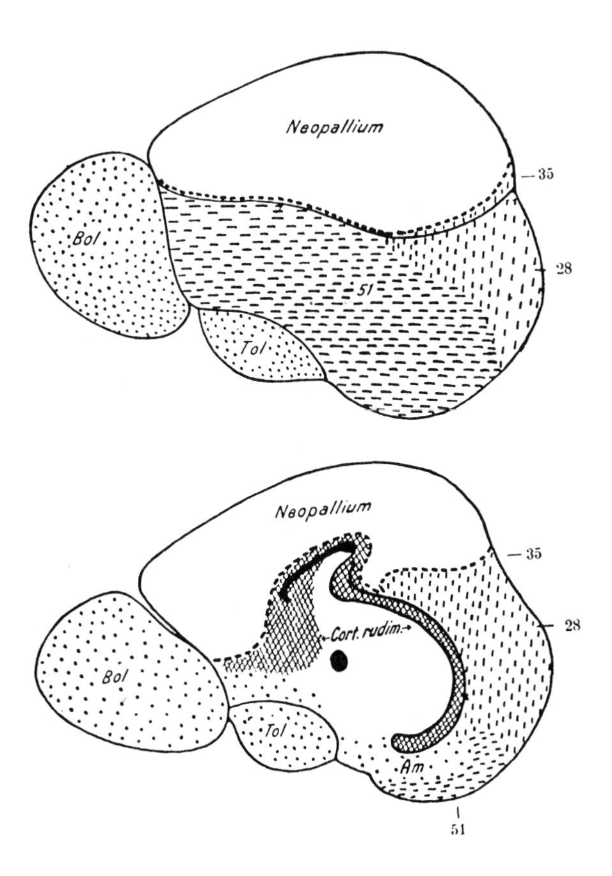

Fig. 145 and 146. Lateral and medial aspects of the hemisphere of the hedgehog. The white surface corresponds to homogenetic cortex, the rest to heterogenetic cortex. Of the latter, the primitive cortex is dotted, the rudimentary cortex is cross-hatched, and the striate cortex is dashed. The striate cortex is further subdivided into an anterior portion (51) and a posterior portion with several individual areas (28, 35). (See Figs. 110 and 111 and Table 7.)

Table 7. (Division of the mammalian cortex according to major architectonic types).

Heterogenetic cortex [1] (Lack of the six-layered pattern in ontogenesis and phylogenesis)		
1. Primitive cortex (no layers)	2. Rudimentary cortex (isolated rudimentary layers)	3. Striate cortex (several secondary, further differentiated layers)
Olfactory bulb Olfactory peduncle Olfactory tubercle Anterior perforated substance Amygdala	Hippocampus Dentate gyrus Subiculum Indusium griseum Septum pellucidum Preterminal area (25)	Presubicular area (27) Retrosubicular area (48) Entorhinal area (28, 34) Perirhinal area (35) Prepyriform area (51)

Homogenetic cortex (All types derived from the basic six-layered variety)		
1. Homotypical cortex (six-layered pattern throughout life)	2. Heterotypical cortex (secondary transformation to six-layered pattern)	
	a) reduction in layers	b) increase in layers
Frontal region (areas 8, 9, 10, 11, 44, 45, 46, 47) Postcentral region (1, 2, 3, 43) Parietal region (5, 7, 39, 40) Temporal region (20, 21, 22, 36, 37, 38, 41, 42, 52) Occipital region outside area 17 (areas 18 & 19) Postcingulate subregion (23, 31)	Precentral region (areas 4 & 6) Precingulate subregion (24, 32, 33) Retrosplenial region (areas 26, 29, 30)	Insular region (areas 13-16) Striate area (area 17)

[1]) The numbers in brackets denote the areas of the brain map.

many animals, they consist of secondarily highly atrophied structures. To this group of primitive cortical structures belong the olfactory bulb, the anterior perforated substance and the amygdaloid nucleus.

b) *Rudimentary cortex* is characterised by the beginnings of a certain laminar pattern, with individual basic layers (I and VI) of the phylogenetically younger homogenetic cortex present, although in only rudimentary form, while other basic layers are still entirely absent. One can include in this category the hippocampus with the dentate gyrus, the subiculum, the indusium griseum (*162) and the preterminal area (*157).

c) The (heterogenetic) *striate cortex*, in contrast to the rudimentary type, consists of several clearly formed layers of the basic architectonic cortex which have often undergone a fuller development than many homogenetic formations by secondary differentiation, sometimes involving a massive development, and the separation of sublayers. Usually layers I, V and VI are affected in this way, while the other basic layers of the homogenetic cortex are not at all developed. To this type of cortex belong the entorhinal, the perirhinal, the prepiriform, the presubicular, and the retrosubicular areas (and perhaps also the ectosplenial area).

As can be seen from the foregoing, the ways in which the different cortical locations have differentiated are very diverse. The major types of differentiation are once again summarised in Table 7 [4]); at the same time a list is given of the main regions and areas of human homogenetic cortex. One can see from it that the heterogenetic cortex belongs predominantly or almost exclusively to those sections of the hemisphere that morphologists have designated "rhinencephalon" or "archipallium", while the homogenetic cortex is restricted essentially to the neopallium [5]).

c) *The principle of divergent development of homologous elements.*

Although a large part of the hemispheric surface differentiates in an essentially similar way either in all mammals or, for some portions, only in a number of closely related species, and has thus developed morphologically equivalent, homologous elements, these homologous zones usually display considerable

---

[4]) Meynert has already attempted a division of the cerebral cortex according to structural variations; he distinguished two main types:

1. Cortex with surface white matter, in which he mainly includes his "defective cortex", that is the hippocampus, uncus, septum pellucidum and the "granular formations" of his olfactory lobe. - In general this group corresponds closely to our heterogenetic cortex.
2. Cortex with surface grey matter, corresponding in the main to our homogenetic group, but of which he only distinguishes two types:
   a) the five-layered cortex (including the common variety and the claustral formation);
   b) the eight-layered cortex (corresponding to our calcarine cortex). (Th. Meynert, Der Bau der Grosshirnrinde. p.58) (*163)

[5]) The division of the cortex by Ariëns Kappers into "*archicortex*", "*paleocortex*" and "*neocortex*" is mainly based on the situation in lower vertebrates, and it seems to me that it is not yet proved that this division can be extrapolated directly to all mammals in the form conceived by the author. - C.U. Ariëns Kappers, Die Phylogenese des Rhinencephalon, des Corpus Striatum und der Vorderhirnkommissuren. Folia neurobiolog., I., 1908. (*164)

variations. These variations frequently have such obvious and consistent characteristics that one can recognise the relevant animal group from the specific structural type.

They are associated on the one hand with extrinsic shape, size and position, and on the other with the intrinsic structure of the particular areas and regions, and are thus partly of a quantitative and partly of a qualitative nature.

Of the quanitative features, the most important are the volume, or more exactly the surface area, of the cortical zone under discussion. We have seen many examples above of how the surface extent of a cortical area or region can vary within wide limits (Chapter VI). I should like to illustrate a particularly striking case in the accompanying Figures 147 and 148. It concerns the hippocampal region of man and hedgehog, represented by two coronal sections through the widest extent of this region. In the hedgehog the zone extends over far more than half, and almost two-thirds, of the section (50 to 66%), whereas in man the figure is about 1/20, only about 5%. The difference in surface area appears even more clearly in the surface view of this region (Figures 144 to 146). The most marked differences in volume emerge in the retrosplenial region of various mammalian orders, as we have seen in the descriptions of the brain maps in Chapter IV, but even the physiologically very important striate area varies over an extraordinarily wide range.

The qualitative divergence between homologous cortical zones is manifested in differences in histological structure, of which the important ones for our purposes concern the laminar cytoarchitectonics. In the preceding chapters we have also tried to give illustrative examples of this. From the outset one must distinguish between two forms of cortical evolution that are related in essence, but the results of which diverge profoundly.

On the one hand, a cortical area of a particular animal group undergoes special development such that its intrinsic structure (its cyto- and myeloarchitectonics) differentiates further than in other groups, and in a similar way over the whole of its surface, laying down larger or more numerous cells of a given type or developing completely new cellular structures, or even introducing new sublayers by the aggregation of certain cell types (eg: the calcarine cortex or area 17 of the capuchin monkey). In doing this the surface area need not necessarily increase; it can remain the same as in related species with an incomplete differentiation of the area in question, or it can even diminish relatively, in spite of the increase in structural complexity (eg: the human entorhinal area, area 28).

The second form of qualitatively divergent development consists of the structural differentiation within a cortical area of distinct loci in different, but specific, ways. Thus regions that are homogeneously built in other species are split into several spatially separate special zones, or areas are divided into several subareas. The retrosplenial area, area 29, of the rabbit is a particularly striking example of the latter process (Figure 107), and another is the postcentral region of primates.

Among these major forms of divergent cortical development are included

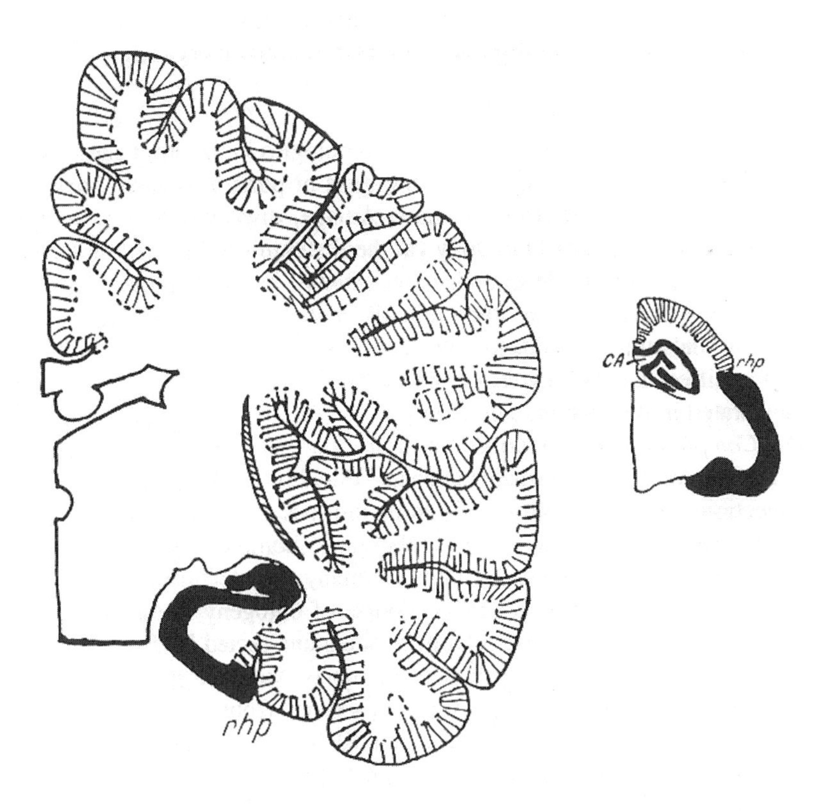

Fig. 147 and 148. Diagrams of coronal sections through the hemispheres of man and hedgehog at the level of the largest extent of the piriform lobe or hippocampal region, drawn in their natural size relations. The neopallium is hatched, the cortex of the hippocampal region is black. rhp = posterior rhinal sulcus, CA = Ammon's horn. Man 1:1, Hedgehog 2:1 (*165). (Note the huge development of the neopallial cortical surface in man!)

all the phenomena that we recognised in Chapters III and VI as polymorphism of cortical development and progressive and regressive modifications of cortical areas. They are the causes of the surprising multiplicity of forms of cortical organisation in the various mammalian species.

d) *Special homologies.*

We describe as special homologies, or homologies in the strict sense according to C. Gegenbaur [6]), the relationship between organs or parts of organs sharing the same origin, that "being derived from the same Anlage, display the same morphological pattern". Various degrees of homology can be distinguished according to the appearance of a particular organ, depending upon whether its morphological state is essentially unaltered compared with other animals or if it has undergone modifications in individual animals through additions or deletions of parts. We have now to determine whether the different subdivisions of special homology defined by morphologists can be demonstrated in the mammalian cortex [7]).

α) *Complete homology* arises when an organ is modified in extrinsic features such as shape, size etc., but remains unchanged and intact in terms of spatial relations and connections, and most importantly in its intrinsic structure. Such cases abound in the cerebral cortex, as we have seen. Naturally, homologies will usually only be complete between closely related groups, as many new features have been added to individual parts and old ones lost in the course of ontogeny through progressive and regressive events: a cortical type seldom remains unaffected by fundamental changes throughout the whole mammalian class. Nevertheless there are enough examples where one can accept such complete homology over a broad front. Among them are many of the *monomorphic types* described in Chapter III (page 87) and found throughout the mammalian class with few exceptions. Examples are cortical types in the hippocampal region (areas 27, 28 and 35), the hippocampal cortex itself together with the dentate gyrus, and also the giant pyramidal cortex of several orders, especially primates, prosimians, many carnivores and ungulates. It is also true of parts of the retrosplenial region in a large number of species, such as area 29 in prosimians, macrochiropterans, carnivores and ungulates on the one hand, and in many rodents on the other. In primates this whole region is completely homologous and represents a regressed organ, as opposed to other orders in which it is defectively homologous. Finally complete homology can be accepted for individual *polymorphic types* in several species, such as the calcarine cortex of many primates, prosimians, carnivores and macropods.

β) *Incomplete homology* (*Gegenbaur*) means that a cortical zone is profoundly modified in its intrinsic characteristics, either by addition or by reduction of

---

[6]) Carl Gegenbaur, Vergleichende Anatomie der Wirbeltiere mit Berücksichtigung der Wirbellosen. Leipzig 1898. (*166)

[7]) In the following groupings I shall follow Gegenbaur's classification strictly. loc. cit., p.24.

component parts, in relation to other zones that are otherwise completely homologous or structurally similar. There are also plenty of examples of this in the above descriptions, and in particular the qualitative divergences of cortical development mentioned earlier belong here, as do all the deviations of the localisational pattern described in Chapter VI as essential variations of cortical areal parcellation. We also find everywhere the two subtypes of incomplete homology that Gegenbaur distinguished: *defective homology* arising from loss of elements, and *augmentative homology* from the addition of new elements.

We have already seen (Chapter I, page 36) that one can speak of defective homology in a narrower sense not only with reference to cortical areas themselves but to the lamination within an area, and that imitative homologies (Fürbringer) also appear in the cerebral cortex in a specific sense.

2. The question of the "organ".

a) *Organ formation by differentiation*. Together with generation (Generatio) and growth (Crescentia), differentiation (Divergentia) is the most important fundamental function responsible for the formation of individual organs, and on it all their secondary development and refinement depends [8]). This process, which involves the emergence of dissimilar elements from similar fundamental bases as an adaptation to differing life styles, or, in physiological terms, to functional differences, has led to an enormous multiplicity of organisational patterns in the cerebral cortex of various animals and also often in one and the same brain, especially in higher species, as we have seen.

The final laminated cortex develops by morphological differentiation from a homogeneous Anlage that is common to all mammals, the primitive unlaminated cortical plate (W. His), and within this cortex emerge numerous extensive regional structural modifications as a result of local heterogeneous differentiation. The result of this developmental process throughout the mammalian class is the emergence of specialised histological complexes that are well demarcated from other parts, that each possess their own specific structure and, as far as we can determine, each subserve a specific function and, in

---

[8]) Ernst Haeckel, Generelle Morphologie, Vol. II, p.72ff. See also: ibid, Gesammelte populäre Vorträge aus dem Gebiet der Entwicklungslehre, Part 1, p.99ff. Über Arbeitsteilung in Natur und Menschenleben. (*167)

[9]) One could object that the term "*organ*" for such histological units within a major organ is out of place. However, I feel that, according to the usual zoological terminology, the concept is justified in its strictest sense. Victor Carus used it in this general sense. He described an "organ" as a "sum of specific elements or tissues with specific interrelations and form". Haeckel also has a very broad morphological concept of an organ; an organ is (without being an individual higher order system) "a consistent homogeneous spatial element of defined form made up of several tissues with consistent relations" (Generelle Morpholog. I, p.291). R. Hertwig gives the definition: "An organ can be called a tissue complex that is dermarcated from other tissues and has adopted an individual, unique form to perform a specific function" (Lehrbuch der Zoolog., p.83). (*168) Finally, Claus states "one means by organ each part of the body that, as a unit subordinate to the higher unity of the organism, displays a specific form and inner structure and performs a corresponding function" (Lehrbuch der Zoologie, pp.38-40). As one can see, all these definitions can be applied without difficulty to our cortical areas.

a word, form organs [9]). What we have determined to be an ensemble of structural areas and regions in the preceding chapters, represents a complex of organic elements subordinate to a higher organism when considered morphogenetically.

From the organic point of view, the mammalian cerebral cortex is thus to be considered as an *organ complex* or, in other words, an ensemble or aggregate of suborgans derived from a common Anlage and subjected to various degrees of progression and regression, sometimes coordinated and sometimes subordinate, that are specifically differentiated and more or less sharply demarcated from each other by their histological structure.

These individual cortical organs represent composite or "*heteroplastic*" organs (Haeckel's third-order organs) as they consist of a combination of several diverse tissues, such as neurons and their processes, including their fibres, neuroglia and connective tissue, in contrast to the simple or "*homoplastic*" organs that consist of only one tissue. In Haeckel's sense one could perhaps call such an organ complex an "*organ system*" or fourth-order organ, a term relating to those morphological units made of a multiplicity of composite or heteroplastic organs [10]). In this sense the cerebral cortex would be a lower order organ system subordinate to the whole central nervous system.

b) *Refinement through differentiation*. Organ formation relies on functional activity which itself depends on adaptation to changing conditions of life. An organ changes according to the conditions that influence it, like the whole organism. If such an influence operates in a particular way for a long time, the organ will change its function; it adapts to new circumstances because this represents an advantage for the organism in the struggle for survival. With the change in function goes a slow, but steady, change in the form of the organ. By an accumulation of very small alterations parts of the organ change permanently and differentiate morphologically. The newly structured elements with their new function develop certain characteristics and become more clearly differentiated from each other with time, and so finally spatially separate parts emerge as new organs. Thus, adaptation and the resultant differentiation determine functional localisation, that is a feature of increased overall performance. Certain functions originally subserved by the whole organ are now accomplished by particular parts and the performance of the organ is therefore divided into various partial functions that, all together, represent the total function.

Refinement of an organ is necessarily related to this process of differentiation, that is expressed physiologically as functional adaptation and morphologically as anatomical complexity. In this way the cerebral cortex develops a wealth of areas, or individual organs. The greater multiplicity and independence of different sections of the cortex means greater freedom of action, and with the larger number of relatively autonomous elements there is more latitude for differentiation. Thus is produced a lasting modification and

---

[10]) Generelle Morphologie, I, p.301.

refinement of the whole cortex, along with a multiplication of functions [11]).

Compared with the multiple organs derived by specialisation, the original single organ can be regarded as "*primitive*" (Gegenbaur). In this sense, one might often consider certain regions of the more simply organised mammals described above as primitive organs from which the multiple cortical areas present in more highly developed species have emerged. In the final analysis, of course, one can also postulate such a primitive condition for the whole cortex somewhere in its ancestral past, that is to say a single relatively poorly differentiated primitive cortex, but this condition has long been surpassed in phylogeny and is no longer demonstrable. It must remain an open question whether the refinement of the cortex through differentiation is always the result of external, physical causes, or whether many of the associated phenomena may be explained in other ways, unrelated to external living conditions and unrelated to the struggle for existence, rather due to a property of the organism itself, an "*energy for refinement*" (R. Hertwig) [12]) or, as Naegli [13]) expresses it, a "*principle of progression*" [14]).

c) *Different degrees of development.*

Rules of divergence, and especially divergence in cortical development, express themselves in two forms, as we have already seen several times: firstly, by the development of extremely different characteristics in a particular part of the cortex of certain individual species (Haeckel's *primary phylogenetic differentiation*), even if the cortex in question is in other ways generally morphologically similar to comparable ones, and of similar origin, so that they are still homologous. They manifest themselves to us as development towards a higher level, or progressive development, in some cases, and as regressive changes towards a lower level in others.

The second form of divergence involves the transformation of different parts of the cortex of the same brain in different ways (Haeckel's *secondary individual* or *ontogenetic differentiation*). Here also one must distinguish between qualitative and quantitative divergence as well as between progressive and regressive development of individual zones as above.

Both processes, phylogenetic differentiation that leads to transmutations of species and thence to gradual development of various parts of the cortex in more highly organised groups, and ontogenetic differentiation that governs the emergence of different cortical organs in individual animals, are the expression of one and the same basic biological phenomenon of physiological functional specialisation based on adaptation and heredity [15]). Just as reciprocal relations

---

[11]) See Gegenbaur, loc cit, p.3ff., on "Adaptation".

[12]) cf. R. Hertwig, Lehrbuch der Zoologie, 1900. Also: idem, Der Kampf um Grundfragen der Biologie, 1909. (*169) See also O. Hertwig, Über die Stellung der vergleichenden Entwicklungslehre zur vergleichenden Anatomie etc. Handbuch d. Entwicklungsgesch., 1906. (*170)

[13]) Carl Naegeli, Entstehung und Begriff der naturhistorischen Art. Munich, 1865. (*171)

[14]) C.E. v. Baer's "*pursuit of a goal*" or Eimer's "*orthogenesis*".

exist between the organs of an individual such that local changes in one region also cause modifications in distant body parts, according to Cuvier's *principle of correlation of parts*, so functional specialisation affects not only anatomical refinement, or development in the strict sense, but also regression of other parts, especially during ontogenetic development. Accordingly, development and regression, anatomical complexity and simplicity, are not only not mutually exclusive, but they proceed together and are mutually dependent in all sites and at all times, according to the correlation principle.

This correlation is visible grossly in the cortex both in the size and shape of its organs (areas and regions), and in their number and position. If a cortical area or a larger cortical zone increases in relative size in an animal, or if the number of the component subfields increases considerably, we frequently see a regression in neighbouring or even distant cortical zones, and this is again expressed partly by a decrease in volume and partly by a reduction of differentiated component parts, that is of regions and areas. Sometimes changes brought about by such correlation lead merely to a spatial displacement, that is to say a relocation of an organ. At a higher level, regression is accompanied by true "atrophy" with the emergence of rudimentary or cataplastic organs.

In earlier chapters we have encountered abundant examples of all these developmental processes. The way in which rudimentary elements arise is obviously the same as that by which new elements are formed. Only the direction of the formation is opposite in each case. "Just as during the genesis of an organ numerous small additions accumulate over many generations thus finally leading to the appearance of a completely new part, so numerous small deletions gradually accumulate during the regression of an organ until finally, after many generations, it disappears completely. In both cases adaptation and heredity work together and reveal natural selection as the underlying cause in the struggle for existence." (Haeckel.) We have already pointed out above that it is not always possible to decide in individual cases what has really regressed. Cortical organs of moderate developmental level within a mainly poorly organised cortex will differ little, if at all, in terms of morphological importance, from organs that have regressed from a higher cortical type.

In certain circumstances it may be particularly difficult to avoid confusion between rudimentary cortical areas and developing or newly formed cortical organs. Anaplastic or newly developing structures may appear morphologically regressed or rudimentary, and thus of little physiological value.

The same applies to unchanging (*172) structures. One has to assume such unchanging elements, or "*stock structures*" (*173), in the cerebral cortex, an organ undergoing continual and progressive development, at least in many animals and particularly in man, as much as those undergoing clearly

---

[15]) Haeckel writes: "When a morphological individual's embryological development causes the ontogenetic differentiation of its characteristics to unfold rapidly before our eyes, we can easily recognise therein the inheritance of the phylogenetic differentiation which the ancestors of the organism in question have undergone during their slow paleontological development" (Generelle Morphologie II, p.256).

demonstrated anaplastic and cataplastic processes. Where they are to be found and what particular morphological features might represent them cannot yet be stated. Only the most detailed knowledge of the progress of an individual's ontogenetic development, together with that of the formation of the same organ in related animals, may perhaps reveal in some cases whether a given section of cortex is progressing, regressing or unchanging. But the relations become even more complicated, and a decision therefore considerably more difficult, in that one must certainly agree with Haeckel and distinguish two types of organisational pattern with regard to cortical organisation: firstly "*monotropic types*", that is those that have become exclusively physiologically and morphologically adapted to particular special conditions, thus forfeiting further capacity for development, and secondly "*polytropic types*", or those that are not particularly exclusively or specially adapted and have thus retained a high capacity to develop in other directions. As we are here dealing with the end product of the development of physiological function, a major part of the work of determining these various types will fall upon comparative physiology.

3. The systematic significance of our results.

"Natural systematics" fulfil the task of revealing the natural family relationships of organisms, that is their lineage, by means of their greater or lesser degree of morphological relationship. I should now like to briefly raise the question as to how much localisational data can contribute to our knowledge of mammalian phylogenetic relations.

a) *Phylogenetic relations in general.*

The principle and most important manifestations of family relations that we know are comparative anatomy and ontogeny. The difficulties in drawing definite conclusions about phylogeny from them are particularly great when dealing with such a highly complex and so profoundly secondarily modified organ as the cerebral cortex. They arise partly from the incompleteness of the material, but also especially from the facts of caenogenesis (Haeckel).

We have constantly had to emphasise that our material is extremely incomplete. This is partly due to our research methodology that involves simply taking occasional samples, and partly on the gaps in the animal series itself. Numerous transitional forms between modern species have certainly died out, and we could not examine many others. Thus, on the one hand we are lacking many important intermediate varieties, while on the other hand morphological relationships are often so changed by caenogenetic "alterations" of palingenetic development taking place during embryonic development that they can no longer be explained phylogenetically. We can, therefore, only point out here the most general phylogenetic relationships.

The most important finding of our study of comparative cortical topography, and the most significant for our investigation, is that the cellular structure of the cerebral cortex of all mammals, placental as well as aplacental, manifests a common architectonic plan, a standardised pattern of cell lamination. First, we were able to distinguish two types of such architectonic cortical formations, derived from (presumably) different primitive cytoarchitectures, the heterogenetic and the homogenetic types. The essential features of each of these structural types is consistently demonstrable throughout all species, with greater or lesser modifications. Further, we ascertained that in all mammals the homogenetic cortex again manifests a regular series of variations (homologous types), that themselves all stem from a common basic histological form, the original six-layered primitive tectogenetic type. Finally, we were able to show that, thanks to these homologous structural formations, it is possible to demonstrate an essentially standardised organisation of the surface of the hemispheres throughout the whole mammalian class into spatially delineated zones, or secondary cortical organs, that we denominate regions and fields (or areas).

These observations can only be explained by the acceptance of a common developmental pattern of all cortical formations. Particularly convincing is the evidence that even those cortical structures that manifest completely novel characteristics in their mature form, such as profoundly modified architecture (through loss or gain of layers), also possess the six-layered basic form during embryonic development, even if only temporarily. Therefore, the six-layered state must be seen as a primordial, atavistic condition, at least for homogenetic cortex (the question remains uncertain for many heterogenetic types), with later patterns representing secondary variations derived from it by differentiation.

With all this, a common origin for at least the whole of the neocortex, with all its modifications, becomes extremely likely, and the accumulated data on architectonics and localisation represent a body of evidence for the development of all mammals from a common stem, that is to say for their monophyletic origin.

This is not the place to discuss the special family relationships of individual members of this stem. For this the available material is hardly sufficient, for in many cases the proof of homology is entirely lacking and in others it is not easy to distinguish homology from analogy. One can, however, state that many systematic groups demonstrate obvious close phylogenetic relationships, even with regard to their cortical organisation. This point will be discussed in more detail elsewhere.

b) *The position of man.*

For Huxley [16]) the "question of questions" is that of man's place in nature. In fact this problem is of such fundamental theoretical importance from

---

[16]) Thomas Henry Huxley, Zeugnisse für die Stellung des Menschen in der Natur (Mans place in nature). German version by Victor Carus, Brunswick 1863.

On the same subject see also: Carl Vogt, Vorlesungen über den Menschen, seine Stellung in der Schöpfung und in der Geschichte der Erde. Giessen 1863. (*174)

the point of view of the development of the cortex, the organ with which we associate all higher intellectual capacities, such as particularly distinguish man, that we do not wish to omit at least a few aphoristic remarks.

Huxley summarises his observations in the famous pithecometric thesis that differences in body build between man and the great apes are smaller than the differences between these great apes and lower monkeys (*175).

Recently objections have been made to this statement from different sides. In particular Johannes Ranke, in accord with the theory of Cuvier and Blumenbach, has maintained "that in spite of the relative closeness of man to the apes, there exists a quite essential and even systematically tangible difference between man and ape", consisting principally of a difference due to the dissimilar structure of the central nervous system. Ranke takes the superiority of the brain over the viscera as crucial and believes that he can contrast man, the "cerebral being", to the whole of the rest of the animal kingdom, the "visceral beings", in contrast to the hitherto generally accepted Linnaean classification whereby man and the apes together form a mammalian order, the primates.

Even Haeckel, that most persistent supporter of the theory of evolution, expressly emphasises certain highly significant differences between human organisation and that of his next lower relation, namely "the higher degree of differentiation of the larynx (speech), the brain (the soul), the extremities, and finally the upright gait" [17]).

How can this problem be resolved from the standpoint of cortical development? From the foregoing arguments, the monophyletic origin of the cortex of all mammals, including man, emerges unequivocally. The intrinsic and essential correlations of the basic features of cortical structure support this idea unmistakably, just as does the general similarity of body structure. But also the close and special relationship of man and primates emerges from their histological localisational relations. From the architectonics of the cerebral cortex and also from the topographic distribution of cortical areas, man is closer to the monkeys, and especially to the great apes, than any other mammal. The cortical pattern of an orang-utan that we studied resembled remarkably that of a young human both in general features and in individual cortical types.

On the other hand, however, in spite of the unmistakable essential relationships, there exist such important quantitative differences between man and great apes that the universal validity of Huxley's statement is subject to a severe limitation. These differences concern the surface extent and volume [18]) of the whole organ, as well as its intrinsic structure and topical organisation.

With respect to the first point, the development of the surface of the

---

[17]) Generelle Morphologie II., p.430.

[18]) In this context it should be noted that H. Friedenthal has recently claimed to have established a confirmation of Huxley's rules for curves of body weight (Über das Wachstum des menschlichen Körpergewichts in den verschiedenen Lebensaltern und über die Volumenmessung von Lebewesen. Med. Klinik. 1909. 19). (*176)

cerebral cortex, there can be no doubt that the total area of the cortex of man largely exceeds that of a great ape. According to H. Wagner [19]) the whole cortical surface measures between 187,000 and 221,000 mm^2; R. Henneberg determined a mean of about 110,000 mm^2 for one hemisphere - according to figures that he obligingly made available to me - using a more precise method [20]). If one compares this with the cortical area of an orang-utan of about 50,000 mm^2, according to Wagner, and that of a lower monkey (macaque) of 30,000 mm^2, one arrives at a different relationship than might be expected from Huxley's statement quoted above.

However, the comparison indisputably leans even more in favour of man if one compares the total cortical mass, that is to say the product of cortical area and depth [21]). Systematic measurements of this are still lacking, but are urgently needed for a deeper understanding of the subject of the cortex [22]). Nevertheless, simple observation reveals that in man, as the extent of the cortex, and also its cross-sectional depth, strongly dominate not only absolutely but also relatively, the difference between man and the great apes is also a considerably greater one in this respect than that between great apes and lower monkeys.

The same is true of the topographic parcellation of the cerebral cortex. The richness of topical differentiation is incomparably greater and more varied in man than in any monkey. This is already obvious from the number of cytoarchitectonic areas. In man, about fifty different areas can be distinguished, in many lower gyrencephalic monkeys up to thirty, and in great apes (orang-utan) about as many or a few less. When one considers the myeloarchitecture, the difference between human and monkey cortex becomes much greater. O. Vogt distinguishes more than fifty different myeloarchitectonic areas

---

[19]) H. Wagner, Massbestimmungen der Oberfläche des grossen Gehirns. Inaugural dissertation, Göttingen 1864. (*177)

[20]) Henneberg's figures will soon be published in the Journal für Psychol. und Neurol. (*178)

[21]) H. Wagner describes the product of surface area and weight "surface development". He made the interesting observation that the surface development of the cortex is less related to brain weight in man than in the orang-utan, and much less than in lower, lissencephalic mammals. This fact can doubtless be explained in large part by the infracortical brain centres, especially the subcortical white matter of man, having attained such a massively dominant volume.

[22]) G. Anton - Gehirnvermessung mittels des Kompensations-Polar-Planimeters (Wien. klin. Rundsch. 1903, 46) (*179) - has carried out similar research on the relationship of the volumes of the cortex and white matter. The measurements also need completing systematically for the main mammalian groups.

No less a problem, that equally needs to be tackled urgently, is the question of the relationship between body size and cerebral or cortical development. As is well known, Bronn (Morphologische Studien über die Gestaltungsgesetze der Naturkörper, 1858) (*158) has postulated the rule that, generally speaking, animals increase in weight progressively in rough proportion to their increase in capability. Haeckel has already spoken out against a general application of Bronn's rule in his Generelle Morphologie, and Fürbringer established that only exceptionally could this postulated parallel between body weight and level of development be observed to reach its full expression (see M. Fürbringer, Untersuchungen zur Morphologie und Systematik der Vögel. Allgemeiner Teil. Amsterdam, 1887. p.155ff.). (*180) In complete agreement with Fürbringer, I can simply add here with regard to the cerebral cortex that indeed, in general, larger animals in many ways manifest a higher cortical development than smaller ones, but that this is only valid for individuals of a restricted group, such as a family or genus, and that in addition many exceptions occur.

in the human frontal lobe alone (about a hundred in the whole cortex), T. Mauss [23]) was able to delimit thirty-two corresponding areas in lower monkeys and, according to personal communications, now distinguishes about forty myeloarchitectonic areas in the orang-utan and gibbon [24]).

Finally, it should be remembered that the intrinsic structure of the human cerebral cortex manifests on the whole a much finer organisation, both in the form and polymorphism of the elements and in the complexity of the connections, and that thus man seems to be further removed from the great apes than they are from the lower apes. In all these facts, then, we see a flagrant contradiction to the content of Huxley's law. Whether qualitatively different and more sophisticated research methods will again bridge this apparent rift, is a question for the future.

As the last point in this summary, we should again just mention briefly the importance of cortical localisation for anthropology. Positive findings are rather few. Elliot Smith [25]) took up the problem first from the point of view of localisation and showed that in the occipital cortex (striate area) of the Egyptian brain features are apparent that, in many ways, recall those in the orang-utan. Then I obtained similar findings, first in the Javanese brain [26]), and later found essentially the same in other races, particularly hereros and hottentots, and at the same time proved that in a large proportion of these foreign races the striate area demonstrates localisational features that are considerably different from those of European brains and often reveal a greater similarity with the anthropoid brain [27]).

With regard to this isolated finding from anthropological investigation of the cerebral cortex, although it relates to a single cortical area, I believe that in the future we shall not have to stand by despondently, as in the days when we simply compared sulci. There are many questions concerning anthropology which, while certainly necessitating the greatest care and long-term critical comparative research, can be tackled from the point of view of cortical localisation, with the prospect of success. (*184)

---

[23]) Th. Mauss, Die faserarchitektonische Gliederung der Grosshirnrinde bei der niederen Affen. Journal f. Psychol. u. Neurol., 1908, 13. (*181)

[24]) The relevant research by O. Vogt and Mauss is almost completed and will soon be published. (*182)

[25]) E. Smith, The morphology of the occipital region of the cerebral hemisphere in Man and Apes. Anat. Anz., 1904, 24.

idem. Studies in the morphology of the human brain. Records of the Egyptian Governement School of medicine, II.

[26]) Brodmann, V. Mitteilung zur histologischen Rindenlokalisation. Journ. für Psych. u. Neurol., 1906, VI.

[27]) For a discussion of the incidence of the "Affenspalte" in foreign races of man: presentation to the Berlin Society for Psychiatry and Neurology. Ref. Zentralbl. f. Nervenheilk., 1908. (*183)

Chapter VIII.

Localisation and histopathology.

In his famous policy paper on the present state of pathological anatomy of the central nervous system, Franz Nissl [1]) acclaims as the greatest step forward of our times the clear recognition "that the next goal to attain in the pathological anatomy of the central nervous system does not consist in the most detailed possible identification of the affected nervous elements, nor an attempt to relate clinical signs to anatomical observations, but in the systematic investigation of individual histopathological processes underlying clinical pathological conditions, and also in the most detailed possible definition of the various histopathological states".

These few words contain a whole programme and signify the renunciation of his own hopes and aspirations cherished for years and the beginning of a new epoch in the histopatholgy of the central nervous system. Until then - undoubtedly under the influence of the neuron theory proposed by Forel, His and Waldeyer - one had placed the emphasis on pathological changes in the "functional network elements", especially of neurons, and the effort of

---

[1]) F. Nissl, Zum gegenwärtigen Stande der pathologischen Anatomie des zentralen Nervensystems. Zentralbl. f. Nervenheilk. und Psychiatr. 26, 1903, p.517ff. - See also F. Nissl, Die Hypothese der spezifischen Nervenzellenfunktion. Allg. Zeitschr. f. Psychiatrie 54, 1898, p.1ff. (*185)

researchers had been directed toward discovering specific neuronal changes for the different pathological states and pathogenic circumstances. Now it was to be suddenly different. Nissl warns explicitly against an overemphasis on neurons, and deviations from their normal state, for estimating pathological conditions. He considers the present-day priority to be the application of practical pathological anatomy; one must turn one's attention more to the overall pathological process, including the function of the supporting tissues such as glia and vascular and connective tissue, rather than to individual components of the neuronal network.

If one examines the literature of the last decade on this question, one must recognise how right Nissl's opinion was. The achievements of this period have been obtained in the way suggested by Nissl's school [2]) and it can hardly be doubted that, for the present, this will remain the case.

Nevertheless, I remain hopeful that the results of histological localisation will not be without influence on the histopathology of the cerebral cortex.

I am however not so optimistic as to believe that field topography, as I have described it in this treatise, will at present lead to the cortical localisation of individual psychiatric disorders or even individual psychological symptoms. But I also know that the question of the seat of pathological changes in tissues has not been entirely neglected and that histopathologists, at least the more critical of them, have always paid attention as to which strictly delimited portions of cortex are involved predominantly (or exclusively) in defined psychological conditions. Meanwhile, many histopathological questions are seen in a quite new light thanks to the results of cortical localisation as described here, and now require consideration from a histotopographic point of view. In this chapter - with the aid of some examples - I wish to expose briefly what direction any benefits to be expected from this approach might take and where one should next direct one's attention in the investigation of specific cortical functions. An exhaustive treatment of all the questions arising from this is however impossible. New problems affront the researcher continuously, so only a few passing references will be made with the aim of elucidating the relationships between histopathology and cortical localisation and which can serve as initial guidelines.

1. There can be little hope that our localisational data will lead to advances in our understanding of the basic or specific cellular pathology of brain diseases. Selective cell pathologies, whether in the form of a regional malfunction or of the exclusive involvement of particular cell types, are extreme rarities and, judging by our present concepts, should not be expected very often, even theoretically. Certain intoxications would have to be considered first, especially experimental ones.

Schröder, in his "Einführung in die Histologie und Histopathologie des

---

[2]) See Nissl's "Histologische und histopathologische Arbeiten über die Grosshirnrinde" (*186), Vol. 1 and 2, in which a wealth of fundamental research on the pathological anatomy of neurological and psychiatric diseases is published. Examples are the contributions of Nissl himself, Alzheimer, Schröder, Spielmeyer, Forster, Ranke, Merzbacher etc.

Nervensystems" [3]), which is excellent both in its critical command of the material and its clarity of expression, has shown, on the basis of the literature and his own broad experience, what our position regarding this question of brain diseases should be. He describes as a useless undertaking today to seek specific neuronal changes in particular clinical conditions or particular lesions. "Experience so far teaches that most known cellular changes in human pathology are probably traceable to general somatic disturbances that accompany the neurological problem (fever, anaemia, malnutrition, exhaustion, oedema etc. together with the particular trauma of a long agony)." Nevertheless, Schröder does not exclude the possibility that there do exist disease entities of the central nervous system that are distinguished by "characteristic changes of the normal appearance of neurons".

Now this seems to me the important point that will in time justify histotopography. If this assumption is true, in my opinion we shall only reach our goal and avoid mistakes by paying attention to normal local cellular paradigms. We saw earlier to what a massive extent even individual cell shapes varied in different regions of the cortex. So, two possibilities can be envisaged. First, a lesion could affect all neurons diffusely and in a similar way, regardless of their histological characteristics and their position. This case is outside the scope of our consideration. However, another possibility - which entirely concurs with Schröder - is quite conceivable, and even probable on the grounds of pathological experience: that a particular disease process might affect only certain cell types or act partially selectively on limited cell varieties in a particular region. The recognition of such localised cell pathology, however, definitely requires a most detailed knowledge of localisation. An example from neuropathology, that has been a very recent subject of discussion, can illustrate this.

It consists of the secondary changes in the cerebral cortex in *amyotrophic lateral sclerosis*. Apart from the ascending pyramidal degeneration in this disease, that has been unequivocally followed into the cortex with the Marchi method in 18 cases published so far, according to Rossi and Roussy [4]), changes in the cortical cells have also been sought for a long time. The results are however extremely contradictory, particularly regarding localisation [5]); some authors claim to have found changes in the large pyramidal cells (Kojewnikoff, Charcot and Pierre Marie, Mott, Spiller and Sarbo), others do not refer to this at all; some speak of changes "*in the central gyri*" or in the "*motor cortex*", without specifying what they mean by the latter and without any topographical description of the extent of the pathological process. Other authors who make

---

[3]) P. Schröder, Einführung in die Histologie und Histopathologie des Nervensystems. G. Fischer, Jena 1908. (*187)

[4]) J. Rossi and G. Roussy, Un cas de sclérose latérale amyotrophique avec dégénération de la voie pyramidale. Revue neurolog. 1906, No.9. (*188)

[5]) J. Rossi and G. Roussy, Contribution anatomo-pathologique à l'étude des localisations motrices corticales à propos de trois cas de sclérose latérale amyotrophique avec dégénération de la voie pyramidale suivie au Marchi de la moelle au cortex. Revue neurol. 1907, 15. (for the literature, see this paper.) (*189)

statements about localisation emphasise that cells in the precentral gyrus are more severely changed than in the postcentral (Czylharz-Marburg, Spiller), and finally a small number maintain that only the precentral gyrus is involved, while the postcentral remains entirely untouched. In particular Probst [6]) and Campbell [7]) emphatically support the last point of view and describe cell atrophy and cell loss, especially of the large pyramids and Betz cells, exclusively anterior to the central sulcus. Rossi and Roussy essentially concur with them concerning the cells, but also claim a limited involvement of the postcentral gyrus ("une participation, bien que très minime").

For me there is no doubt that these divergent views are due in large part to a lack of knowledge of normal structural criteria in the central region. Certainly, the magnitude of the disease process will influence the histopathological appearance; in some cases there will be atrophy and decreased density of certain cell types, in others there will be complete cell loss, depending on the severity of the disease. I observed this myself in two parallel cases; in one specimen the giant pyramids [8]) were totally absent, in the other they were merely shrunken and fewer in number. However, I take it as quite impossible, in one and the same disease, and one that is so well characterised in its chronic and localised action on the motor system as amyotrophic lateral sclerosis, that sometimes the whole central region is involved, sometimes the precentral gyrus exclusively, but sometimes only predominantly, while sometimes even the postcentral gyrus is involved. It is indeed quite significant that with changed physiological views about the central gyri and with the replacement of the theory of uniformity by one of histological localisation, opinions have multiplied that relate the amyotrophic process exclusively to the precentral gyrus.

Finally, Campbell, Rossi and Roussy, and Schröder [9]) have based their investigations from the beginning on our new knowledge of localisation with results that are in complete agreement with them.

2. More important for the histopathologist than a knowledge of individual cell types and their distribution over the hemispheric surface is a knowledge of laminar relations in general, and their regional variations in particular, or in other words of cytoarchitectonics (and myeloarchitectonics) [10]) in the broadest sense, with all their local extremely variable characteristics.

The changes in cortical histology described above were either largely

---

[6]) Probst, Zu den fortschreitenden Erkrankungen der motorischen Leitungsbahnen. Arch. f. Psychiat. 30, 1898, p.766. (*190) Also, Sitzungsbericht der Kais. Ak. d. W. Wien, 112, 1903, p.683.

[7]) Campbell, Histological studies, p.85ff.

[8]) In the precentral gyrus, of course! I refer again to what was said on page 66. One should also become accustomed in histopathology to use the expression "giant pyramids" only in the narrow sense formulated there.

[9]) My colleague Schröder was kind enough to communicate to me that in three typical cases he found gross cell loss in the precentral gyrus corresponding entirely to the extent of my giant pyramidal area, with disappearance of all giant pyramids, while the postcentral gyrus remained intact.

[10]) I freely admit that myeloarchitectonics, provided that their localisational aspects are considered, promise to give more practically comprehensible results on many problems of human brain pathology than cytoarchitectonics, on technical grounds alone.

unknown, or at least little heeded, until very recently. Further, the overall "histopathological appearance" of a condition may also be marked by alterations in architectonics, even if a major role is played by vascular and connective tissues. These alterations are not always merely secondary ones but quite frequently also primary ones involving degeneration of functional cellular elements, and such architectonic changes will be different in different localities of the cortical surface. When one realises all this, it becomes immediately clear how important it is to consider the normal structural differences within the cortex in order to understand and assess a pathological process.

Everywhere in articles in the domain of the histopathology of the cerebral cortex one can find arguments about cell shrinkage, loss of neurons, overall cell increase or cell loss, disappearance of layers, thinning of the cortex, and many other criteria, in short details of architectonic relations. Frequently these data are produced without defining the site they refer to, even to a rough approximation. It is obvious from the descriptions in Chapters I and II that such results can only be accepted with reserve. Cell density, cell size, cortical thickness, size and composition of individual layers, are all factors that in man regularly undergo very great regional variations. Figures 7, 12, 16, 27, 28 and 32 show such differences at a standard magnification of 25:1 in areas 3, 4, 5, 6, 17 and 18 of the human brain map, that is from the cortex of the post- and precentral gyri, the superior frontal gyrus, the most anterior section of the superior parietal lobe, the occipital lobe and the calcarine region. Three extreme examples of different human structural formations are also represented at higher magnification in Figures 42, 43 and 53. One must admit that it is difficult to imagine bigger differences in the histological composition of different parts of the body than those seen here in one and the same organ. It therefore follows that if one does not know the changing structural features of cells in different regions, or does not pay enough attention to them, one will easily succumb to serious errors when dealing with estimation of cell density, the absence of normal cell types or the presence of pathological ones, the grouping and the orientation of cells in a section, the cortical thickness, the density and size of individual layers or, in short, problems of normal cytoarchitecture of the cerebral cortex [11]). One will even run the risk, under certain circumstances, of being mistaken as to whether normal or pathological signs are present in a given case.

In this context, an example from the recent literature can be illustrative:

In his great monograph "Histological Studies on the Localisation of Cerebral Function", Campbell described particular pathological changes of the laminar organisation in a cortical area - his postcentral area - in three cases of tabes dorsalis. He found a consistent marked reduction of cortical thickness,

---

[11]) In any case all these questions still need the most careful, detailed investigation, especially with regard to cell density and specific cell types limited to individual regions. There is still insufficient comparative material and evaluation of sections; particularly needed are reliable quantitative data on cell numbers and cell sizes, and also on laminar and cortical thicknesses in different areas and at different developmental stages.

considerable cell loss, especially of large pyramidal cells of the supra- and infragranular layers, blurring of individual layers and atypical arrangements and orientations of cells. In none of his cases did these presumed changes affect the whole area, but only the anterior edge of his postcentral area on the anterior bank of the postcentral gyrus, deep in the central sulcus. As he admitted that the rest of the cortex was intact, Campbell concluded that here was the localised, anatomopathological basis of tabes in the cerebral cortex, and he even went so far as to maintain that, as tabes was a disease of the sensory system, this pathological strip of cortex must be regarded as the primary cortical end-station of the nerve fibres for general bodily sensation [12]).

One must admit that if this finding was correct it would have the greatest significance for our theoretical understanding of the nature and pathogenesis of certain disease entities, not only of the so-called "systematic" diseases, such as tabes, but also for our whole perception of the localisation of central disease processes. However, unfortunately - almost regretfully - Campbell's results have not been confirmed. From the outset they gave rise to serious theoretical misgivings. If the changes in the particular circumscribed cortical zones were secondary in nature, as Campbell considers, a consequence of degeneration of primary sensory neurons, the ascending disease process would have to experience at least two interruptions, once in the dorsal horns or the dorsal nuclei of the spinal cord, and once in the thalamus (*192). However, consistent changes in the secondary sensory pathways have never been found in tabes. Furthermore, it is difficult to imagine, and hardly in agreement with our experimental pathological findings, that a cortical centre should manifest such fundamental disturbances when the peripheral end of its afferent system is diseased [13]). In addition, Campbell's results have been directly shown to be erroneous by specific investigation. As a result of my description of the cytoarchitectonic structure of the cortex in question (my first communication, 1903, on the central region), Gordon Holmes [14]) has determined that the structural peculiarities that Campbell mistakenly attributed to tabes occur normally in all healthy brains and that they represent regional structural variations within the postcentral gyrus. There were absolutely no detectable pathological features in the cortex of the postcentral gyrus of any of four cases of tabes investigated by Holmes. He found all structural details just as as I had described them and illustrated them by microphotographs in healthy humans years earlier - and refers specifically to this.

I can but confirm these observations from my own experience. I have been no more able to identify specific changes in the cortex of the postcentral gyrus, and particularly in its anterior portion, in tabes dorsalis than Holmes. What

---

[12]) To quote Campbell: "the primary terminus or arrival platform for nerve fibres conveying impulses having to do with "common sensation"" (*191)

[13]) Campbell counters this objection with the equally poorly valid hypothesis that it is the effect of a "complete interruption of physiological impulses" occasioned by the destruction of the spinal dorsal roots (*193).

[14]) Gordon Holmes, A note on the condition of the postcentral cortex in Tabes dorsalis. Review of Neurology and Psychiatry, 1908.

Campbell took to be pathological represents a quite normal situation. The anterior bank of the postcentral gyrus, as I demonstrated in 1903, normally has an extremely thin layer of cortex, it possesses mainly small, often irregularly oriented cell types, and an overall laminar pattern (even myeloarchitectonically) that can be found nowhere else. For orientation, Figures 149 and 150 again show two sections of cortex from directly neighbouring sites in the pre- and postcentral gyri.

An insufficient knowledge of localisational and architectonic criteria misled Campbell into interpreting pathological changes, atrophy of the cortex, and cell loss when in fact conditions were quite normal [15]). Similar situations must arise in other cases. In particular, findings of pathognomonic changes in hippocampal cortex in epilepsy, originally derived from Meynert and later supported from many quarters, may be explicable in this way.

3. We come to a third point, that of the significance of cortical architectonics and topographic localisation for the study of the pathogenesis of certain brain diseases related to abnormal structure, especially the idiocies and hereditary and familial diseases.

That the clinical picture of *idiocy* can be resolved histopathologically into a series of different entities was especially due to the pioneering investigations of Alzheimer [16]). According to this, one must distinguish initially three types of idiocy, according to their origins. They are: a) caused by focal, usually inflammatory lesions of the brain, b) based on diffuse cortical processes and c) governed by primary developmental disturbances (agenesis, aplasia etc.). We are only interested here in the last group.

H. Vogt [17]) has shown that in those forms of idiocy due to developmental disturbances (including those beginning only postnatally) there is retardation of ontogenesis. Now we know from the discussion in Chapter I that, during embryonic development, the cerebral cortex undergoes what can be seen as a stage when its laminar pattern is uniform throughout, and that the later layers and areas of the mature brain represent specifically differentiated local transformations arising by variously directed local growth processes. In many forms of idiocy one can recognise an arrest at an embryonic developmental stage. In this connection it is of the greatest importance for the understanding of congenital and early acquired forms of mental retardation to know the approximate time at which the arrest of ontogenetic development occurs. As in these cases the architectonic structure in a way mirrors a more or less undifferentiated

---

[15]) Campbell's mistake is excusable, for in one of today's most popular textbooks of the anatomy of the central nervous system it is still said that the pre- and postcentral gyri represent the thickest cortex of the whole hemispheric surface, while Kaes, in his large atlas gives the cortical thickness for both gyri as about the same, some 4 to 5mm, whereas I find 4mm in the middle of the precentral gyrus and about 2mm in the postcentral, especially its anterior bank.

[16]) A. Alzheimer, Einiges über die anatomischen Grundlagen der Idiotie. Zentralbl. fr Nervenheilkunde und Psychiatrie, 1904, p. 497ff. (*194)

[17]) H. Vogt, Über die Anatomie, das Wesen und die Entstehung mikrocephaler Missbildungen. Arbeiten aus dem hirnanatomischen Institut in Zürich. Vol. I, Wiesbaden 1905. - H. Vogt and P. Rondoni, Zum Aufbau der Hirnrinde. Deutsche med. Wochenschrift 34, 1908, p.1886. (*195)

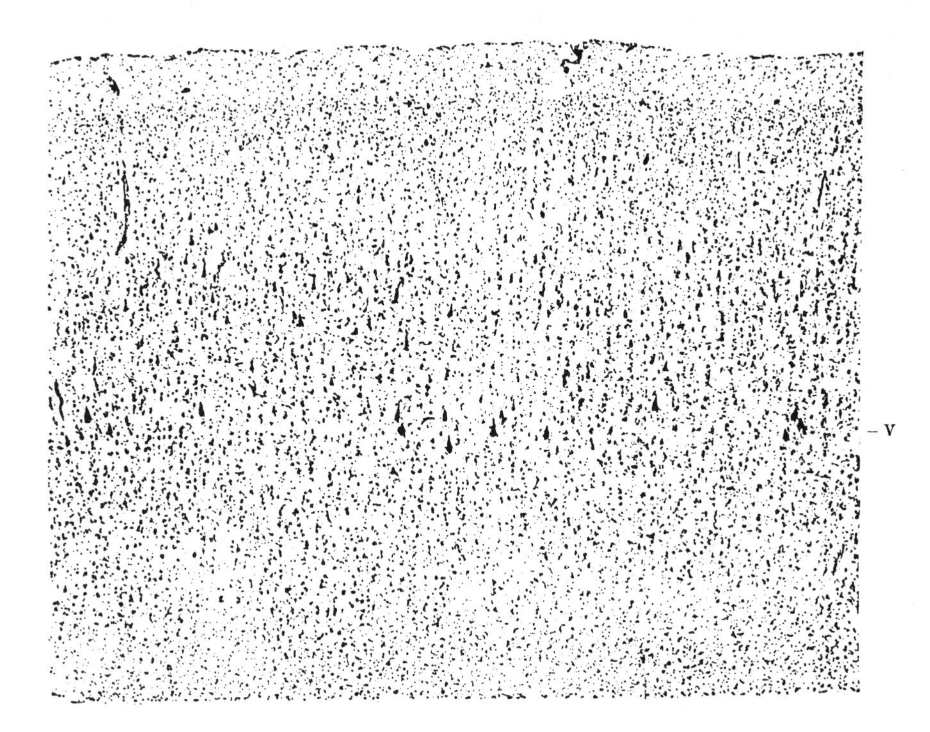

Fig. 149. Cortical section of the posterior bank of the precentral gyrus (area 4) from the immediate neighbourhood of the following figure. 25:1, 10μm.

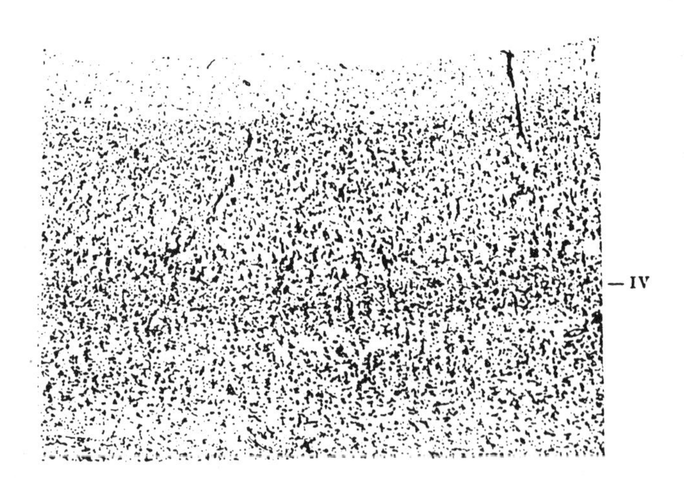

Fig. 150. Cortical section from the anterior bank of the postcentral gyrus (area 3), in the immediate vicinity of the previous figure. 25:1, 10μm.

ontogenetic transitional stage, it provides an indicator for the assessment of the problem. In any case, it should be taken into consideration that the architectonic differentiation of different regions and different layers does not occur at the same time during ontogenesis, and above all not at the same speed, a factor that has certainly not been sufficiently emphasised until now. Only the most detailed knowledge of the regional structural differences that already arise during embryonic development can help one to avoid errors. Many important questions, such as which aspects of a case are primary pathology and which are secondary degeneration, and the further question as to whether one is dealing with a simple arrest at a particular foetal stage or with concurrent overproduction, that is to say increased growth of histological elements or whole layers - for example the granular layer, as stated by H. Vogt - will only be decided with certainty by consideration of topical localisation in the embryonic or foetal cortex, a subject that, admittedly, still remains to be developed.

The same applies to another clearly defined syndrome, Tay-Sachs *familial amaurotic idiocy*, within which Spielmeyer [18]) and H. Vogt [19]) have recently distinguished two particular sub-varieties, one infantile, the other juvenile. The valuable research of Schaffer [20]) has determined that in this disease consistent severe changes occur in all neurons of the whole central nervous system, consisting predominantly of a peculiar swelling of the cell body and dendrites, as well as dissociation of the intracellular fibrils, whereas in his opinion the overall cortical structure, especially its laminar architecture, is unchanged. Schaffer concludes from this, in essential agreement with Spielmeyer and Vogt, that it represents a generalised, selective, primary cellular disease based on an abnormal structural development (a congenital weakness and tendency to precocious degeneration of the cytoplasm) within a central organ that is otherwise normal from the developmental and organogenetic point of view. Without wishing to cast the slightest doubt on the positive findings of these authors, I think I must assume that here also a more detailed knowledge of regional architectonic criteria in the juvenile cerebral cortex will later reveal local or general anomalies of gross laminar structure and may thus be able to explain the differences between the clinical varieties (cell number, cell size, size of various layers etc.).

I am particularly motivated in this conclusion by the recently published work by Kölpin [21]) describing changes in cortical lamination in *Huntington's*

---

[18]) W. Spielmayer, Klinische und anatomische Untersuchungen über eine besondere Form von familiärer amaurotischer Idiotie. Gotha 1907. (*196)

[19]) H. Vogt, Über familiäre amaurotische Idiotie und verwandte Krankheitsbilder. Monatschrift für Psychiat. und Neurol. 18, 1905. (*197)

See also the review by the same author: Zur Pathologie und pathologischen Anatomie der verschiedenen Idiotieformen. Monatschrift für Psychiat. und Neurol. 22, 1908. (*198)

[20]) K. Schaffer, Über die Anatomie und Klinik der Tay-Sachsschen amaurotisch-familiären Idiotie mit Rücksicht auf verwandten Formen. Zeitschrift für die Erforschung und Behandlung des jugendlichen Schwachsinns 3, 1909. (*199)

[21]) Kölpin, Zur pathologischen Anatomie der Huntingtonschen Chorea. Journal für Psychologie und Neurologie 12, 1908, p.57. (*200) (Jelgersma has recently referred to changes in subcortical centres, and especially certain fibre systems.)

chorea. There is undoubtedly an exaggerated development of certain layers, especially the inner granular layer, as I have now been able to determine in a third case, kindly referred to me by Kölpin. Because of this, and because of the dense concentration of small granular, neuroblastoma-like elements, the cortical cross-section strikingly reminds one in many ways of the laminar pattern found at immature cortical stages. I have found this especially characteristically in the occipital lobe and the calcarine cortex. The resemblance is even closer when one sees a distinct laminar accumulation of granules in certain cortical regions in which there is normally no inner granular layer in the adult, such as in the giant pyramidal type, just as we have observed during an ontogenetic transitional period in this region. The coincidence of these factors suggests a partial persistence of infantile or foetal patterns of lamination. Thus, in chronic hereditary chorea, we could be dealing with abnormal structural development expressed in histotectonic terms, a *vitium primae formationis* (*201) or a partial developmental defect of the cerebral cortex, upon which a precocious atrophy of the nervous system is superimposed leading to degeneration during the course of life.

This is not the place to engage in further theoretical discussion on this question which is also of great importance for the problem of aging. I am well aware of the many objections that could be brought against such a concept [22]). I should not wish to omit to point out that Gowers has already expressed a similar opinion concerning spinal cord sclerosis, suggesting that this pathology is due to an inborn vital nutritional deficiency, a so-called "*abiotrophy*", that is first apparent at puberty. Jendrassik also proposes that various hereditary or familial disease entities, at whatever age they appear, have their origin in differently localised defects of the central nervous system (for instance, muscular dystrophy is a local defect of certain neurons). Perhaps we should consider the anomalies of lamination arising from developmental errors found in Huntington's chorea as the visible histological expression of such a hereditary defect.

---

[22]) I must not omit to mention that our colleague Schröder has recently found a similar distinct development of a small-celled cortical layer within the otherwise agranular precentral gyrus in three cases of amyotrophic lateral sclerosis, similar to that described as characteristic of Huntington's chorea. Schröder, who was kind enough to demonstrate his slides to me, considers these indeed quite remarkable laminar cell aggregates as hyperplastic glial cells, and I must admit that, in view of the broad similarities, the concept cannot be dismissed that in our cases we may also be dealing with such secondary hyperplasia. In any case, it must not be overlooked that the overall architectonic pattern of the cortex in Huntington's disease is entirely different; the cortex of the occipital lobe, and especially the calcarine type, has a surprisingly similar laminar pattern to that of juvenile cortex, which is not the case in lateral sclerosis. What is more, it would be difficult to understand how, in a diffuse syndrome with mental retardation, a single layer (the inner granular layer) could have undergone a selective exuberant development over the whole cortical surface through secondary glial hyperplasia while in the remaining cortical layers the glial cells are not at all or only insignificantly hyperplastic. Such an explanation seems to me, at the very least, rather forced, compared with the above mentioned view, especially when one considers that the question is still open and must await further research for a conclusion to be reached.

[23]) Bourneville, Idiotie et épilepsie symptomatique de sclérose tubéreuse ou hypertrophique. (*202)

Similar considerations apply for other similar syndromes; I might simply mention Bourneville's [23]) *tuberous sclerosis*. These brief remarks must suffice here.

4. Finally I should like to draw attention in a few words to those well known, clearly delimited and localised cortical changes described after disease or loss of circumscribed peripheral parts of the body or regions of innervation. There can be no doubt that when the normal architectonics of the different regions is known more precisely and the details of the parcellation of structural areas are revealed, many controversial questions about the extent and degree of such secondary processes in the cerebral cortex will be resolvable. One should just remember what was said above about tabes dorsalis on the one hand and amyotrophic lateral sclerosis on the other. The first to be considered are cases of amputation of suitable duration (members, tongue, larnyx etc.), then encapsulated foci (*203), experimental section of certain pathways and, last but not least, congenital agenesis of body parts [24]).

A particularly profitable field for the study of such histotopographic questions is represented by certain diseases, lesions and developmental disorders of the sensory organs, especially those of the special senses. This includes particularly congenital or early acquired blindness, long standing labyrinthine deafness, and deaf-mutism.

As representative of this approach one could mention the research of Bolton [25]) on the alterations in cortical architectonics in the occipital lobe, especially in his visuo-sensory area, our striate area, in anophthalmic and congenitally blind patients. He showed for the first time for a sensory organ and a cortical area how such pathotopographic problems can be tackled. Von Monakow [26]), Fürstner [27]), Henschen [28]), Cramer [29]), Moeli [30]), H. Berger [31]), Leonowa [32]) and others have presented similar, for the most part older, research

---

[24]) Gudden has already observed atrophy of the large pyramidal cells in the cortex after section of the internal capsule and quite rightly speculated that one will be able to obtain details of localisation in this way; he has also already posed the question whether different functions are related to different cell types. - Gudden, Über die Frage der Lokalisation der Funktionen der Grosshirnrinde. Allgem. Zeitschr. für Psychiat. 42. (*204)

[25]) S. Bolton, The exact histological localisation of the visual area of the human cerebral cortex. Philosoph. Transact. 193, 1900.

[26]) von Monakow, Über einige durch Exstirpation zirkumskripter Hirnrindenregionen bedingte Entwicklungshemmungen. Arch. f. Psychiatr. 12, p.143, 1882. idem, Experimentelle und patholog. anat. Untersuchungen über die Beziehung der sog. Sehsphäre zu den infrakortikalen Optikuszentren und zum Nervus opticus. Arch. f. Psychiatr. 14, p.699, 1883. (*205)

[27]) Fürstner, Weitere Mitteilungen über den Einfluss einseitiger Bulbuszerstörung auf die Entwicklung der Hirnsphären. Arch. f. Psychiatr. 12, p.612, 1882. (*206)

[28]) Henschen, On the visual path and centre. Brain, 16, p.170.

[29]) Cramer, Beitrag zur Kenntnis der optischen Kreuzung im Chiasma. Anatom. Hefte, X, p.416. (*207)

[30]) Moeli, Veränderungen des Tractus und Nervus opticus bei Erkrankungen des Occipitallappens. Arch. f. Psychiatr. 22, p.73, 1891. (*208)

[31]) H. Berger, Beiträge zur feineren Anatomie der Grosshirnrinde. Monatsschr. f. Psychiatr. u. Neurol. 6, 1899, p.405. - idem, Experimentellanatomische Studien über die durch den Mangel optischer Reize veranlassten Entwicklungshemmungen in Occipitallappen des Hundes und der Katze. Arch. f. Psychiatr. 33, 1900. (*209). Berger describes localisation in the visual cortex of the dog that corresponds with that of Munk.

on the "visual cortex", but with very contradictory results. H. Berger, in particular, has tackled the problem experimentally from a strictly topographical standpoint, and it seems that when normal localisational features relating to the animals in question are understood in even more detail a potentially successful future is promised for experimental studies (*211). Observations also exist concerning cortical changes after destruction or lack of the organ of hearing, or in deaf-mutism (Waldschmidt [33]), Eberstaller [34]), Probst [35]), Strohmayer [36]), Brouwer [37]), and the results and conclusions of these studies are no less in disagreement than the corresponding results on the visual cortex. This is not the place for entering into detail; it does, however, seem to me that the blame for this divergence can be attributed not only to interpretation, but to a large extent, if not entirely, to a lack of sufficiently solid localisational or histotopographic criteria.

From all this we see the following: topographical localisation forms the surest basis and starting point for solving a considerable number of histopathological problems concerning the cerebral cortex. One must accept the principle that if we wish to investigate pathological changes in a particular part of the cortex in the way described above, we must first understand the normal structure of precisely the same cortical area and use it as a comparison. An approximate definition of a region, such as "occipital cortex", "parietal lobe", "frontal lobe", or similar terms, is not sufficient to this end. In the future it will be more important in such studies to consider as an equally significant factor alongside the "*specific cellular appearance*" (*217) of Nissl the "*specific architectonic appearance*" (*218) of each regional cortical area.

However, in doing this it will not suffice to merely limit oneself to overall gross architectonics of an area; rather, it will be essential for histopathological problems to know the extent and the delimitation of areas. Further, one must become accustomed to taking into account within each area specific local structural differences and especially individual variations in cellular density and size, cortical and laminar thickness etc.

Indeed, one must not forget in this respect that at every stage much desir-

---

[32]) v. Leonowa, Über das Verhalten der Neuroblasten des Occipitallappens bei Anophthalmie und Bulbusatrophie und seine Beziehungen zum Sehakt. Archiv f. Anat. von His 1893, p.308. - idem, Beiträge zur Kenntnis der sekundären Veränderungen der primären optischen Zentren und Bahnen in Fällen von kongenitaler Anophthalmie und Bulbusatrophie bei neugeborenen Kindern. Arch. f. Psychiatr. 28, p.53. 1896. (*210)

[33]) Waldschmidt, Beitrag zur Anatomie des Taubstummengehirns. Zeitschr. f. Psychiatr. 43, 373. (*212)

[34]) Eberstaller, Das Stirnhirn. Wien-Leipzig 1890. (*213)

[35]) Probst, Über das Gehirn der Taubstummen. Arch. f. Psychiatr. 34, 584, 1901. (*214)

[36]) W. Strohmayer, Anatomische Untersuchung der Höhsphäre beim Menschen. Monatschr. f. Psychiatr. u. Neurol. 10. 172, 1901. (*215)

[37]) B. Brouwer, Over Doofstomheid en de acustische Banen. Akademisch Proefschrift. Amsterdam 1909. (*216)

able work is lacking. We are at the threshold of a new field of research, for which we can only give a sketchy outline here. What is needed now is a systematic and thorough, broad collaborative study, based on what we have learned about normal histological localisation. Only then will we ultimately progress to an understanding of the pathology of the cortex as an organ.

Chapter IX.

Physiology of the cortex as an organ.

It is a basic biological principle that the function of an organ is correlated with its elementary histological structure. Just as every organic structure is a product of its developmental history and, further, as development represents the sum of diverse vital processes, so in the last analysis the explanation of the emergence of organic structure and its genesis by histological differentiation is a physiological problem [1]). Function creates organs. The recognition of this seems to me to enable those who would not otherwise be inclined, to deduce the function of an organ from its structure, and to draw conclusions about the activity of the whole organ, as well as its component parts, from its overall histological composition and the structural similarities or dissimilarities of individual components, in short from its intrinsic structural differentiation.

Although my studies of localisation are based on purely anatomical considerations and were initially conceived to resolve only anatomical problems, from the outset my ultimate goal was the advancement of a theory of function and its pathological deviations. Now the question arises as to what we can deduce from our histotopographical findings in terms of the physiology of the cerebral cortex.

---

[1]) See R. Hertwig, Lehrbuch der Zoologie, 1900, p.48. (*219)

1. Localisation by elements. (*220)

Our comparative anatomical studies have furnished such fundamentally new data concerning the histological structure of the elements of the cerebral cortex, and especially about homologies of individual elements, that I believe we may even succeed in providing new directions for many physiological concepts.

a) Previous interpretations.

Previous interpretations were based on the concept of the morphological, and therefore physiological, uniformity of all elements in the central nervous system, as derived from older data. Even Meynert - the very same researcher from whom came the first thrust toward a histotopographic parcellation of the cerebral cortex and who equally has exercised a lasting influence on physiological theory to this very day - adheres to the uniformity of functional elements and emphatically supports the view that all neurons have the same intrinsic function, the differences in activity of individual components of the central nervous system arising simply through differences in the connections between cell complexes. He even goes so far as to state that the spinal grey matter and its neurons has the same basic functional role (formation of the "ego") (*221) as the cerebral cortex, but it is inhibited by the relative poverty of the connections between its elements [2]). He thus also refuses to restrict the modality of sensation primarily to the cortex, referring to Pflüger's well known investigations, and ascribing the same properties to deeper brain structures.

Exner's theory is essentially founded on the same concepts. Like Meynert he bases his studies of central nervous activity mainly on pathways, that is the conducting elements of the central nervous system; he argues that nerve cells play a quite subordinate role. The grey matter forming the cerebral cortex is merely an organ for processing qualitatively similar impulses arriving in quantitatively varying numbers. He thus endeavours to explain all physiological activity and the most important psychic phenomena to variations

---

[2]) He qualifies "as the only natural interpretation that the neurons of the cerebral cortex are in no way different from other neurons in terms of specific function, such as their capacity for sensitivity, but only differ in their connections with centres related to abundant forms of sensory impressions, that we know by experience are involved in the formation of the ego". (Der Bau, p.1 and 2).

Elsewhere he writes: "It is absolutely unnecessary to accept that this graded difference in duration of excitation is governed by different intrinsic properties of the neurons; it can also result from extrinsic properties." (ibid., p.3).

[3]) S. Exner, Entwurf zu einer physiologischen Erklärung der psychischen Erscheinungen. 1894. p.3. (*222)

Exner writes: "All manifestations of quality and quantity of conscious sensations, percepts and concepts can be traced back to quantitatively variable excitation of various parts of the nervous system as a whole. Two sensations have the same conscious effect if the same cortical pathways are equally excited by the sensory stimulus. Two sensations are similar if at least parts of the excited cortical pathways are identical in the two cases. The quality of the sensation and its local effect are thus the result of stimulation of different cortical pathways" (p.225).

in excitation, and in particular "to ascribe everything that seems to us like variations in consciousness to quantitative factors and to different cerebral connections of otherwise essentially similar nerves and centres" [3]).

Similarly Wundt, in his "Grundzügen der physiologischen Psychologie" [4]), maintains that differences in form and organisation of central nervous elements are "so insignificant that they also suggest an equivalent similarity in central processing". In this he does not in any way seek to exclude functional differences in individual brain regions, but only means that "they must be attributed in all probability [5]) not to specific functional differences in brain elements, but simply to the different connectivity of the latter between themselves and with peripheral organs" (p.148).

As one can see, all these theories are in agreement in their basic concepts and depend on the supposition that the neurons are structurally equivalent, independent of their origin, their position and their external form. In this respect we should recall the anatomical facts!

b) The histological facts and their consequences.

It is well established that there are numerous extremely different types of neuron in the cerebral cortex in terms of their outward shape and internal structure. It is also clear that certain of these cell types are grouped in distinct, laminar histological formations, that we recognise as the layers, either over the entire hemispheric surface or over a large part of it. It has been further demonstrated, through new localisational studies, that the occurrence of many particularly distinctive cellular features is limited exclusively to a spatially more or less sharply delimited portion of the cortical surface. Moreover, it has finally been proved that such regionally restricted or localised cell types of similar structure are regularly found in particular (identical) positions in the cerebral cortex, not only in man but throughout the whole mammalian class.

Although our present knowledge in no way suffices to propose an exclusive or preferential function for given unique, local, individual cell types, the very great morphological variation of cell shape, both in terms of external form and internal structure, obliges one to accept that these elements reflect different functions. This is not only because they are related to different nervous pathways or because they belong to different architectonic formations - which, incidentally, is usually also the case - but because they have differentiated in specific morphological directions.

The morphological process of differentiation depends, as explained above, on the "construction of dissimilar forms from a similar basis" (Haeckel); physiologically speaking, histological specificity corresponds to functional specialisation. These specialised cell elements begin to differentiate very early in ontogeny in many regions. This differentiation manifests itself in a

---

[4]) W. Wundt, Grundzügen der physiologischen Psychologie. IVth edition. Vol. 1. p.147ff. (*223)

[5]) Wundt expressly speaks here of a mere probability.

congruous manner in animals and man for homologous cells in similar positions and often, for given cell types, preferentially at a time when all other cells are still "dormant". In many cases this differentiation attains a higher degree than we can find in any other organ or even than exists between cells of different organs. Now, whenever we find new cell types elsewhere in an organism or in an organ, we anticipate a corresponding qualitative specialisation. Why should the cells of the cerebral cortex constitute an exception? The cells of two glands such as the liver and the kidney do not differ from each other histologically in any way as much as, for instance, a giant pyramidal "motor" neuron and a small "granule cell" of the calcarine cortex in one and the same central nervous system. Apart from differences in external form and size, many cortical neurons differ from each other in the internal structure of their cell body in fibrillar preparations (according to Bielschowsky or Cajal), as well as in corresponding Nissl preparations, to such an extent that we must certainly speak of a high degree of specific histological differentiation. It is difficult to imagine that such morphological specificity is merely the expression of quantitative factors, such as gradations of one and the same qualitatively similar form of excitation (Exner) or a different number of excitatory connections (Meynert). How can the same nervous activity that excites two cells differently only in quantitative terms cause this high degree of histological specialisation; how can it be explained if one accepts that these cells differ physiologically only in the amount, and not in the manner, of excitation? Thus it seems to be more correct, and from the histological point of view even unavoidable, to ascribe qualitatively different functions to specific, different cortical cells. In spite of the objections that have been made, this is not contradicted by these cells being spatially located within the same organ, nor by their all being of the same histogenetic origin.

However, the following argument seems the most important to me. In my opinion, the quantitative hypothesis does not suffice to explain why certain cell types begin to differentiate very early in ontogeny, long before all other cells and also long before there is any question of functional activity. We cannot avoid the conclusion that there are inherited qualitative differences already present in the cerebral Anlage. There are also histogenetic, as well as morphological, reasons that compel us to reject Wundt's "*principle of indifferent function*" for the cells of the cerebral cortex, according to which no elements fulfill specific functions but are entirely dependent on their connections and relations with other elements for their functional activity.

The polymorphism of cells, their asynchronous histogenetic differentiation, the strict regional separation of certain cell types, and the regular appearance of similar (homologous) cell types in identical positions on the cortical surface in all mammals, all justify the view that in the cerebral cortex a broad division of function has occurred among the cellular elements, in other words that functional specificity is divided between cells of different individual morphology, different localisation and different capacity [6]).

c) Newer hypotheses.

It is now the time to approach the question as to what we might localise in these morphologically differentiated elementary components or, in other words, how we are to imagine the relationship between specific morphological elements and various psychic phenomena we may experience.

From the outset all hypotheses according to which individual psychic phenomena such as percepts and concepts (*224) are embodied in individual morphological elements of the cerebral cortex, that is in neurons, are to be rejected on basic psychological grounds. As is well known, H. Munk [7]), who has developed this theory most substantially, accepts that certain categories of concepts can be attributed to specific cells, and he comes to the logical conclusion that in the central sensory system all concepts related to a given sense that an individual disposes of are situated alongside the corresponding percept, for example in the visual system all visual concepts, and in the auditory system all auditory concepts.

The cardinal error in this view, as Wundt has demonstrated in detail, is the fact that it attributes complex functions to a single histological structure. A concept, that is here taken to be a psychic element, whether related to visual, auditory or verbal percepts, is really a highly complex psychic function. A real concept is composed of psychic elements of the widest variety, and in addition every concept possesses spatial and temporal qualities; it is quite impossible to imagine that all these properties are located precisely in one cell. A neuron is always a relatively simple entity, in comparison with the infinitely complicated structure of the cerebral cortex, or compared with the histological complex making up a cortical area, or even with a single cell layer within an area. It is impossible to conceive that such complex psychic functions could operate within so simple a component of the whole organ as a cell. Only physiological events and functions of the most elementary kind, such as a primtive motor impulse or an elementary sensory event [8]), can be associated with histological elements, but never complex psychic images such as concepts.

Even less plausible are the views recently propagated over-enthusiastically in several quarters that consider specific familiar psychic phenomena as linked to specific well-defined cell types. In particular, Ariëns Kappers has edified an extensive functional hypothesis for individual cell types or major laminae of the cerebral cortex. He believes himself justified on the grounds of comparative

---

[6]) Whether this functional diversity of specific cell types can be traced back in the last analysis - as many physiologists maintain - to quantitative differences in a single chemokinetic principle, must remain an open question, that cannot be decided either physiologically or anatomically at present.

[7]) H. Munk, Über die Funktionen der Grosshirnrinde. Gesammelte Abhandlungen. 2nd edition. Berlin 1890. (*225)

[8]) In this respect it remains an open question whether a simple sensory event, such as the visual percept corresponding to a light stimulus of a given frequency, really represents the absolute "psychic element", just as must the other question whether cells are the ultimate histological element of the cerebral cortex.

anatomical observations to postulate that the specific cellular elements of the individual cortical layers control specific psychic phenomena throughout the whole mammalian class, from high to low; in particular the granule cells have receptor functions, the elements of the infragranular layer lower-level intraregional associative functions, and the supragranular pyramids higher-order interregional associative functions [9]).

Quite apart from the fact that in this interpretation a too schematic view of physiological brain activity is expressed, with above all a completely unacceptable generalisation of human psychophysiological theories to the whole mammalian class, on closer inspection Kappers' actual material includes no adequate support for such a far-reaching functional hypothesis. It is unacceptable to conclude from a simple increase in volume of a layer or in number and size of individual cells in higher animals that these should be signs of higher psychic order, as if higher physiological function should be expressed merely in volume or size rather than as depending more on internal organisation (chemical, dynamic or even histological, although not amenable to microscopic proof). Moreover, most of what appears in the literature concerning the extent, depth and cell density of the major layers in different animal groups, including just those layers considered by Kappers in his theory, is partially incomplete or contradictory, and partially absolutely wrong. For instance, I have demonstrated elsewhere [10]) that cortical areas characterised by an especially distinct granular layer that have been described as visual cortex, or as the "visuosensory area" by English authors, have often been mistakenly homologised. Indeed, in various animals the most heterogeneous architectonic formations have been described as such, that have nothing to do with the visual cortex of man (our calcarine cortex). In this respect I cannot make an exception of the otherwise very meritorious works of Mott and Watson, in which appear the most disastrous errors. When Mott claims to have found a parallel between progressive improvement of visual function (the development of binocular vision) and the development of the "visual cortex", I must object that in a number of his experimental animals what he described is absolutely not "visual cortex" and could never be, for the cortex in question stems demonstrably from the cingulate gyrus, specifically from my retrosplenial

---

[9]) Ariëns Kappers (The phylogenesis of the Palaeo-cortex and Archi-cortex compared with the evolution of the visual Neo-cortex. Archives of Neurology and Psychiatry, Vol. IV, 1909) writes literally:

1. "The granular layer in the cortex is primary in character, and has originally receptory functions."
2. "The infra-granular layer, as already pointed out by Watson, has projection and intra-regional associative functions."
3. "The supra-granular pyramids - as already was proved by Dr. Mott are the latest to appear - have chiefly associative functions of a higher order."

- See also: A. Kappers and W. Theunissen, Zur vergleich. Anatomie des Vorderhirns der Vertebraten. Anat. Anz. 30, p.496. (*226)

[10]) See my fifth communication on "Histological localisation"; also Edinger and Wallenberg's "anatomical reports" in Schmidt's "Jahrbücher der Medizin", 1907 and 1909. (*227)

region, thus not corresponding in any way to the calcarine cortex of primates. Thus Mott compares quite heterologous cortices and relates them to vision, making his conclusions untenable.

Kappers, however, leans heavily on this particular author. This proves that his functional localisation according to cell layer is unacceptable, for to a great extent it is based on obvious anatomical errors.

In particular, notions that claim that peripheral sensory modalities are reflected in the layers of specific cortical zones are based on complete fantasy. There is, for example, Flechsig's [11]) of a copy of the retinal layers in the layers of the visual cortex, or the theory of Ewens (*229) [12]) who supposes that the individual layers of the occipital cortex are devoted to the perception of individual colours, or even that of Leonowa [13]) (based on research on anophthalmics) which arrives at the conclusion that the fourth layer of the occipital lobe carries the representation of visual percepts and is also the seat of visual concepts and ideas, or finally that of the audacious psychiatrist who determined in brains of paralysed patients that the various levels of consciousness are accommodated in the various deep cortical layers. One will agree with Wundt when he speaks of "fanatical brain anatomists". They are without exception speculative errors that far exceed our genuine knowledge and do not have the remotest support from either anatomical or physiological data.

In contrast to the above-mentioned older hypotheses, all that we wish to conclude from our observations is simply the acceptance of a functional difference between the various cellular elements of specific structures that constitute the cerebral cortex, and indeed this is an inevitable conclusion. Our present-day knowledge of the elementary structural features of the cerebral cortex is insufficient for further physiological interpretations; in particular the specific significance of individual cell types remains completely obscure, however strictly localised they may be and however constantly they appear throughout the mammalian class [14]).

2. Regional functional localisation in general.

a) Historical retrospect.

In order to understand the progress represented by a purely anatomical approach to the problem of regional or areal localisation it is useful to provide a brief retrospect of the history of the question. However, it is impossible to discuss in detail all the various modifications of the localisational hypothesis.

---

[11]) Flechsig, Gehirn und Seele. 2nd edition, 1896. (*228)

[12]) Evens, A theory of cortical visual representation. Brain 16, p.475.

[13]) v. Leonowa, Beiträge zur Kenntnis der sekundären Veränderungen der primären optischen Zentren und Bahnen in Fällen von congenitaler Anophthalmie und Bulbusatrophie bei neugeborenen Kindern. Arch. f. Psychiatr. 28. p.53, 1896. (*210)

[14]) Sherrington has drafted a very ingenious and certainly fruitful hypothesis about central nervous processes, using general physical concepts as a basis, in "The integrative action of the nervous system" (London 1906).

Monakow [15]), in his critical report "Über den gegenwärtigen Stand der Frage nach der Lokalisation im Grosshirn" (*231) reviewed the literature comprehensively; Wundt [16]) discussed the principal points of view. For our purposes it suffices to recall that, since the principles of the theory of functional localisation emerged, we must distinguish two main groups of researchers: the "strict localisers" and the "half localisers" (*232).

The older functional hypothesis, introduced to science by the physiologist Flourens [17]), according to which all parts of the surface of the cerebral hemispheres have the same physiological function, like the parts of a gland, is only comprehensible for us today in the context of the anatomical knowledge of brain structure at the time. It could only be tenable as long as one possessed no knowledge, or only a very fragmentary knowledge, of the structural features of this organ, and it is significant that long before physiological studies led to the acceptance of spatial localisation of central nervous activity, neuroanatomists had constant recourse to the concept of a circumscribed representation of the parts of the body within the cerebral cortex [18]). The physiological theory of the unity and indivisibility of the central nervous system came into disfavour when, anatomically speaking, one began to convert from gross morphological dissection to microscopic study of the basic histological structure of the cerebral cortex. Mainly thanks to the anatomical data so gathered, the "*principle of functional localisation*" eventually triumphed and today, in spite of considerable divergences in details, differences of opinion no longer concern major points, but only matters of degree. The discussion hinges on the extent and nature of what can be localised, and how, and no more on the principal question as to whether localisation exists.

Among the "half localisers", Friedrich Goltz [19]) takes pride of place. Originally completely refractory, he later made substantial concessions so that

---

[15]) Asher and Spiro, Ergebnisse der Physiologie, I-VI. cf. also von Monakow, Gehirnpathologie, 2nd edition. (*230)

[16]) cf. W. Wundt, Physiologische Psychologie, 4th edition, Vol. 1, p.236. (*223)

[17]) Flourens, Recherches expérimentales sur les propriétés et les fonctions du système nerveux. 2nd edition. Paris 1842. (*233)

This extreme anitlocalisational theory stemmed from the same experiments - partial lesions of the cerebral hemispheres and observation of the resulting deficits - as later strict localisational theories that gained acceptance. It culminates in the proposition that any part of the cerebrum can be functionally replaced by any other part of the same organ. Flourens even went so far on the grounds of his experimental observations as to state that if one extirpates all but the smallest fraction of the cerebrum of an animal, the remnant is still capable of carrying out the entire role fulfilled until then by the whole cerebrum.

Among the supporters of this major theory of representation one might mention Carville and Duret, who propose mutual functional replacement by all cortical areas of the same side; Soltmann, in contrast, defends the idea of substitution by symmetrical cortical areas of the opposite hemisphere. (cited by Goltz, Über die Verrichtungen.) (*234)

[18]) In this context, I look upon it as a pious duty to stress that the much maligned and, as Möbius rightly emphasises, highly meritorious research neuroanatomist, Gall, was the first to introduce a practical system of physiological cerebral localisation. On the basis of anatomical data he elaborated a concept of the brain being composed of a large number (27 in all) of intrinsic organs, analogous to the extrinsic sensory organs and mediating an interpretation of the inner self just as the latter interpret the environment.

[19]) F. Goltz, Über die Verrichtungen des Grosshirns. Gesammelte Abhandlungen. Bonn 1881.

the difference between him and his opponents (*235) finally became only a quantitative one. In his later works he expressly conceded that "the lobes of the cerebrum certainly do not have the same function" and that, in particular, the anterior and posterior quadrants of the hemisphere are functionally different. Thus Goltz recognises the principle of localisation, but he has always denied the existence of specific sensory zones and profoundly opposed the correspondence of circumscribed sections of the hemisphere to specific sensory modalities [20]). Among the older clinicians, Bernhard Gudden [21]) adopts a related point of view. He also does not believe "that circumscribed regions exist in the cerebral cortex that exercise a specific function exclusively and in all circumstances". On the other hand, on the basis on the anatomically demonstrated interdependence of different cerebral regions, he comes to the conclusion that there remains "no alternative than to admit the view, quite resolutely, that with normal development and experience the cerebral cortex localises functions in at least two main regions, one for movement and the other for sensory concepts", the former having its seat in the anterior and the latter in the posterior quadrants of the hemisphere [22]).

The most extreme localisational hypothesis states that the cerebral cortex is divided into a large number of sharply separated zones, rather like a topographical map, that correspond exactly to the different sensory organs and motor apparatuses of the body. Today its most influential representative must still be Hermann Munk [23]). He maintains not only the principle of strictly circumscribed localisation but also the matching of cortical centres with the various peripheral sensory surfaces, and therefore divides the whole hemispheric surface (in dogs and monkeys) into his well-known sensory and motor centres (*240) (centres for somatic sensation) (*241). Flechsig's theory [24]), based on myelogenesis, surpasses even this interpretation; according to it, not only do the individual sensory organs project strictly topographically to the cortical surface, but also the highest psychic phenomena project spatially to circumscribed loci, the so-called "centres of understanding".

---

[20]) Goltz writes expressly: "While I consider as proved a large influence of the posterior part of the brain on visual function, on the basis of my recent studies, I cannot accept it in the sense of a circumscribed visual area, such as Ferrier, Munk and Luciani have postulated" (loc. cit. p.169).

Also: "The idea of centres within the cerebral cortex subserving circumscribed specialised functions is unacceptable. Thus there is no part of the cerebral cortex that is devoted exclusively to vision, none exclusively to hearing, smell, taste or feeling ... The manifestations of life from which we infer intelligence, emotions, suffering, instinct, do not depend on functionally specialised sections of cortex" (loc. cit. p.173).

Also: "There are no so-called motor centres (*236) on the surface of the brain that form obligatory and exclusive relay stations for a voluntary movement" (p.114).

[21]) H. Grashey, Bernhard v. Guddens gesammelte und hinterlassene Abhandlungen. Wiesbaden, 1889. (*237)

[22]) B. von Gudden, Über die Frage der Lokalisation der Funktionen der Grosshirnrinde. Allgemeine Zeitschrift für Psychiatrie 42. (*238) See also his collected and posthumous works, p.200ff (*239)

[23]) H. Munk, Über die Funktionen der Grosshirnrinde. Gesammelte Mitteilungen, 2nd edition. Berlin, 1890. (*225)

[24]) Flechsig, Gehirn und Seele. (*228) Rectoral address. Leipzig 1894.

The majority of other researchers who have become involved experimentally with questions of localisation in the meantime, such as Hitzig, Ferrier, Luciani and Seppilli, Luciani and Tamburini, Horsley and Schäfer, Monakow, Bianchi, Tonnini, Christiani, J. Loeb, Lussana and Lemoigne, and many others [25]), fall between these two extremely contradictory interpretations, and one must say that there are almost as many variants and shades of localisational theory as there are supporters of it. In the practical disciplines the stricter tendencies have predominated recently, especially under the influence of the theory of aphasia and the work of Flechsig. Meanwhile one will certainly agree with Monakow [26]) that "there lies an error in the way that the majority of today's clinicians and physiologists consider cerebral localisation theoretically, an error in the formulation of the question and in the conclusions drawn from observation".

b) The principle of functional localisation from the morphological point of view.

Let us now try to establish and substantiate a personal point of view on the basis of the anatomical data, and decide which physiological interpretation agrees best with the results of histological localisation.

We mentioned earlier that Gudden, who always strived to reconcile anatomical data with physiological activity, was convinced that he had to reject the concept of circumscribed, localised, functional centres because the largely homogeneous histological structure of the cortex, including its cellular components and their lamination, as well as the arrangement of its fibres, contradicted this view [27]). These interpretations are demonstrative of how erroneous views about the anatomical features of an organ lead to faulty

---

[25]) I shall cite here only a few of the most important older works, in addition to those already mentioned earlier:

Hitzig, Physiologische und klinische Untersuchungen über das Gehirn. Gesammelte Abhandlungen, Parts I and II. Berlin 1904. (*242)

Horsley and Schäfer, A Record of Experiments upon the Functions of the Cerebral Cortex. Philos. Transact. 1888.

Ferrier, Vorlesungen über Hirnlokalisation. (German edition by Dr. Weiss, Vienna, Leipzig 1892). (*243)

Luciani and Sepilli, Die Funktionslokalisation auf der Grosshirnrinde. German edition by O. Fränkel. Leipzig 1886. (*244)

Luciani and Tamburini, Sui centri psico-sensori corticali. Revist. speriment. di Freniat. 1879, p.1. (*245)

J. Loeb, Die Sehstörungen nach Verletzung der Grosshirnrinde. Pflügers Archiv, Vol. 34. p.18. (*246)

J. Loeb, Beiträge zur Physiologie des Grosshirns. Pflügers Archiv, Vol. 39 p.31. (*247)

H. Christiani, Zur Physiologie des Gehirns. Berlin, Enslin, 1885. (*248)

See also the very pertinent textbooks of physiology by Schiff, Hermann, Ranke, and others. Also: v. Monakow, Gehirnpathologie. Also: Exner, Untersuchungen über die Lokalisation der Funktionen in der Grosshirnrinde des Menschen. Vienna 1881. (*249)

[26]) v. Monakow, Neue Gesichtspunkte in der Frage nach der Lokalisation im Grosshirn. Correspondenzblatt für Schweizer Ärzte, 1909, 12. (*250)

[27]) Gudden, Gesammelte Abhandlungen, p.206. (*251)

concepts about its function. The intimate interdependence of anatomy and physiology is reflected in this. Everyone who used data about the intrinsic structure of the cortex as it was known then was bound to arrive at similar conclusions.

In contrast we have now been able to determine that the cerebral cortex consists of a number of individual histologically highly differentiated organs, each of which has a clearly determined position and its own specific structure, not only in terms of the arrangement and connections of its cellular elements and its fibre architecture, but also, and most important, due to the variety of individual cell types. We have also seen that such histological "centres" can be delimited in all other mammals as well as in man, often equally sharply even if in smaller numbers, and that individual homologous structural zones adopt the same, clearly determined positions on the cortical surface in all mammalian brains - with only minor variations.

Even if one does not wish to go so far as to consider these circumscribed histological units as absolutely independent physiological units and attribute complete functional autonomy to them, their demonstration is still of decisive importance for the question of localisation. There is an undisputed axiom: physiologically dissimilar elements have dissimilar structures. Reversing this statement one may equally justifiably conclude: parts of organs that are structurally different must serve different purposes. The genesis of histological specialisation depends on the transformation of similar rudiments into dissimilar forms, and can be traced back to physiological functional specialisation. Even if this last point were not firmly established for the cerebral cortex and if today localisation were not beyond all doubt from the standpoint of clinical experience and experimental data, it would of necessity have to be inferred from the anatomical facts. They alone compel us to formally advance the "*principle of functional localisation*" as an irrefutable rule, according to which every specific function corresponds to a specific region of the central nervous system with given conditions of conduction or, if the function is a composite one, to a complex of regions. The acceptance that all parts of the cerebral cortex participate uniformly in all its functions must forthwith be considered as obsolete by anatomists, physiologists and clinicians alike.

But one can go even further. Our findings do not simply justify the acceptance of the principle of a spatial functional specialisation across the cortical surface, they also prove with unassailable certainty the existence of a strictly circumscribed regional localisation of specific functions. The zones described above are sometimes sharply delimited from their differently structured surroundings. In other words, all around them sharp boundaries mark the transition to their specific histological structure, so that they stand out clearly from the rest of the cortex as a completely circumscribed area of tissue, like a histological organ *sui generis*. As examples, one may recall areas 4, 17, 27, 28 and the insula as a whole. In many mammalian orders abrupt histological transitions exist, just as we find between different sections of other organ systems, such as many parts of the gastrointestinal tract, for which no one doubts distinct local functions. The meaning, and even more so the origin, of such histological transformations, and their appearance throughout the whole mammalian class, would be absolutely inexplicable if one did not accept that this was a matter of an anatomical substrate, preformed in Anlage and spatially strictly circumscribed, for fixed, specific functions, functions associated exclusively with the particular zone and which specifically differ in some way from the functions of all other zones.

The question now arises as to what we can generally localise in these anatomically demarcated cortical regions. Naturally, this question cannot be treated here in detail for the various regions, but only its principal aspects considered. It is hardly necessary to mention that, in the present state of our knowledge, much still remains hypothetical.

1. Total or collective functions.

The first thing to say is that just as untenable as the idea of a "concept cell" or an "association layer" is the assumption of specific "higher order psychic centres". Indeed recently theories have abounded which, like phrenology, attempt to localise complex mental activity such as memory, will [28]), fantasy, intelligence or spatial qualities such as appreciation of shape and position to circumscribed cortical zones. Older authors such as Goltz, Rieger, Wundt, and recently, particularly outspokenly, Semon [29]), have already quite rightly expressed their opposition to such a "naive view" and pleaded simple psychological facts against it.

These mental faculties are notions used to designate extraordinarily involved complexes of elementary functions. What was said above about percepts, with regard to the cortical physiological processes underlying such complex functions, is thus even more appropriate in relation to such universal "faculties". One cannot think of their taking place in any other way than through an infinitely complex and involved interaction and cooperation of numerous elementary activities, with the simultaneous functioning of just as many cortical zones, and probably of the whole cortex [30]), and perhaps also including even subcortical centres [31]). Thus we are dealing with a physiological

---

[28]) The abstract notion of the will is certainly not a real process, but a general concept derived from a number of concrete realities. The individual concrete act of will, that is the only aspect to exist in reality, is however always a composite process stemming from numerous sensations and feelings and that therefore always includes numerous physiological processes of various types and localisations. cf W. Wundt, Physiolog. Psycholog. I, p.261ff, 1908.

[29]) R. Semon, Die Mneme als erhaltendes Prinzip im Wechsel des organischen Geschehens. Leipzig 1906. (*252)

[30]) Von Monakow goes so far as to state that the wave of physiological excitation necessary just to recognise an individual sound must spread over the whole cerebral cortex.

[31]) The idea that sensory processes, especially those of a composite nature such as emotions or passions, can only be manifested with the functional support of deep (*253), subcortical portions of the central nervous system, arises from the consistent systemic phenomena that accompany emotions (pulse, respiration, muscular tone). Without this, the "effect of emotion on enhancing sensation" (Wundt) would be inexplicable.

process extending widely over the whole cortical surface and not a localised function within a specific region. We must therefore reject as a quite impossible psychological concept the idea that an intellectual faculty or a mental event or a spatial or temporal quality or any other complex, higher psychic function should be represented in a single circumscribed cortical zone, whether one calls this an "association centre" or "thought organ" or anything similar.

In particular, to speak of a multitude of "psychic centres" one must invoke hypotheses that are equally unfounded both physiologically and psychologically [32]). In reality there is only one psychic centre: the brain as a whole with all its organs activated for every complex psychic event, either all together or most at the same time, and so widespread over the different parts of the cortical surface that one can never justify any separate specially differentiated "psychic" centres within this whole.

Naturally this does not mean that all the individual organs make equal physiological contributions to higher psychic processes. With regard to the infinite variety and richness of the psyche one should rather envisage the situation that in each particular case supposed "elementary functional loci" are active in differing numbers, in differing degrees and in differing combinations. One must further accept that specific complex processes occur mainly in one locality and others mainly in another. One must therefore also assume a certain regional preference for higher activities, sometimes more in occipital and temporal areas, sometimes more in frontal. Such activities are, however, always the result (and not merely the sum) of the function of a large number of sub-organs distributed more or less widely over the cortical surface; they can never be the product of a morphologically or physiologically independent "centre". The variety and the gradations of form and degree of higher intellectual activity are thus merely the expression of the infinite variability of functional combinations of individual cortical organs. The possibility of such variable and diverse complexity is supported by the evidence that I have given that the cortical surface is composed of numerous such specific morphological organs. As the possibilities for complexity are potentiated immeasurably with each new organ, we must recognise in the richer topical organisation of the cortical surface in certain mammals, and especially in man, one of the bases for their higher intellectual potential and thus for their perfection.

---

[32]) As is well known, Flechsig distinguishes four such "thought organs" or, as he calls them by contrast to the "inner senses", "mental" centres, one frontal, one insular, one parietal and one temporal, and he believes that his anatomical division, based on the asynchronous myelinisation of different sections of the cortex, also points to a differentiation of organs for individual "spiritual powers in the older psychological sense", without, of course, giving further detailed explanations as to the localisation of such powers (Gehirn und Seele, p.97, 1896).

2) Localised functions.

If the considerations developed here make the localisation of higher psychic activity impossible in the sense of its underlying physiological processes being restricted to spatially demarcated cortical centres, on the other hand our histotopographic findings, that agree from many points of view with connectional data and clinicopathological experience, demonstrate emphatically that nevertheless certain central nervous activities must be considered to have a spatially circumscribed localisation within the cerebral cortex.

The specific histological differentiation of cortical areas provides irrefutable proof of their specific functional differentiation - for, as we have seen, it is based on functional specialisation. The large number of distinct structural zones suggests a spatial specialisation of various individual functions, and finally the all-round sharp demarcation of many areas indicates inexorably a strictly circumscribed localisation of their corresponding physiological functions.

α) *The principle of absolute localisation.*

Although psychologists have often spoken out against this concept, one is nevertheless obliged to adopt the principle of absolute localisation for many such cortical processes on account of the anatomical data outlined above.

The sharply demarcated structural zones that are identifiable as specific morphological organs within the cortex can only be explained by assuming the localisation of equally sharply delimited specific functions and, in other words, that each such organ represents an exclusive individual function that is different from those of all the other organs. It does not necessarily follow from this that such a "centre" is the locus of only a single function or that a sensory centre is only activated by the sensory organ corresponding to a given peripheral stimulus (visual, auditory, somatosensory etc.). It is more correct that within a given organ an association of these elements also occurs and thus in one and the same place elementary activity is already synthesised into higher, more complex functions, and at the same time other modalities are added to specific sensory elements in these cortical sensory centres.

Thus we should not consider such a sensory centre as simply a repetition or a copy of a peripheral sensory organ, such as the retina, but rather characterise it as a "centre" in the truest sense of the word, along with Wundt [33]), that is as an organ within which various peripheral activities that participate in the sensory function in question are centralised or concentrated. To make this clear with an example, in the visual centres one can localise not only the functions associated with actual light sensitivity, but also the control of eye movements and of certain ocular reflexes etc. [34]). The essential point of our theory is that

---

[33]) Grundzüge der physiol. Psychol. 1908. (*223)

those elementary cortical processes strictly related to a peripheral sensory organ remain restricted to a circumscribed cortical zone. Their physiological limits are just as sharp and unchanging as the morphological ones, and the related pathways also have their origin or termination immutably within these absolutely constant boundaries.

β) *The principle of relative localisation.*

Alongside absolutely sharply demarcated organs, we have also discovered structural regions in the cerebral cortex whose borders are indistinct and where the architectonic features merge and overlap more or less completely those of neighbouring zones. This fact suggests a more or less extensive superimposition of functions in the cerebral cortex. Certain pathophysiological observations also support this. However, one should not consider as being superimposed in this way those elementary functions mentioned above related to a single sensory zone and which are inseparable and in any case are situated within a single centre. We mean, rather, cortical activities related to different locations in the periphery of the body that partially overlap in the cortex. In this case one may speak of relative localisation whereby a physiological process is not sharply limited to an area devoted exclusively to it, but rather this cortical area can partially also subserve other functions (*the principle of multiple functional cortical representation*).

But one should also maintain the concept of relative localisation in another sense. Human and animal clinical experience convincingly demonstrates again and again an extensive capacity for recovery of lost function after destruction of parts of the cortex, not only in acute cases but also frequently when the lesion occured in the distant past and has not in itself resolved. Recovery of function can sometimes appear years later and may be relatively complete, as one can observe particularly in cases of speech disturbance. Such recovery from physiological deficits [35]) cannot occur without other parts of the cortex gradually replacing the lost functional zone and taking over its activity (*the principle of functional replacement*). Indeed clinical pathology also recognises such cortical plasticity (*254).

To explain this undisputable phenomenon one can propose two hypotheses. On the one hand it is conceivable that there are elements or organs in the cerebral cortex that in normal circumstances remain in a state of physiological inactivity throughout life, like nonfunctioning reserves, destined only to function as replacement organs for any element that is put out of action. This proposition is improbable in that we know that totally functionless organs atrophy in the long term. The second, much more plausible, hypothesis is that

---

[34]) That subcortical centres are also involved in such sharply circumscribed cortical functions should however not cause us to include them in the corresponding "sensory centre". This would be a weakening of the concept of a cortical centre. If this were the case one would also logically have to include the peripheral organ.

[35]) We are, of course, not including here any transitory deficits such as temporary diaschisis.

elements that have a role in replacing a function already normally play a part in that lost function, either by serving as one active component of it and subsequently taking over the entire function after the loss of the complementary element, or by their normally being responsible for a relatively subsidiary or gradually attenuated contribution to the total function, but progressively increasing in importance in case of need. In the first case it is a form of secondary acquisition of functions related to the original activity, in a way a partial change of function in an organ; in the second case the substitution depends only on an activity-related increase in a function that is normally present.

Only further research can clarify which morphological components of the cerebral cortex can replace each other mutually in this way; it could be symmetrical parts of both hemispheres, as sometimes appears probable in aphasia, or spatially contiguous areas or, finally, normally closely functionally related, but spatially separated, elements. But it is in no way admissable to presuppose the same replacement mechanism for all activities, from quite primitive basic functions to higher psychic phenomena; it is even much less possible, as is so often done nowadays, to simply conclude from observations on humans the existence of precisely the same processes in animals, and vice versa.

3. Special functional localisation.

Having so far examined the basic principles of the problem of functional localisation in the cerebral cortex, we shall now briefly touch upon the question as to whether it is already possible today to localise certain specific physiological activities in our individual "anatomical centres". This question is related to the other extremely important question for the further development of the localisation theory as to whether, and to what extent, histological structure corresponds to the currently accepted physiological subdivisions of the cortical surface in man and various animals or, in other words, the question as to the relationship between anatomical and physiological localisation in general.

As any sure and unequivocal localisation on the cortical surface is still lacking for the majority of cortical functions, even the very simplest such as the cortical process corresponding to general skin sensation or visual or auditory perception, our investigation must be restricted to correlation of relevant data from a few examples. When considering the many cases of contradictory observations that still exist between physiologists and clinicians, it would represent progress if, in the case of two divergent opinions, one agreed with the results of histological localisation thus favouring its acceptance on anatomical grounds.

a) *The motor cortex.* (*255)

A striking example exists in the localisation of the *electrically excitable (motor) area* of the cerebral cortex. Ferrier, Beevor and Horsley, Munk and others proposed the theory which prevailed exclusively for a long time that the "motor" cortex, and especially the electrically excitable zone, included the whole Rolandic region, that is to say the entire precentral and postcentral gyri and part of the parietal lobe, both in man and in monkeys. In contrast, more recently the opposing view, originally suggested by Hitzig (as early as 1874), that only the section of the cortex lying anterior to the central sulcus is excitable by weak currents, ie that it is the seat of electromotor activity, has achieved wide acceptance, especially under the influence of the stimulation experiments of Sherrington, Grünbaum and C. and O. Vogt on monkeys, and by F. Krause and others on man. Today there is a general tendency to abandon the earlier theory according to which the precentral and postcentral gyri represented equal components of a single functional zone, Munk's so-called "*somatic sensory zone*". Rather each of these gyri forms a special functional centre and subserves specific activities. Only Munk and a few of his pupils adhere tenaciously to the old unitary theory.

What is the position of anatomical localisation with regard to this question now? The answer is implicitly put forward in Part II and I have also repeatedly exposed the same arguments in my earlier communications. According to these it is clear that in terms of major architectonic features the central sulcus forms the boundary between two fundamentally different structural regions, the agranular precentral region and the granular postcentral region, and that that this structural borderline can be traced throughout the whole mammalian class and even occurs when the sulcus itself is absent. From this one must necessarily conclude, as we have done in our earlier discussions, that there exists an essential fundamental physiological difference between the pre- and postcentral gyrus. Indeed this essential difference now emerges clearly and unequivocally from the new data on physiological excitation from the above-named authors [36]). We have consequently in this respect a gratifying and, in terms of further research, an encouraging agreement between the results

---

[36]) If myelogenetic studies do not reveal at all such a marked difference and, moreso, if judging by the temporal sequence of myelinisation the two central gyri represent a homogeneous and unseparable entity, it is evidence, like so many other findings, that myelinisation can only be considered an indicator of functional localisation with considerable reservations and with critical prudence. From this example, there can be no question of the functional selectivity of a centre being demonstrated absolutely systematically by myelogenetics.

[37]) It should just be noted that the concept described here is also supported by results with other methods. Thus Monakow (Gehirnpathologie, 2nd Edition), on the basis of new clinical and pathological data, accepts the view that the motor pathway originates, if not exclusively at least to a major extent, from the precentral gyrus, and O. Vogt (Über strukturelle Hirnzentra, mit besonderer Berücksichtigung der strukturellen Felder des Cortex pallii. Anat. Anz. Verh. der Anat. Ges. XX, 1906) (*256) was able to demonstrate positively with experimental degeneration techniques that the two central gyri have different projections from the thalamus, namely that the precentral gyrus is innervated directly from the area of termination of the tegmental radiation in the thalamus (*257), while the postcentral gyrus is related to the terminal region of the medial leminiscus (*258). Thus a fundamental systematic difference in fibre connections is demonstrated and therefore also essentially different functions. Some years ago, at the Berlin Society for Psychiatry and Neurology, I myself called attention to unpublished lesioning experiments by O. Vogt, that in lower monkeys also essentially different deficits emerge according to whether one destroys the pre- or postcentral gyrus (Neurologisches Zentralblatt, 1905, p.1158ff)

of anatomical, and especially cytoarchitectonic, localisation studies [37]) and a strictly physiological view. The practical importance of this finding cannot be ignored.

But the agreement goes even further. An unprejudiced demonstration is provided not only by the correspondence of detailed anatomical and physiological borders but also, and much more conclusively, one can establish a very extensive, if not absolute, overall regional spatial organisation. The precentral region that we delimited in cercopithicids (Figures 92 and 93) coincides with the excitable zone determined by C. and O. Vogt [38]) in the same family of monkeys in that the latter is entirely enclosed within the anatomical region. The congruence is not absolute in that the electrically excitable motor zone only encroaches a little onto the medial surface, if at all, whereas the histological field extends widely medially. On the other hand, however, there is an astonishing agreement in that several of the different physiologically excitable subdivisions of C. and O. Vogt equally correspond exactly to individual cytoarchitectonic (and myeloarchitectonic) areas within the precentral region.

The same is true for prosimians, especially the lemur family, as a comparison of the parallel anatomical and physiological studies undertaken by Vogt and myself shows. Here also the "excitatory" field coincides almost exactly with my precentral region, the two cytoarchitectonic areas 4 and 6 (thus not just the giant pyramidal area) [39]). This interpretation is again confirmed by Mott and Halliburton, who independently from us have determined in lemurs that there is a total agreement between the results of their anatomical and physiological experiments [40]).

Finally, if one compares my other histological brain maps with the surface localisations described by C. and O. Vogt on the basis of stimulation [41]), one can also recognise a tolerable correspondence between anatomical and physiological borders even in other inferior mammals.

Thus we arrive at the conclusion that, in very different animals, there exists a considerable correspondence between a physiologically very important cortical zone, the electrically excitable motor area, and an anatomically defined zone. This fact merits our consideration from the beginning. It forms a valid criterion for the judgement of the techniques in question. The minor disagreements will certainly be clarified in time. They could firstly be due to deficiencies in technique and observation, but secondly it is also possible that the more extensive anatomical region might represent a higher element within which the smaller "excitatory zone" represents only a partial function, as it is absolutely not necessary to think that the electrically excitable zone must

---

[38]) C. and O. Vogt, Zur Kenntnis der elektrisch erregbaren Hirnrindengebiete bei den Säugetieren. Journal f. Psych. u. Neurol., VIII Ergänzungsheft, 1907. (*259)

[39]) Precise details can be found in the published works. Perhaps our small area 8 lying anterior to the precentral region could also coincide with Vogt's "eye field".

[40]) Mott and Halliburton, Localisation of Function in the Lemurs Brain. Proc. Roy. Soc. B. Vol. 80, 1908. They speak of a "close correspondence of the results", but this seems to me to be going rather too far, especially in view of their anatomical observations.

[41]) Elektrisch erregbare Hirnrindengebiete, loc. cit., Plates 12 and 13.

be identical with the "motor region" in the strictest sense, that is the centre for voluntary movement.

If we attempt to study other functional domains we encounter greater difficulties. As we have said, there is still a general lack of adequate data concerning the localisation of the most fundamental functions. In particular for man we still do not possess even approximately certain and unequivocally proven localisational data for the main senses.

b) *The human visual cortex.*

Observations relating to localisation within the human visual system are very divergent. While certain localise vision exclusively to the medial surface and assume it to be limited to the "calcarine" cortex in the narrow sense (Henschen [42]), others propose a much greater extent for the cortical visual area, covering almost the whole occipital cortex, including the lateral surface (von Monakow [43]), Bernheimer [44]) Förster [45]) etc). It seems to me that the time has come for histological localisation to make its voice heard. Our striate area represents such a histologically characteristic and topically so sharply delimited zone, that is moreover so constant and readily demonstrable throughout the mammalian class, that one must suppose it to have a major function that is highly specific and essential in all mammals. In addition, clinicopathological observations are consistent with the idea that there exists a close relationship between this cortical area and the major sensory modality that is usually localised in this region, that is to say the perception of visual stimuli. Now also Henschen's localisation of the "visual area", based on simultaneous clinical and anatomopathological observations, coincides overwhelmingly with the striate area even in detail. This agreement should be quite capable of tipping the balance between the two opposing localisational hypotheses in favour of the concept represented by Henschen that is in better agreement with the anatomy [46]). The neuropathologists will henceforth have to turn their attention to the question.

The features of the other cortical sensory areas of man are even less clear. The localisation of taste and smell has not yet reached beyond the stage of conjecture. Any attempt to ascribe these physiological centres to a cytoarchitectonic zone (region or area) must therefore provisionally seem hopeless from the outset. What is clear is that the cortical zone that is usually referred to as the olfactory centre, or in a wider sense as the "rhinencephalon" (*263), is always composed of a number of very distinct anatomical areas. Which of these is the true olfactory cortex cannot be decided as yet [47]).

---

[42]) Henschen, Klinische und anatomische Beiträge zur Pathologie des Gehirns. Upsala, 1890-1903. Vol. 1-4. Also idem: La projection de la rétine sur la cort. calc. Sém. méd., 1903. (*260)

[43]) v. Monakow, Gehirnpathologie.

[44]) Bernheimer, Die kortikalen Sehzentren. W. kl. W., 1900. (*261)

[45]) Förster, Unorientiertheit, Rindenblindheit, Seelenblindheit. Arch. f. Oph., 1890. (*262)

[46]) English authors also do not hesitate to refer to the striate area in short-hand form as the "visuo-sensory area" (Bolton, Mott, Watson).

c) *The human auditory cortex.*

The situation of the so-called auditory cortex in man is not much clearer. Here also diametrically opposed views confront each other. Most authors, exemplified by von Monakow on the basis of secondary degeneration and Munk on the basis of physiological experiments involving deprivation in animals, adhere to the theory put forward by Wernicke, representing the opinion that the major part of the temporal lobe should be considered to be the central organ of hearing, or in any case at least the whole of the superior temporal gyrus. However, Flechsig [48]), on the basis of myeloarchitectonic observations, has recently proposed that the auditory cortex occupies only a tiny cortical area mainly limited to the anterior transverse gyrus and comprising scarcely two square centimeters of the exposed surface of the superior temporal gyrus.

Against this last view - apart from basic objections to such an interpretation of physiological localisational based on developmental, and particularly of myelogenetic, processes - the following three facts about histological localisation [49]) should now be emphasised. Firstly the anterior transverse gyrus is not an independent anatomical entity. As was already described in detail above (page 121), it can be divided on cytoarchitectonic and myeloarchitectonic grounds into several structural areas that possess common architectonic features within themselves and with neighbouring cortex of the rest of the superior temporal gyrus and, together with these represent a larger, homogeneous structural zone. Further, the zone in question is not just limited to the anterior transverse gyrus but climbs significantly beyond it anteriorly and equally surrounds the whole posterior transverse gyrus and a considerable part of the free surface of the superior temporal gyrus (Figure 89). These facts alone would be enough to prove conclusively that the anterior transverse gyrus does not in itself form an independent functional centre and that therefore it cannot represent alone and exclusively the "auditory cortex". If one wishes to define a sensory centre within this region, one will have to admit to its having a considerably greater

---

[47]) Retzius has already stressed that even anosmatic animals possess a "rhinencephalon" and that therefore this anatomical formation undoubtedly serves more than an exclusively olfactory function.

[48]) Flechsig, Bemerkungen über die Hörsphäre des menschlichen Gehirns. Neur. Cbl., 1908, p.1 & 50ff. (*264)

[49]) Indeed Flechsig (loc. cit.) seeks to diminish the significance of histological localisational studies by maintaining that they had contributed absolutely nothing new apart from the finding long since introduced by himself, and secondly that histological localisation was not a useful heuristic principle as the aim of brain research was not the demarcation of "anatomical cortical fields" but functional organisation. As to the first point, I prefer to pass over it in silence; as to the second the following should be noted. Using the same argument, one must qualify all morphological studies as useless, for the goal of all biological investigation consists not in recognising criteria of shape but rather life processes. Further, one must confront Herr Flechsig with the question whether his own myelogenetic technique is also not merely "anatomical". Or is he today trying in all seriousness to make us believe that, with his myelogenetic studies, he is pursuing pure physiology?

[50]) If one accepts Flechsig's localisation, only a few square centimetres of the whole auditory cortex would be involved, that is to say approximately not much more than 1/200 of the total surface of the hemisphere; it is, however, quite unthinkable that such an important cortical function as hearing should be restricted to so small a part of the whole cortex.

expanse [50]).

In addition, there is a third very important observation from comparative localisation, that was also already mentioned above, the observation that the extremely characteristic cell and fibre architecture, typical of both transverse gyri in man, is lacking in all other animals. In other words, a human structural zone in which Flechsig places the cortical end-station of the auditory pathway, the auditory cortex, is completely absent in animals, and even in monkeys that otherwise possess a very similar cortical structure to that of man [51]) thus facilitating the homologising of cortical areas between man and monkeys. There are only two explanations for this: either the other mammals do not possess an equivalent of the human "auditory cortex", or, as is the more likely interpretation, this cortical zone serves other functions in man and is, at least not exclusivley, or in the form and extent proposed by Flechsig, the "hearing centre". What is more, its composition from several, histologically widely different, individual areas (41, 42 and 52 of the brain map) suggests a physiologically more complex function (*265) of the region in question.

The solution to this question will only be furnished by an impartial examination of pathological cases with reference to anatomical localisational data, as well as by a detailed comparative study of cortex, using both anatomical (especially myeloarchitectonic) and physiological methods.

d) *Localisation of speech and aphasia.*

It would be particularly tempting in this connection, considering the controversy recently engendered by Pierre Marie about aphasia, to also engage in a discussion of the specific localisation of speech. However, it seems to us that the time is hardly ripe for this for most of the necessary physiological preparatory work is lacking. What is more, it is in no way to be seen as definite that the cortical localisation of speech coincides with that of aphasia. In relation to aphasia, however, one can already immediately conclude two things from the psychophysiological considerations described above. Firstly, an aphasia, regardless of whether it belongs to the motor or sensory subcategory, can never be linked to a single structural centre, and therefore to an individual one of our cytoarchitectonic areas, but rather it always includes a complex of such areas, forming a larger region. Secondly, the "aphasia centre" covers a much greater expanse [52]) than one was formerly accustomed to believe. I have already made brief reference elsewhere [53]) to the fact that in particular, according to all that can be concluded from anatomical localisational data, the seat of motor aphasia must extend much further anteriorly than appears from Broca's classic

---

[51]) One may simply compare the great similarities between the calcarine cortex (the "visual cortex") and the giant pyramidal cortex (the "motor cortex") in man and monkey.

[52]) Liepmann and Knauer have recently firmly expressed the same opinion, the former on the basis of pathological findings, the latter on the basis of concepts in the domain of general physiolgy. See H. Liepmann, Zum Stand der Aphasiefrage. Neur. Cbl., 1909, p.449. A. Knauer, Zur Pathologie des linken Schläfenlappens. Klinik. f. psych. u. nervöse Krankheiten, IV, 1909. (*266)

[53]) Neur. Cbl., 1909, p.720.

theory, and that at least the anterior sections of the inferior frontal gyrus, and perhaps even part of the actual orbital surface, must be included in it (thus, apart from area 44, also areas 45 and 47 of the brain map, Figure 85). That the insular cortex cannot be the "speech centre" in a strict sense can be concluded with certainty from the fact that regions of similar structure appear throughout the whole mammalian class, and are even much more extensive in lower species than in man. But within which specific histological regions, and within which individual boundaries, we must localise such complex functional zones as must be proposed to explain the synthesis of human speech, only the future will reveal.

e) *Individual functional centres in animals.*

Functional subdivision of the cerebral cortex in animals has only been attempted so far in isolated cases; for the most part, this has only been concerned with the verification of electrically excitable zones. Now we possess a physiological parcellation of the whole cortical surface in two mammals, the dog and the monkey, at least on the lateral aspect. It is the abiding merit of H. Munk to have attempted a first outline of the physiological structure of the whole cortical surface in these two animals with experimental methods. Thus the way was first paved for the real theory of "centres" (*267), the principle of localisation in the strict sense.

If we now compare Munk's brain maps of the monkey and dog with the results of histological localisation, one cannot fail to recognise that a high degree of agreement is attained between a large number of data obtained with both localisational methods, whereas in other points equally great and not always insubstantial divergences exist in the results. We cannot enter into a detailed analysis of all areas here, but must limit ourselves to selecting a few particularly instructive examples.

To begin with, our two brain maps of the monkey show extensive agreement in relation to two regions: firstly, Munk's area A, the visual cortex of the lateral surface [54]), corresponds entirely with the extent of my striate area, or area 17 (Figure 90, page 128), and secondly area B, or the auditory cortex, of Munk's monkey map coincides with my temporal region (but not with any single area in it). It may also be noted that within the frontal lobe directly anterior to the arcuate sulcus lies a small zone that coincide absolutely in both physiological and anatomical maps (Munk's area H, my area 8), and finally this sulcus represents in both maps a strict boundary between different regions (Figure 90) (*269).

On the other hand, as we have already seen above, there exists a substantial divergence between anatomical and physiological localisation in that, according to Munk, the region of the two central gyri of the monkey represents

---

[54]) Physiological localisation has unfortunately failed almost completely on the very important medial side (*268) because of technical difficulties.

a functionally homogeneous sensory zone, the somatic sensory cortex, whereas anatomically the pre- and postcentral gyri must be separated into two essentially different centres strictly divided by the central sulcus. According to this, Munk's extremities zone (*270), areas C and D, on either side of the central sulcus, is composed anatomically of an (electromotor) precentral region anterior to the sulcus, and a posteriorly situated (sensory?) postcentral region, structurally different from the former from all points of view.

The structural divergences are even greater in the brain of the dog. It is less easy to make even the physiological "somatic sensory cortex" agree with one of our anatomical regions than in the monkey. Only the extremities zone placed by Munk mainly in the posterior sigmoid gyrus coincides approximately with the dorsalmost part of our giant pyramidal area, or area 4; in the ventral section, however, all correspondence is lacking. A physiological separation into two major divisions corresponding to the agranular precentral region and the granular postcentral region, which must certainly be postulated on comparative anatomical grounds, is absent in Munk's map of the brain of the dog.

The physiological visual cortex of the dog, Munk's area A, stretches very widely laterally and includes an extensive zone that occupies nearly a fifth of the whole lateral surface. On the other hand, the anatomical striate area in the dog, that is certainly closely concerned with vision and of which the lateral borders correspond with Munk's visual area in the monkey, lies almost exclusively on the medial surface and covers only the dorsalmost part of the marginal gyrus near the posterosuperior edge of the cortex. The extensive sections of the suprasylvian and ectosylvian gyri that Munk includes in his visual cortex, and especially the real centre of the visual cortex (*271), area A1, lie quite outside the striate area. This divergence can only be explained in two ways: either the anatomical area in question is not the "visual area" (but then how can one explain the obvious coincidence with Munk's visual area in the monkey?), or the extent of the physiological area has been traced much too far laterally in the dog (*272). I am inclined toward the latter view and am convinced that, provided it was surgically technically feasible, if one succeeded in destroying the whole of the lateral area A, including area A1, without damaging the medial surface and above all without a lesion in the optic radiation, there would be no visual disturbance, but that on the contrary the destruction of exclusivley the medial surface that coincides with the striate area, leaving the lateral cortical surface fully intact, would always cause a hemianopsia (or total blindness if the lesion was bilateral). In other words, anatomical data, and especially comparative localisational information, make it very highly likely that the real visual area of the dog lies almost completely on the medial surface of the hemisphere and that Munk's localisation needs correction in this respect. Monakow has already determined that the anterior extent along the cortical margin must be considered to be greater than given by Munk, based on anatomical examination of Munk's dogs. This is fully supported by histological localisation, for the dorsal part of the striate area extends much further rostrally than Munk's map suggests.

Just a word about the "auditory cortex" of the dog, Munk's area B! As is well known, Kalischer has recently absolutely disputed the existence of an auditory centre in the dog in the position described by Munk and his interpretation of it, on the basis of his behavioural experiments. From other quarters it has been defended with dogmatic assuredness using the same methods. We do not wish to enter into this dispute here, that is really mainly one of method. Once again anatomical, and especially myeloarchitectonic, data and, above all, comparative histotopography, indicate firstly that the auditory centre as a whole is situated more anteriorly, mainly toward the superior end of the sylvian sulcus, probably partly even anterior to it [55]), and secondly that the ventral portions of Munk's auditory area are in no way connected with the "auditory cortex". The latter zone, lying directly lateral to the posterior rhinal sulcus, certainly corresponds, judging from its structure, to my areas 35 and 36 in man, which lie extremely medially along the parahippocampal gyrus; it therefore cannot be really auditory cortex, and one must exclude at least this part from the auditory area, and doubtless also the more caudal portions.

I have come to the end of my reflections. As can be seen, the results of anatomical and physiological localisation agree well on the whole. The principle of both is the delimitation of superficial zones, such that the surface of the cerebral hemisphere is divided topographically into different organs. In many points there is even satisfying agreement with respect to the specific localisation of individual "centres". In other points, certainly, physiological views will have to undergo revision in the light of irrefutable anatomical data. For most histological organs there is still a total lack of functional localisation. A wide field of fruitful activity is opened up to physiology here through newly acquired anatomical localisational information.

However, one thing must be stressed quite firmly: henceforth functional localisation of the cerebral cortex without the lead of anatomy is utterly impossible in man as in animals. In all domains, physiology has its firmest foundations in anatomy. Anyone wishing to undertake physiological localisational studies will thus have to base his research on the results of histological localisation. And today with greater reason than ever, one must recall the words of the past master of brain research, Bernhard Gudden, spoken three decades ago in the face of a partial and dangerous tendency to specialise in extirpation experiments: "Faced with an anatomical fact proven beyond doubt, any physiological result that stands in contradiction to it loses all its meaning... So, first anatomy and then physiology; but if first physiology, then not without anatomy."

---

[55]) Also compare especially the position of the early myelinated temporal cortex according to C. Vogt (Etude sur la myelinisation des hémisphères cérébraux. Paris 1900). (*273)

Literature.

Only publications are listed here that are directly related to the question of cortical parcellation and histological localisation. All other papers are cited in the main text (*274).

R. Arndt, Studien über die Architektonik der Großhirnrinde des Menschen. Arch. f. mikrosk. Anat. Bd. 3-5, 1867-69.

Berlin, Beitrag zur Strukturlehre der Großhirnwindungen. Inaug.-Dissert. Erlangen 1858.

W. Betz, Anatomischer Nachweis zweier Gehirnzentra. Centralbl. f. d. mediz. Wiss. 1874, Nr. 37 und 38.

–, Über die feinere Struktur der Großhirnrinde des Menschen. Ebenda 1881, Nr. 11, 12, 13.

V. Bianchi, Il Mantello cerebrale del Delphino (Delphinus delphys). Napoli 1905. Tipografia d. R. Akadem. d. Science.

S. Bolton, The exact histological Localisation of the visual area of the human cerebral cortex. Philos. Transact. **193**, 1900.

K. Brodmann, Beiträge zur histologischen Lokalisation der Großhirnrinde.
I. Mitteilung: Die Regio rolandica. Journ. f. Psych. u. Neurol. 2, 1903.
II. Mitteilung: Der Calcarinatypus. Ebenda 2, 1903.
III. Mitteilung: Die Rindenfelder der niederen Affen. Ebenda 4, 1905.
IV. Mitteilung: Die Riesenpyramidentypus und sein Verhalten zu den Furchen bei der Karnivoren. Ebenda 6, 1905.

V. Mitteilung: Über den allgemeinen Bauplan des Cortex pallii bei den Mammaliern und zwei homologen Rindenfelder im besonderen. Zugleich ein Beitrag zur Furchenlehre. Ebenda 6, 1906.

VI. Mitteilung: Die Cortexgliederung des Menschen. Ebenda 10, 1907.

VII. Mitteilung: Die cytoarchitektonische Cortexgliederung der Halbaffen. Ebenda 12, 1908 (Ergänzungsheft).

–, Über Rindenmessungen. Zentralb. f. Nervenh. u. Psychiatr. 19, 1908.

–, Demonstrationen zur Cytoarchitektonik der Großhirnrinde mit besonderer Berücksichtigung der histologischen Lokalisation bei einigen Säugetieren. (Vortrag auf der Versammlung deutscher Psychiater, 1904.) Allg. Zeitschr. f. Psychiatr. 61, H. 5, 1904.

Campbell, Histological studies on the Localisation of cerebral function. Cambridge 1905.

Farrar, On the motor Cortex. Americ. Journ. of insan. 59, 1903.

Golgi, Sulla fina Anatomia degli organi centrali del sistema nervose. Milano 1886.

B. **Haller**, Die phyletische Entfaltung der Großhirnrinde. Arch. f. mikr. Anat. und Entwicklungsgesch. 71, 1908.

C. **Hammarberg**, Studien über Klinik und Pathologie der Idiotie, nebst Untersuchungen über die normal Anatomie der Hirnrinde. Upsala 1895.

Hermanides und **Köppen**, Über die Furchen und über den Bau der Großhirnrinde bei den Lissencephalen, insbesondere über die Lokalisation des motorischen Zentrums und der Sehregion. Arch. f. Psychiatr. **37**, 1903.

G. **Holmes**, A note on the condition of the postcentral cortex in tabes dorsalis. Rev. of Neurol. and Psychiatr. 1908.

Ariëns Kappers und **W.F. Theunissen**, Die Phylogenese des Rhinencephalon, des Corpus striatum und der Vorderhirnkommissur. Fol. neurobiol. I. 1908.

A. **Kappers**, The phylogenesis of the Palaeo-cortex and Archi-cortex, compared with the Evolution of the Visual Neo-cortex. Arch. of Neurol. and Psych., Vol. IV, 1909.

Th. **Kaes**, Die Großhirnrinde des Menschen in ihren Massen und in ihrem Fasergehalt. Ein gehirnanatomischer Atlas. Jena 1907.

Köppen und **Löwenstein**, Studien über den Zellenbau der Großhirnrinde bei den Ungulaten und Karnivoren und über die Bedeutung einiger Furchen. Monatsschr. f. Psychiatr. und Neurol. **18**, H. 6, 1905.

W. **Kolmer**, Beitrag zur Kenntnis der "motorischen" Hirnrindenregion. Arch. f. mikr. Anat. u. Entwicklungsgesch. **57**, 1901.

B. **Lewis**, On the comparative structur of the cortex cerebri. Brain 1, 1878. (Vgl. auch Lewis, "Textbook of mental Diseases".)

–, and Clarke, The cortical lamination of the motor area of the brain. Proc. Roy. Soc. **28**, 185, 1878.

B. **Lewis**, Researches on the comparative structure of the cortex cerebri. Philos. Transact. **171**, I, 1880.

Th. **Mauss**, Die faserarchitektonische Gliederung der Großhirnrinde bei niederen Affen. Journ. f. Psych. u. Neurol. Bd. XIII, 1908.

O. **Marburg**, Beiträge zur Kenntnis der Großhirnrinde der Affen. Obersteiners Arbeiten 1908.

Th. Meynert, Studien über das pathologisch-anatomische Material der Wiener Irrenanstalt. Vierteljahresschr. f. Psychiatr. **1**, 1866.

–, Der Bau der Großhirnrinde und seine örtlichen Verschiedenheiten, nebst einem pathologischanatomischen Corollarium. Leipzig 1868. (Siehe auch Vierteljahresschr. f. Psychiatrie, 1868.)

–, Kap. 31 in Strickers "Handbuch der Lehre von den Geweben". Leipzig 1872.

F.W. Mott, The progressive evolution of the structure and functions of the visual cortex in mammalia. Arch. of Neurol. from the patholog. Labor. of the London county asylums. Vol. III, 1907.

F.W. Mott and **A.M. Kelley**, Complete survey of the cell lamination of the cerebral cortex of Lemur. Proceed. Roy. Soc. B., Vol. 80, 1908.

F.W. Mott and **W.D. Halliburton**, Localisation of function in the Lemurs brain. Proceed. Roy. Soc. B., Vol. 80, 1908.

Ramon y Cajal, Studien über die Hirnrinde des Menschen. J.A. Barth, Leipzig. 1900-1906. Deutsch von Bresler.

1. Heft. Die Sehrinde, 1900.
2. Heft. Die Bewegungsrinde, 1900.
3. Heft. Die Hörrinde, 1902.
4. Heft. Die Riechrinde beim Menschen und Säugetier, 1903.
5. Heft. Vergleichende Strukturbeschreibung und Histogenesis der Hirnrinde, 1906.

L. Roncoroni, La fine morfologia del cervello degli epilettici. Arch. di Psich. 17, 1896.

–, Sul tipo fundamentals di stratificazione della cortecia cerebrale. Anat. Anz. 34, 1909.

L. Rosenberg, Über die Cytoarchitektonik der ersten Schläfenwindung und der Heschlschen Windungen. Monatsschr. f. Psych. u. Neurol. **23**, 1908.

Schlapp, Der Zellenbau der Großhirnrinde der Affen. Arch. f. Psychiatr. **30**, 1889.

–, The microscopic structure of cortical areas in man and some mammals. Publicat. of Cornell University II, 1902.

E. Smith, The so-called "Affenspalte" in the human (Egyptian) brain. Anat. Anz. **24**, 1904, S. 74.

–, The Morphology of the occipital region of the cerebral hemisphere in Man and the Apes. Anat. Anz. **24**, 1904, S. 436.

–, Studies in the morphology of the human brain. Nr. 1. The occipital region. Records of the Egyptian Government School of medicine. Vol. **II**. (Separat.)

–, A new topographical survey of the human cerebral cortex. Journ. of Anat. and Physiol. **41**, 1907.

Taalman Kip, Over den bouw van den Cortex cerebri bij mol en egel. Psychiatr. en Neurol. Bladen 1905.

–, De Phylogenie van de Cortex cerebri. Verhandelingen Natur- en Geneeskunde Congress 1905.

H. Vogt und **P. Rondoni**, Zum Aufbau der Hirnrinde. Deutsche med. Wochenschrift **34**, 1908, S. 1886.

O. Vogt, Zur anatomischen Gliederung des Cortex cerebri. Journ. f. Psychol. und Neurol. **2**, 1903.

–, Über strukturelle Hirnzentra, mit besonderer Berücksichtigung der strukturellen Felder des Cortex pallii. Anat. Anz. 1906, Verh. der Anat. Gesellschaft, S. 74ff.

–, Der Wert der myelogenetischen Felder der Großhirnrinde (Cortex pallii). Anat. Anz. **29**, 1906, S. 273.

C. und O. **Vogt**, Zur Kenntnis der elektrisch erregbaren Hirnrinden-Gebiete bei den Säugetieren. Journ. f. Psycholog. u. Neurolog. **8**, 1906, Ergänzungsheft, S. 277.

A. **Watson**, The mammalian cerebral cortex, with special reference to its comparative Histology. I. Ordre Insectivora. Arch. of Neurol. from the pathol. Labor. of the London county asylums. Vol. III, 1907.

Th. **Ziehen**, Das Zentralnervensystem der Monotremen und Marsupialier. II. Teil: Mikroskop. Anat. II. Abschn.: Der Faserverlauf im Gehirn von Echidna und Ornithorhynchus. (Aus Semon, Zool. Forschungsreisen III, Band II, 2, 1909.)

G. **Zunino**, Die myelarchitektonische Differenzierung der Großhirnrinde beim Kaninchen (Lepus cuniculus). Journ. f. Psych. u. Neur. Bd. 14, 1909.

Translator's References

Brodmann was inconsistent and often inaccurate in his referencing. In the following list I include the references used in Brodmann's original text (including those appearing as footnotes) and in his Bibliography, as well as all those used in the Translator's Introduction, in order to provide a single source of references. Brodmann also often gives just an author's name without a bibliographic reference. Where possible, I have suggested a suitable reference in such cases, but sometimes I have not been able to identify the exact source. Some older journals omitted the authors' initial(s) and I have indicated this where relevant by the use of brackets.

Alzheimer A (1904) Einiges über die anatomischen Grundlagen der Idiotie. Zentralblatt für Nervenheilkunde und Psychiatrie 27 497-505

Alzheimer A (1904) Histologische Studien zur Differentialdiagnose der progressiven Paralyse. In: Histologische und histopathologische Arbeiten über die Grosshirnrinde, ed F Nissl Vol. 1. pp.18-314. Fischer, Jena

Anton GT (1903) Gehirnvermessung mittelst des Kompensations-Polar-Planimeters. Wiener klinische Wochenschrift 16 1263-1267

Arndt R (1867, 1869) Studien über die Architektonik der Grosshirnrinde des Menschen. Archiv für mikroskopische Anatomie 3 441-476, 5 317-331

Asher L, Spiro K (1902-1907): see Monakow (1902, 1904, 1907)

Baer KE von (1828, 1837) Über Entwickelungsgeschichte der Thiere. Beobachtung und Reflexion. Vols. 1, 2. Bornträger, Königsberg

Bailey P, von Bonin G (1951) The isocortex of man. University of Illinois Press, Urbana

Baillarger J(GF) (1840) Recherches sur la structure de la couche corticale des circonvolutions du cerveau. Mémoires de l'Académie Royale de Médecine 8 149-183

Bechterew W von (1894) Die Leitungsbahnen in Gehirn und Rückenmark. Translated by J Weinburg. Besold, Leipzig

Berger H (1899) Beiträge zur feineren Anatomie der Grosshirnrinde. Monatschrift für Psychiatrie und Neurologie 6 405-420

Berger H (1900) Experimentell-anatomische Studien über die durch den Mangel optischer Reize veranlassten Entwicklunghemmungen im Occipitallappen des Hundes und der Katze. Archiv für Psychiatrie und Nervenkrankheiten 33 521-567

Berlin R (1858) Beitrag zur Structurlehre der Grosshirnwindungen. Inaugural dissertation. Junge, Erlangen

Bernheimer S (1900) Die corticalen Sehcentren. Wiener klinische Wochenschrift 13 955-963

Betz (W) (1874) Anatomischer Nachweis zweier Gehirncentra. Centralblatt für die medicinischen Wissenschaften 12 578-580, 595-599

Betz W (1881) Ueber die feinere Structur der Gehirnrinde des Menschen. Centralblatt für die medicinischen Wissenschaften 19 193-195, 209-213, 231-234

Bianchi L (1894) Ueber die Function der Stirnlappen. Berliner klinische Wochenschrift 31 309-310

Bianchi V (1904) Il mantello cerebrale del delfino (Delphinus Delphis). Annali di Nevrologia 22 521-542

Bianchi V (1905) Il mantello cerebrale del delfino (Delphinus Delphis). Accademia di scienze fisiche e matematiche di Napoli 17

Biedermann W (1895) Elektrophysiologie. Part 2. Fischer, Jena

Bielschowsky M (1905) Die histologische Seite der Neuronenlehre. Journal für Psychologie und Neurologie 5 128-150

Bielschowsky M, Brodmann K (1905) Zur feineren Histologie und Histopathologie der Grosshirnrinde mit besonderer Berücksichtigung der Dementia paralytica, Dementia senilis und Idiotie. Journal für Psychologie und Neurologie 5 173-199

Bolton JS (1900) The exact histological localisation of the visual area of the human cerebral cortex. Philosophical Transactions of the Royal Society of London, Series B 193 165-222

Bonin G von (1953) Alfred Walter Campbell. In "The Founders of Neurology", ed W Haymaker. Thomas, Springfield. pp 16-18

Bonin G von (1960) Some papers on the cerebral cortex. Thomas, Springfield. pp 201-230

Boulder Committee (1970) Embryonic vertebrate central nervous system: revised terminology. Anatomical Record. 166 257-261

Bourneville (DM) (1900) Idiotie et épilepsie symptomatiques de sclérose tubérose ou hypertrophique. Archives de Neurologie 10 29-39

Brodmann K (1903a) Beiträge zur histologischen Lokalisation der Grosshirnrinde. Erste Mitteilung: Die Regio Rolandica. Journal für Psychologie und Neurologie 2 79-107

Brodmann K (1903b) Beiträge zur histologischen Lokalisation der Grosshirnrinde. Zweite Mitteilung: Der Calcarinatypus. Journal für Psychologie und Neurologie 2 133-159

Brodmann (K) (1904a) Demonstrationen zur Cytoarchitektonik der Grosshirnrinde mit besonderer Berücksichtigung der histologischen Localisation bei einigen Säugethieren. (Proceedings of the German Psychiatry Association - Deutscher Verein für Psychiatrie - 25-27 April 1904 in Göttingen) Neurologisches Centralblatt 23 489

Brodmann K (1904b) Demonstrationen zur Cytoarchitektonik der Grosshirnrinde mit besonderer Berücksichtigung der histologischen Lokalisation bei einigen Säugetieren. (Proceedings of the German Psychiatry Association - Deutscher Verein für Psychiatrie - 25-27 April 1904 in Göttingen) Allgemeine Zeitschrift für Psychiatrie 61 765-767

Brodmann K (1905a) Beiträge zur histologischen Lokalisation der Grosshirnrinde. Dritte Mitteilung: Die Rindenfelder der niederen Affen. Journal für Psychologie und Neurologie 4 177-226

Brodmann K (1905b) Beiträge zur histologischen Lokalisation der Grosshirnrinde. Vierte Mitteilung: Die Riesenpyramidentypus und sein Verhalten zu den Furchen bei den Karnivoren. Journal für Psychologie und Neurologie 6 108-120

Brodmann K (1905c) Physiologische Differenzen der vorderen und hinteren Centralwindung. (Proceedings of the Berlin Society for Psychiatry and Neurology - Berliner Gesellschaft für Psychiatrie und Nervenkrankheiten - 6 December 1905) Neurologisches Centralblatt 24 1158-1160

Brodmann K (1906) Beiträge zur histologischen Lokalisation der Grosshirnrinde. Fünfte Mitteilung: Über den allgemeinen Bauplan des Cortex pallii bei den Mammalieren und zwei homologe Rindenfelder im besonderen. Zugleich ein Beitrag zur Furchenlehre. Journal für Psychologie und Neurologie 6 275-400

Brodmann K (1907) Bemerkungen über die Fibrillogenie und ihre Beziehungen zur Myelogenie mit besonderer Berücksichtigung des Cortex cerebri. Neurologisches Centralblatt 26 338-349

Brodmann K (1908a) Beiträge zur histologischen Lokalisation der Grosshirnrinde. VI. Mitteilung: Die Cortexgliederung des Menschen. Journal für Psychologie und Neurologie 10 231-246

Brodmann K (1908b) Beiträge zur histologischen Lokalisation der Grosshirnrinde. VII. Mitteilung: Die cytoarchitektonische Cortexgliederung der Halbaffen (Lemuriden). Journal für Psychologie und Neurologie 10 287-334

Brodmann (K) (1908c) Zur histologischen Lokalisation des Scheitellappens. (Proceedings of the Berlin Society for Psychiatry and Neurology - Berliner Gesellschaft für Psychiatrie und Nervenkrankheiten - 11 November 1907). Zentralblatt für Nervenheilkunde und Psychiatrie 19 26-28

Brodmann K (1908d) Über den gegenwärtigen Stand der histologischen Lokalisation der Grosshirnrinde. (Proceedings of the German Psychiatry Association - Deutscher Verein für Psychiatrie - 24-25 April in Berlin). Zentralblatt für Nervenheilkunde und Psychiatrie 19 695-696

Brodmann K (1908e) Über Rindenmessungen. Zentralblatt für Nervenheilkunde und Psychiatrie 31 781-798

Brodmann K (1909a) Über Rindenmessungen. Neurologisches Centralblatt 28 129, 635-639

Brodmann (K) (1909b) (Discussion on aphasia, Proceedings of the Berlin Society for Psychiatry and Neurology - Berliner Gesellschaft für Psychiatrie und Nervenkrankheiten - 14 June 1909) Neurologisches Centralblatt 1909 28 720-722

Brodmann, K (1913) Neue Forschungsergebnisse der Grosshirnrindenanatomie mit besonderer Berücksichtigung anthropologischer Fragen. Gesellschaft deutscher Naturforscher und Ärtze 85 200-240 (Translated by Elston and Garey (2004) qv)

Bronn HG (1858) Morphologische Studien über die Gestaltungsgesetze der Naturkörper überhaupt, und der organischen insbesondere. Winter, Leipzig

Brouwer B (1909) Over doofstomheid en de acustische banen. Dissertation, Amsterdam

Cajal S Ramón y (1893) Beiträge zu feineren Anatomie des grossen Hirns. II: Ueber den Bau der Rinde des unteren Hinterhauptlappens der kleinen Säugethiere. Zeitschrift für wissenschaftliche Zoologie 56 664-672

Cajal S Ramón y (1900-1906) Studien über die Hirnrinde des Menschen. JA Barth, Leipzig. Translated by J Bresler.
Volume 1. Die Sehrinde, 1900.
Volume 2. Die Bewegungsrinde, 1900.
Volume 3. Die Hörrinde, 1902.
Volume 4. Die Riechrinde beim Menschen und Säugetier, 1903.
Volume 5. Vergleichende Strukturbeschreibung und Histogenesis der Hirnrinde, 1906.

Campbell AW (1905) Histological studies on the localisation of cerebral function. Cambridge University Press

Carus JV (1853) System der thierischen Morpologie. Englemann, Leipzig

Changeux JP (1985) Neuronal man. The biology of mind. Translated by L Garey. Pantheon, New York

Charcot JM, Marie P (1885) Deux nouveaux cas de sclérose latérale amyotrophique suivis d'autopsie. Archives de Neurologie 10 1-35, 168-186

Christiani A (1885) Zur Physiologie des Gehirnes. Enslin, Berlin

Claus C (1891) Lehrbuch der Zoologie. Elwert, Marburg

Cramer A (1898) Beitrag zur Kenntnis der Optikuskreutzung im Chiasma und des Verhaltens der optischen Centren bei einseitiger Bulbusatrophie. Anatomische Hefte (Section 1, Arbeiten des anatomischen Institut Wiesbaden) 10 415-484

Cuvier G (1800) Leçons d'anatomie comparée. Vol. 1. Baudoin, Paris

Czylharz E von, Marburg O (1901) Beitrag zur Histologie und Pathogenese der amyotrophischen Lateralsclerose. Zeitschrift für klinische Medizin 43 59-74

Doinikow B (1908) Beitrag zur vergleichenden Histologie des Ammonshorns. Journal für Psychologie und Neurologie 13 166-202

Eberstaller O (1890) Das Stirnhirn. Ein Beitrag zur Anatomie der Oberfläche des Grosshirns. Urban & Schwarzenberg, Vienna & Leipzig

Economo C von, Koskinas GN (1925) Die Cytoarchitektonik der Hirnrinde des erwachsenen Menschen. Springer, Vienna, Berlin

Edinger L (1866) Vorlesungen über den Bau der nervösen Zentralorgane des Menschen und der Tiere. 6th edition. Vogel, Leipzig
Edinger L, Wallenberg A (1907) Bericht über die Leistungen auf dem Gebiete der Anatomie des Centralnervensystems in den Jahren 1905 und 1906. Schmidt's "Jahrbücher der in- und ausländischen Medicin" 295 1-33, 113-152
Edinger L, Wallenberg A (1907, 1909) Bericht über die Leistungen auf dem Gebiete der Anatomie des Centralnervensystems im Laufe der Jahre 1907 und 1908. Schmidt's "Jahrbücher der in- und ausländischen Medicin" 303 113-151, 225-231
Eimer GHT (1897) Orthogenesis der Schmetterlinge. Engelmann, Leipzig
Elston GN, Garey LJ (2004) New research findings on the anatomy of the cerebral cortex of special relevance to anthropological questions. University of Queensland Printery, Brisbane
Ewens GFW (1893) A theory of cortical visual representation. Brain 16 475-491
Exner S (1881) Untersuchungen über die Localisation der Functionen in der Grosshirnrinde des Menschen. Braumüller, Vienna
Exner S (1894) Entwurf zu einer physiologischen Erklärung der psychischen Erscheinungen. Deuticke, Vienna
Farrar CB (1903) On the motor cortex. American Journal of Insanity 59 477-514
Ferrier D (1890) The Croonian Lectures on cerebral localisation. Smith, Elder, London
Flechsig PE (1894) Gehirn und Seele, Rectoral address, Leipzig
Flechsig PE (1896) Gehirn und Seele, Rectoral address, 2nd edition, Veit, Leipzig
Flechsig P (1898) Neue Untersuchungen über die Markbildung in den menschlichen Grosshirnlappen. Neurologisches Centralblatt 17 977-996
Flechsig P (1904) Einige Bemerkungen über die Untersuchungsmethoden der Grosshirnrinde, insbesondere des Menschen. Berichten der königlichen sächsischen Gesellschaft der Wissenschaften, mathematisch-physikalsche Klasse 56 177-214
Flechsig P (1905) Einige Bemerkungen über die Untersuchungsmethoden der Grosshirnrinde, insbesondere des Menschen. Archiv für Anatomie und Entwickelungsgeschichte 337-444
Flechsig P (1908) Bemerkungen über die Hörsphäre des menschlichen Gehirns. Neurologisches Centralblatt 27 2-7, 50-57
Flechsig P (1920) Anatomie des menschlichen Gehirns und Rückenmarks auf myelogenetischer Grundlage. Thieme, Leipzig
Flourens P (1842) Recherches expérimentales sur les propriétés et les fonctions du système nerveux dans les animaux vertébrés. 2nd edition. Baillière, Paris
Foerster O (1926) Die Pathogenese des epileptischen Krampfanfalles. Deutsche Zeitschrift für Nervenheilkunde 94 15-53
Forster E (1908) Experimentelle Beiträge zur Lehre der Phagozytose der Hirnrindelemente. In: Histologische und histopathologische Arbeiten über die Grosshirnrinde, ed F Nissl Vol. 2. pp.173-192. Fischer, Jena
Förster (R) (1890) Ueber Rindenblindheit. Archiv für Ophthalmologie 36 94-108

Friedenthal H (1909) Ueber das Wachstum des menschlichen Körpergewichtes in den verschiedenen Lebensaltern und über dir Volumenmessung von Lebewesen. Medizinische Klinik 5 700-703

Fritsch G, Hitzig E (1870) Ueber die elektrische Erregbarkeit des Grosshirns. Archiv für Anatomie, Physiologie und wissenschaftliche Medicin. 300-332

Fürbringer M (1887) Untersuchungen zur Morphologie und Systematik der Vögel. Allgemeine Teil. Amsterdam

Fürstner (C) (1882) Weitere Mitteilungen über den Einfluss einseitiger Bulbuszerstörung auf die Entwicklung der Hirnhemisphären. Archiv für Psychiatrie und Nervenkrankheiten 12 611-615

Gall FJ (1822-1825) Sur les functions du cerveau et sur celles de chacune de ses parties. Vols. 1-6. Baillière, Paris

Gaskell WH (1886?) On the structure, distribution and function of the nerves which innervate the visceral and vascular systems. Journal of Physiology 7 1-80

Gegenbaur C (1898) Vergleichende Anatomie der Wirbelthiere mit Berücksichtigung der Wirbellosen. Vol. 1. Engelmann, Leipzig

Golgi C (1886) Sulla fina anatomia degli organi centrali del sistema nervoso. Hoepli, Milan

Goltz FC (1881) Ueber die Verrichtungen des Grosshirns. Strauss, Bonn

Gowers WR (1892-1893) A manual of diseases of the nervous system. 2nd edition. Churchill, London

Grashey H (1889) Bernhard von Guddens gesammelte und hinterlassene Abhandlungen. Bergmann, Wiesbaden

Grünbaum ASF, Sherrington CS (1901) Observations on the physiology of the cerebral cortex of some of the higher apes. Proceedings of the Royal Society of London 69 206-208

Gudden B von (1886) Über die Frage der Localisation der Functionen der Grosshirnrinde. Allgemeine Zeitschrift für Psychiatrie 42 478-499

Haeckel EH (1866) Generelle Morphologie der Organismen. Vol. I: Allgemeine Anatomie, Vol. II: Allgemeine Entwickelungs-Geschichte. Reimer, Berlin

Haeckel E (1898) The last link. Our present knowledge of the descent of man. Black, London.

Haeckel E, Gesammelte populäre Vorträge aus dem Gebiet der Entwicklungslehre. Part 1. p99ff. Über Arbeitsteilung in Natur und Menschenleben

Haller B (1908) Die phyletische Entfaltung der Grosshirnrinde. Archiv für mikroskopische Anatomie und Entwicklungsgeschichte 71 350-466

Hammarberg C (1895) Studien über Klinik und Pathologie der Idiotie, nebst Untersuchungen über die normale Anatomie der Hirnrinde. Berling, Akademische Buchdruckerie, Upsala

Hassler R (1962) Die Entwicklung der Architektonik seit Brodmann und ihre Bedeutung für die moderne Hirnforschung. Deutsche Medizinische Wochenschrift 87 1180-1185

Henneberg R (1910) Messung der Oberflächenausdehnung der Grosshirnrinde. Journal für Psychologie und Neurologie 17 144-158

Henschen SE (1890-1911) Klinische und anatomische Beiträge zur Pathologie des Gehirns. Vol. 1-4. Almqvist & Wiksell, Upsala

Henschen SE (1893) On the visual path and centre. Brain 16 170-180

Henschen SE (1903) La projection de la rétine sur la corticalité calcarine. Semaine Médical 23 125-127

Hermanides SR, Köppen M (1903) Ueber die Furchen und über den Bau der Grosshirnrinde bei den Lissencephalen insbesondere über die Localisation des motorischen Centrums und der Sehregion. Archiv für Psychiatrie und Nervenkrankheiten 37 616-634

Hertwig O (1906) Lehrbuch der Entwicklungsgeschichte des Menschen und der Wirbeltiere. Fischer, Jena (3rd edition 1890)

Hertwig R (1900) Lehrbuch der Zoologie. Fischer, Jena

Hertwig R (1909) Der Kampf um Grundfragen der Biologie

Heschl RL (1878) Ueber die vordere quere Schläfenwindung des menschlichen Grosshirns. Braumüller, Vienna

His W (1904) Die Entwickelung des menschlichen Gehirns während der ersten Monate. Hirzel, Leipzig

Hitzig E (1904) Physiologische und klinische Untersuchungen über das Gehirn. Gesammelte Abhandlungen. Parts 1 and 2. Hirschwald, Berlin

Holmes G (1908) A note on the condition of the postcentral cortex in tabes dorsalis. Review of Neurology and Psychiatry 6 5-11

Horsley V, Schäfer EA (1888) A record of experiments upon the functions of the cerebral cortex. Philosophical Transactions of the Royal Society of London (B.) 179 1-45

Hubel DH, Wiesel TN (1962) Receptive fields, binocular interaction and functional architecture in the cat's visual cortex. Journal of Physiology 160 106-154

Hubel DH, Wiesel TN (1977) Functional architecture of macaque monkey visual cortex. Proceedings of the Royal Society of London Series B 198 1-59

Huxley TH (1863) Evidence as to man's place in nature. Williams and Norgate, London

Jackson J Hughlings (1890) The Lumleian lectures on convulsive seizures. Lancet 1 735-738

Jendrássik E (1897) Ueber Paralysis spastica, - und über die vererbten Nervenkrankheiten in Allgemeinen. Deutsches Archiv für klinische Medecin 58 137-164

Jendrássik E (1898) Zweite Beitrag zur Lehre von den vererbten Nervenkrankheiten. Deutsches Archiv für klinische Medecin 61 187-205

Kaes T (1893) Beiträge zur Kenntnis des Reichtums der Grosshirnrinde des Menschens an markhaltigen Nervenfasern. Archiv für Psychiatrie und Nervenkrankheiten 25 695-758

Kaes T (1907) Die Grosshirnrinde des Menschen in ihren Massen und in ihrem Fasergehalt. Ein gehirnanatomischer Atlas. Fischer, Jena

Kalischer O (1909) Weitere Mitteilungen über die Ergebnisse der Dressur als physiologischer Untersuchungsmethode auf dem Gebiete des Gehör-, Geruchs- und Farbensinnes. Archiv für Anatomie und Physiolgie 303-322?

Kappers CU Ariëns (1909) The phylogenesis of the palaeo-cortex and archicortex, compared with the evolution of the visual neo-cortex. Archives of Neurology and Psychiatry 4 161-173

Kappers CU Ariëns, Theunissen WF (1907) Zur vergleichenden Anatomie des Vorderhirnes der Vertebraten. Anatomischer Anzeiger 30 496-509

Kappers CU Ariëns, Theunissen WF (1908) Die Phylogenese des Rhinencephalons, des Corpus striatum und der Vorderhirnkommissuren. Folia Neuro-biologica 1 173-288

Kleist K (1934) Gehirnpathologie vornehmlich auf Grund der Kriegserfahrungen. In O von Schjerning, Handbuch der aerztlichen Erfahrungen im Weltkriege 1914-18, vol. 4, p.343 Barth, Leipzig

Knauer A (1909) Zur Pathologie des linken Schläfenlappens. Klinik für psychische und nervöse Krankheiten 4 115-194

Kohnstamm O (1898) Zur Anatomie und Physiologie des Phrenikuskernes. Fortschritte der Medicin 16 643-653

Kohnstamm (O) (1899) Ueber Ursprungskerne spinaler Bahnen im Hirnstamm, speciell über das Athemcentrum. Archiv für Psychiatrie und Nervenkrankheiten 32 681-684

Kohnstamm O (1910) Studien zur physiologischen Anatomie des Hirnstammes. Journal für Psychologie und Neurologie 17 33-57

Kojewnikoff A (1883) Cas de sclérose latérale amyotrophique, la dégéneresence des faisceaux pyramidaux se propageant à travers tout l'encéphale. Archives de Neurologie 5 356-376

Kolmer W (1901) Beitrag zur Kenntniss der "motorischen" Hirnrindenregion. Archiv für mikroskopische Anatomie und Entwicklungsgeschichte 57 151-183

Kölpin-Andernach (1908) Zur pathologischen Anatomie der Huntingtonschen Chorea. Journal für Psychologie und Neurologie 12 57-68

Köppen M, Loewenstein S (1905) Studien über den Zellenbau der Grosshirnrinde bei den Ungulaten und Carnivoren und über die Bedeutung einiger Furchen. Monatsschrift für Psychiatrie und Neurologie 18 481-509

Krieg WJS (1963) Connections of the cerebral cortex. Brain Books, Evanston

Lashley KS, Clark G (1946) The cytoarchitecture of the cerebral cortex of ateles: a critical examination of architectonic studies. Journal of Comparative Neurology 85 223-305

Leonowa O von (1893) Ueber das Verhalten der Neuroblasten des Occipitallappens bei Anophthalmie und Bulbusatrophie und seine Beziehungen zum Sehact. Archiv für Anatomie und Physiologie 308-318

Leonowa O von (1896) Beiträge zur Kenntniss der secundären Veränderungen der primären optischen Centren und Bahnen in Fällen von congenitaler Anophthalmie und Bulbusatrophie bei neugeborenen Kindern. Archiv für Psychiatrie und Nervenkrankheiten 28 53-96

Lewandowsky M (1907) Die Funktionen des zentralen Nervensystems: ein Lehrbuch. Fischer, Jena

Lewis Bevan (W) (1878) On the comparative structure of the cortex cerebri. Brain 1 79-96

Lewis Bevan W (1881) Researches on the comparative structure of the cortex cerebri. Philosophical Transactions of the Royal Society of London 171 35-64

Lewis Bevan W (1889) A text-book of mental diseases: with special reference to the pathological aspects of insanity. Griffin, London

Lewis Bevan (W), Clarke H (1878) The cortical lamination of the motor area of the brain. Proceedings of the Royal Society of London 27 38-49

Liepmann H (1909) Zum Stande der Aphasiefrage. Neurologisches Centralblatt 28 449-484

Loeb J (1884) Die Sehstörungen nach Verletzung der Grosshirnrinde. Archiv für die gesammte Physiologie 34 67-172

Loeb J (1886) Beiträge zur Physiologie des Grosshirns. Archiv für die gesammte Physiologie 39 265-346

Luciani L (1884) (Translated as:) On the sensorial localisations in the cortex cerebri. Brain 7 145-160

Luciani L, Seppilli G (1886) Die Functions-Localisation auf der Grosshirnrinde an Thierexperimenten und klinischen Fällen nachgewiesen. German edition by O Fränkel. Denickes, Leipzig

Luciani L, Tamburini A (1879) Richerche sperimentale sulla funzioni del cervello. Rivista sperimentale di freniatria. 5 1-76

Lussana F, Lemoigne A (1871) Fisiologia dei centri nervosi encefalici. Properini, Padua

Major HC (1876a) The histology of the island of Reil. The West Riding Lunatic Asylum Medical Reports 6 1-10

Major HC (1876b) Observations on the brain of the Chacma baboon. Journal of Mental Science 21 498-512

Marburg O (1907) Beiträge zur Kenntnis der Grosshirnrinde der Affen. Arbeiten aus dem neurologischen Institute an der Wiener Universität 16 581-602

Marie P (1888) De l'aphasie en général et de l'agraphie en particulier. Le progrès médical 7 81-84

Mauss T (1908) Die faserarchitektonische Gliederung der Grosshirnrinde bei den niederen Affen. Journal für Psychologie und Neurologie 13 263-325

Mauss T (1911) Die faserarchitektonische Gliederung des Cortex cerebri der antropomorphen Affen. Journal für Psychologie und Neurologie 18 410-467

Merzbacher L (1910) Untersuchungen über die Morphologie und Biologie der Abräumzellen im Zentralnervensystem. In: Histologische und histopathologische Arbeiten über die Grosshirnrinde, ed F Nissl Vol. 3. pp.1-142. Fischer, Jena

Meynert T (1867a) Der Bau der Gross-Hirnrinde und seine örtlichen Verschiedenheiten, nebst einem pathologisch-anatomischen Corollarium. Vierteljahresschrift für Psychiatrie 1 77-93

Meynert T (1867b) Das Gesammtgewicht und die Theilgewichte des Gehirnes in ihren Beziehungen zum Geschlechte, dem Lebensalter und dem Irrsinn, untersucht nach einer neuren Wägungsmethode an den Gehirnen der in der Wiener Irrenanstalt im Jahre 1866 Verstorbenen. Vierteljahresschrift für Psychiatrie 1 126-170

Meynert T (1867c) Der Bau der Grosshirnrinde und seine örtlichen Verschiedenheiten, nebst einem pathologisch-anatomischen Corollarium. Vierteljahresschrift für Psychiatrie 1 198-217

Meynert T (1868a) Studien über das pathologisch-anatomische Material der Wiener Irrenanstalt. Vierteljahresschrift für Psychiatrie, 1 381-402

Meynert T (1868b) Der Bau der Gross-Hirnrinde und seine örtlichen Verschiedenheiten, nebst einem pathologisch-anatomischen Corollarium. Vierteljahresschrift für Psychiatrie 2 88-113

Meynert T (1872) Vom Gehirne der Säugetiere. In: Handbuch der Lehre von den Geweben des Menschen und der Thiere, ed. S Stricker, Vol. 2, Chapter 31, pp.694-808. Engelmann, Leipzig

Mierzejewsky J (1875) Etudes sur les lésions cérébrales dans la paralysie générale. Archives de Physiologie 7 195-235

Möbius PJ (1905) Franz Joseph Gall, in Möbius PJ, Ausgewählte Werke, vol. 7. Barth, Leipzig

Moeli C (1891) Veränderungen des Tractus und Nervus opticus bei Erkrankungen des Occipitalhirns. Archiv für Psychiatrie und Nervenkrankheiten 22 73-120

Monokow C von (1882) Ueber einige durch Exstirpation circumscripter Hirnrindenregionen bedingte Entwickelungshemmungen des Kaninchengehirns. Archiv für Psychiatrie und Nervenkrankheiten 12 141-156

Monokow (C) von (1883) Experimentelle und pathologisch-anatomische Untersuchungen über die Beziehung der sogenannten Sehsphäre zu den infracorticalen Opticuscentren und zum N. opticus. Archiv für Psychiatrie und Nervenkrankheiten 14 699-751

Monokow C von (1902, 1904, 1907) Über den gegenwärtigen Stand der Frage nach der Lokalisation im Grosshirn. Ergebnisse der Physiologie 1 534-665, 3 100-122, 6 334-605

Monokow C von (1905) Gehirnpathologie, 2nd edition. In H Nothnagel, Spezielle Pathologie und Therapie, vol. 9. Hölder, Vienna

Monakow (C) von (1909) Neue Gesichtspunkte in der Frage nach der Lokalisation im Grosshirn. Correspondenz-blatt für Schweizer Aerzte 39 401-415, 558-560

Mott FW (1895) A case of amyotrophic lateral sclerosis with degeneration of the motor path from the cortex to the periphery. Brain 18 21-36

Mott FW (1907) The progressive evolution of the structure and functions of the visual cortex in mammalia. Archives of Neurology of the Pathological Laboratory of the London County Asylums 3 1-48

Mott FW, Halliburton WD (1908) Localisation of function in the lemur's brain. Proceedings of the Royal Society of London Series B 80 136-147

Mott FW, Kelley AM (1908) Complete survey of the cell lamination of the cerebral cortex of the lemur. Proceedings of the Royal Society of London Series B 80 488-506

Mott FW, Schaefer EA (1890) On associated eye-movements produced by cortical faradization of the monkey's brain. Brain 13 165-173

Munk H (1890) Über die Functionen der Grosshirnrinde. Gesammelte Mittheilungen mit Anmerkungen. 2nd edition. Hirschwald, Berlin

Naegeli C von (1865) Entstehung und Begriff der naturhistorischen Art. Munich

Niessl von Mayendorf (1908) Über die physiologische Bedeutung der Hörwindung. Neurologisches Centralblatt 27 545-546

Nissl F (1898) Die Hypothese der specifischen Nervenzellenfunction. Allgemeine Zeitschrift für Psychiatrie 54 1-107

Nissl F (1903) Zum gegenwärtigen Stande der pathologischen Anatomie des zentralen Nervensystems. Zentralblatt für Nervenheilkunde und Psychiatrie 26 517-528

Nissl F (1904a) Zur Histopathologie der paralytischen Rindenerkrankungen. In: Histologische und histopathologische Arbeiten über die Grosshirnrinde, ed F Nissl Vol. 1. pp.315-492. Fischer, Jena

Nissl F (1904b, 1908) Histologische und histopathologische Arbeiten über die Grosshirnrinde. Vol. 1, 2

Otsuka, R Hassler R (1962) Über Aufbau und Gliederung der corticalen Sehsphäre bei der Katze. Archiv für Psychiatrie und Nervenkrankheiten 203 212-234

Peters A, Jones EG (1984) Cerebral Cortex. Vol. 1. Plenum, New York

Powell TPS, Mountcastle VB (1959) Some aspects of the functional organization of the cortex of the postcentral gyrus of the monkey: a correlation of findings obtained in a single unit analysis with cytoarchitecture. Bulletin of the Johns Hopkins Hospital 105 133-162

Probst M (1898) Zu den fortschreitenden Erkrankungen der motorischen Leitungsbahnen. Archiv für Psychiatrie und Nervenkrankheiten 30 766-844

Probst M (1901) Ueber das Gehirn der Taubstummen. Archiv für Psychiatrie und Nervenkrankheiten 34 584-590

Probst M (1903) Zur Kenntnis der Grosshirnfaserung und der cerebralen Hemiplegie. Sitzungsberichte der kaiserlichen Akademie der Wissenschaften. Mathematisch-naturwissenschaftliche Klasse 112 581-682

Probst M (1903) Zur Kenntnis der amyotrophischen Lateralsklerose. Sitzungsberichte der kaiserlichen Akademie der Wissenschaften. Mathematisch-naturwissenschaftliche Klasse 112 683-824

Ranke O (1908) Beiträge zur Lehre von der Meningitis tuberculosa. In: Histologische und histopathologische Arbeiten über die Grosshirnrinde, ed F Nissl Vol. 2. pp. 252-347. Fischer, Jena

Retzius G (1896) Das Menschenhirn. Studien in der makroskopischen Morphologie. Norstedt, Stockholm

Retzius G (1898) Zur äusseren Morphologie des Riechhirns der Säugerthiere und des Menschen. Biologische Untersuchungen 8 23-48

Roncoroni L (1896) La fine morfologia del cervello degli epilettici. Archivio di Psichitria 17 92-116

Roncoroni L (1909) Sul tipo fondamentale di stratificazione della corteccia cerebrale. Anatomischer Anzeiger 34 58-62

Rosenberg L (1908) Ueber die Cytoarchitektonik der ersten Schläfenwindung und der Heschlschen Windungen. Monatschrift für Psychiatrie und Neurologie 23 52-68

Rossi I, Roussy G (1906) Un cas de sclérose latérale amyotrophique avec dégénération de la voie pyramidale suivie au Marchi de la moelle jusqu'au cortex. Revue neurologique 14 393-406

Rossi I, Roussy G (1907) Contribution anatomo-pathologique à l'étude des localisations motrices corticales. A propos de trois cas de sclérose latérale amyotrophique avec dégénération de la voie pyramidale suivie au Marchi de la moelle au cortex. Revue neurologique 15 785-810

Sarkissov SA, Filimonoff IN, Kononowa EP, Preobraschenskaja IS, Kukuew LA (1955) Atlas of the cytoarchitectonics of the human cerebral cortex. Medgiz, Moscow

Schaffer K (1909) Über die Anatomie und Klinik der Tay-Sachs'schen amaurotisch-familiären Idiotie mit Rücksicht auf verwandte Formen. Zeitschrift für die Erforschung und Behandlung des jugendlichen Schwachsinns 2 19-73

Schiff JM (1858-1859) Lehrbuch der Physiologie des Menschen. Lahr

Schlapp M (1898) Der Zellenbau der Grosshirnrinde des Affen Macacus Cynomolgus. Archiv für Psychiatrie und Nervenkrankheiten 30 583-607

Schlapp MG (1902) The microscopic structure of cortical areas in man and some mammals. American Journal of Anatomy 2 259-281

Schoenemann PT, Sheehan M J, Glotzer LD (2005) Prefrontal white matter volume is disproportionately larger in humans than in other primates. Nature Neuroscience 8 242-252

Schröder P (1908a) Einführung in die Histologie und Histopathologie des Nervensystems. Fischer, Jena

Schröder P (1908b) Zur Lehre von der akuten hämorrhagischen Poliencephalitis superior (Wernicke). In: Histologische und histopathologische Arbeiten über die Grosshirnrinde, ed F Nissl Vol. 2. pp.145-172. Fischer, Jena

Schwalbe G (1881) Lehrbuch der Neurologie, in Lehrbuch der Anatomie des Menschen, CEE Hoffmann, Vol. 2. Erlangen

Schwalbe G (1882) Ueber die Kaliberverhältnisse der Nervenfasern, Vogel, Leipzig

Semendeferi K, Lu A, Schenker N, Damasio, H (2002) Humans and great apes share a large frontal cortex. Nature Neuroscience 5, 272-276

Semon RW (1904) Die Mneme als erhaltendes Prinzip im Wechsel des organischen Geschehens. Engelmann, Leipzig

Sherrington CS (1906) The integrative action of the nervous system. Constable, London

Smith G Elliot (1901) Notes upon the natural subdivision of the cerebral hemisphere. Journal of Anatomy and Physiology 35 431-454

Smith G Elliot (1904a) Studies in the morphology of the human brain, with special reference to that of the Egyptians. No. 1. The occipital region. Records of the Egyptian Government School of Medicine 2 123-173

Smith G Elliot (1904b) The so-called "Affenspalte" in the human (Egyptian) brain. Anatomischer Anzeiger 24 74-83

Smith G Elliot (1904c) The morphology of the occipital region of the cerebral hemisphere in man and the apes, Anatomischer Anzeiger 24 436-451

Smith G Elliot (1907a) New studies on the folding of the visual cortex and the significance of the occipital sulci in the human brain. Journal of Anatomy and Physiology 41 198-207

Smith G Elliot (1907b) A new topographical survey of the human cerebral cortex, being an account of the distribution of the anatomically distinct cortical areas and their relationship to the cerebral sulci. Journal of Anatomy and Physiology 41 237-254

Spielmayer W (1907) Klinische und anatomische Untersuchungen über eine besondere Form von familiärer amaurotischer Idiotie. Gotha

Spielmayer W (1908) Klinische und anatomische Untersuchungen über eine besondere Form von familiärer amaurotischer Idiotie. In: Histologische und histopathologische Arbeiten über die Grosshirnrinde, ed F Nissl Vol. 2. pp.193-251. Fischer, Jena

Spielmayer W (1924) Korbinian Brodmann, 1868-1918, Lebenslauf. In "Deutsche Irrenärzte". Ed T Kirchhoff. Vol 2. Springer, Berlin
Spiller WG (1900) A case of amyotrophic lateral sclerosis in which degeneration was traced from the cerebral cortex to the muscles. Contributions from the William Pepper Laboratory of Clinical Medicine 4 63-106
Strohmayer W (1901) Anatomische Untersuchung der Höhsphäre beim Menschen. Monatsschrift für Psychiatrie und Neurologie 10 172-185
Taalman Kip MJ van Erp (1905a) De phylogenie van de cortex cerebri. Verhandelingen van het negen-de Vlaamsche Congres voor Natuur- en Geneeskunde
Taalman Kip MJ van Erp (1905b) Over den bouw van den cortex cerebri bij mol en egel. Psychiatrische en Neurologische Bladen 9 113-119
Tonnini S (1898) I fenomeni residuali e la loro natura psichica nelle relative localizzazioni dirette e comparate, in rapporto con le diverse mutilazioni corticali del cane. Rivista sperimentale di freniatrii 24 700-744
Vogt Carl (1863) Vorlesungen über den Menschen: seine Stellung in der Schöpfung und in der Geschichte der Erde. Giessen
Vogt Cécile (1900) Etude sur la myélinisation des hémisphères cérébraux. Steinheil, Paris
Vogt Cécile, Vogt O (1906) Zur Kenntnis der elektrisch erregbaren Hirnrindengebiete bei den Säugetieren. Journal für Psychologie und Neurologie 8 277-456
Vogt H (1905a) Über die Anatomie, das Wesen und die Entstehung mikrocephaler Missbildungen. Arbeiten aus dem hirnanatomischen Institut in Zürich 1 1-203
Vogt H (1905b) Über familiäre amaurotische Idiotie und verwandte Krankheitsbilder. Monatschrift für Psychiatrie und Neurologie 18 161-171, 310-357
Vogt H (1907) Zur Pathologie und pathologischen Anatomie der verschiedenen Idiotie-Formen. Monatschrift für Psychiatrie und Neurologie 22 403-418, 490-508
Vogt H, Rondoni P (1908) Zum Aufbau der Hirnrinde. Deutsche medizinische Wochenschrift, 34, 1886-1887
Vogt O (1903) Zur anatomischen Gliederung des Cortex cerebri. Journal für Psychologie und Neurologie 2 160-180
Vogt O (1906a) Der Wert der myelogenetischen Felder der Grosshirnrinde (Cortex pallii). Anatomischer Anzeiger 29 273-287
Vogt O (1906) Ueber strukturelle Hirncentra, mit besonderer Berücksichtigung der strukturellen Felder des Cortex pallii. Proceedings of the Anatomische Gesellschaft, 1-5.6.06. Anatomischer Anzeiger, Ergänzungsheft, 29 74-114
Vogt O (1919) Korbinian Brodmann. Journal für Psychologie und Neurologie 24 I-X
Vogt O (1959) Korbinian Brodmann, Lebenslauf. In "Grosse Nervenärzte" Vol 2, pp 40-44
Wagner H (1864) Massbestimmungen der Oberfläche des grossen Gehirns. Inaugural dissertation, Göttingen
Waldschmidt J (1887) Beitrag zur Anatomie des Taubstummengehirns. Zeitschrift für Psychiatrie 43 373-379

Watson GA (1907) The mammalian cerebral cortex, with special reference to its comparative histology. I.- Order Insectivora. Archives of Neurology of the Pathological Laboratory of the London County Asylums 3 49-122

Wiesel TN, Hubel DH (1963) Single-cell responses in striate cortex of kittens deprived of vision in one eye. Journal of Neurophysiology 26 1003-1017

Wundt W (1880) Grundzüge der physiologischen Psychologie. 2nd edition. Vol 1. Engelmann, Leipzig

Ziehen GT Das Centralnervensystem der Monotremen und Marsupialier. II Teil: Mikroskopische Anatomie. II. Abschnitt: Der Faserverlauf im Gehirn von Echidna und Ornithorhynchus. (From Semon, Zoologische Forschungsreisen in Australien III, Volume II, 2, 1909.) Denkschriften der medizinisch-naturwissenschaftlichen Gesellschaft Jena 1901 6 Part 1: 1-188; Part 2: 677-728 Fischer, Jena Vol 3 1, 229, 677, 789

Zilles K (1990) Cortex. In "The human nervous system" Ed G Paxinos. Academic Press, San Diego. pp 757-802

Zuckerkandl E (1887) Über das Riechzentrum. Eine vergleichend-anatomische Studie. Enke, Stuttgart

Zunino G (1909) Die myeloarchitektonische Differenzierung der Grosshirnrinde beim Kaninchen (Lepus cuniculus). Journal für Psychologie und Neurologie 14 38-70

Glossary of Species Names

Glossary of species names as used by Brodmann, with their modern equivalent in brackets, and their common English name

(Letters in brackets in some of Brodmann's species names indicate variations in his spelling)

Anthropopithecus troglodytes (*Pan troglodytes*) chimpanzee
Ateles ater (*Ateles paniscus*) spider monkey
Bradypus tridactylus (*Bradypus tridactylus*) three-toed sloth
Canis familiaris (*Canis familiaris*) dog
Canis lupus (*Canis lupus*) wolf
Canis vulpes (*Vulpes vulpes*) fox
Capra hircus (*Capra hircus*) goat
Cebus capucinus (*Cebus capucinus*) capuchin monkey
Centetes ecaudatus (*Tenrec ecaudatus*) tail-less tenrec
Cercocebus fulginosus (*Cercocebus torquatus*) mangabey
Cercoleptes caudivolvulus (*Potos flavus*) kinkajou
Cercopithecus fulginosus, probably a mistake for *Cercocebus fulginosus* qv
Cercopithecus mona (*Cercopithecus mona*) mona guenon
Chrysochloris (*Chrysochloris*) golden mole
Ctenomys (*Ctenomys*) tucu tucu

Cynocephalus mormon (?*Papio cynocephalus*, yellow baboon, or *Papio leucophaeus*, mormon drill)
Dicotyles torquatus (*Tayassu tajacu*) collared peccary
Didelphys marsupialis (*Didelphis marsupialis*) opossum
Echidna aculeata (*Tachyglossus aculeatus*) echidna
Elephas africanus (*Loxodonta africana*) African elephant
Erinaceus europaeus (*Erinaceus europaeus*) hedgehog
Felis domestica (*Felis catus*) domestic cat
Felis leo (*Panthera leo*) lion
Felis tigris (*Panthera tigris*) tiger
Hapale jacchus (*Callithrix jacchus*) common marmoset
Hapale pen(n)icil(l)ata (*Callithrix pennicillata*) black-eared marmoset
Hapale ursula (*Saguinus midas*) negro tamarin
Herpestes griseus (*Herpestes edwardsi* = grey mongoose) mongoose
Homo (*Homo sapiens*) man
Hypsiprymnus (*Hypsiprymnodon moschatus*) musk kangaroo
Hyrax capensis (*Procavia capiensis*) rock hyrax
Indris brevicaudatus (*Indri indri*) indris
Lagothrix lagothrica (*Lagothrix lagotricha*) woolly monkey
Lemur macaco (*Lemur macaco*) black lemur
Lepus cuniculus (*Oryctolagus cuniculus*) rabbit
Macacus rhesus (*Macaca mulatta*) rhesus macaque
Macropus dorsalis (*Macropus dorsalis*) black-striped wallaby
Macropus pennicillatus (?*Petrogale penicillata*, rock wallaby)
Macropus rufus (*Macropus rufus*) red kangaroo
Microcebus minimus (*Microcebus murinus*) mouse lemur
Mus musculus (*Mus musculus*) mouse
Mus rattus (*Rattus norvegicus*) rat
Mustela foina (*Martes foina*) stone marten
Nycticebus tardigradus (*Nycticebus coucang*) slow loris
Onychogale frenata (*Onychogalea fraenata*) bridled nail-tailed wallaby
Paradoxurus hermaphrodytus (*Paradoxurus hermaphroditus*) civet
Phalangista vulpina (*Trichosurus vulpecula*) possum
Phoca vitulina (*Phoca vitulina*) common seal
Pithecia satanas (*Chiropotes satanas*) black saki
Propithecus coronatus (*Propithecus verreauxi*) crowned or Verreaux's sifaka
Pterodicticus potto, spelling mistake for *Perodicticus potto*, potto
Pteropus edulis (*Pteropus vampyrus*) large flying fox
Pteropus edwardsi (*Pteropus rufus*) Madagascar flying fox
Saimiris sciurea (*Saimiri sciureus*) squirrel monkey
Sciurus indicus squirrel (This species was not identified. *Sciurus vulgaris* = red squirrel; *Sciurus carolinensis* = grey squirrel)
Semnopithecus leucoprymnus langur (*Cercopithecus leucoprymnus* = *Presbytis senex*, purple-faced langur)
Simia satyrus (*Pongo pygmaeus*) orang-utan
Spalax (*Spalax*) mole-rat
Spermophilus citillus (*Spermophilus citellus*) ground squirrel
Sus scropha (*Sus scrofa*) pig
Talpa europaea (*Talpa europaea*) common mole

Tragu(a)lus minima (?*Tragulus meminna*) chevrotain
Ursus syriacus (*Ursus arctos*) brown bear
Ve(r)sperugo pipistrellus (*Pipistrellus pipistrellus*) pipistrelle or common bat

Translator's Notes

*1 Brodmann usually refers to it as "Laboratorium", but sometimes as "Institut".

*2 Brodmann refers frequently to his seven communications - "Mitteilungen" - in the text, for they represent the initial publications of much of the material of his book. He was an editor of this journal for a number of years.

*3 "Stiftungsdeputation der Stadt Berlin"

*4 "Jagorstiftung"

*5 See the Translator's Introduction, p.X on Brodmann's life to understand the considerable animosity shown him by Berlin University.

*6 ie between white and grey matter

*7 The last two refer respectively to studying axons by staining their myelin sheath or directly their neurofibrils.

*8 Brodmann's references in the text, in his footnotes or in his own list of Literature are not consistent in terms of bibliographic style and abbreviations, or even of accuracy. I have standardised and completed them in my list of Translator's References, but have left them in their original form in the text, footnotes and the original Literature list. However, I have provided translations of the titles of books and papers, but not of journal titles. This first one is: "Histology and histopathology of the cerebral cortex with particular reference to paralytic dementia, senile dementia and idiocy".

*9 "Contribution to the comparative histology of Ammon's horn" (ie: the hippocampus)

*10 Brodmann uses the expression "man and animals" here, as he does mainly in the rest of the text, although he sometimes writes "man and other animals".
*11 The central sulcus.
*12 "kommemorative"
*13 Brodmann does not give a specific reference to Lewandowsky's work, but I suggest his 1907 textbook in the Translator's References.
*14 Again, no precise references are given, but see the Translator's References.
*15 Brodmann here seems to foresee such modern developments as electron microscopy and immunocytochemistry.
*16 No specific reference is given to Mott's work here, but see my list of Translator's References.
*17 Better known in English as the stria of Gennari
*18 Brodmann does not state who is the author of this phrase; in fact, the paper is by Niessl von Mayendorf (1908).
*19 "*Areae anatomicae*"
*20 See Lewis and Clarke.
*21 Does Brodmann mean "Berlin"?
*22 Presumably S. Loewenstein is meant; see Köppen and Loewenstein, 1905.
*23 Given here as "E. Smith", as Brodmann often does.
*24 In his Tables 1 and 2, Brodmann gives his own Latin nomenclature, and German names for all other authors. I have retained Brodmann's Latin, and anglicised the others.
*25 Brodmann only begins his own bibliography at 1903; it is not clear what this reference is.
*26 Meynert used "Körnerschicht" for "granular layer", whereas Betz used "Kernschicht".
*27 Lewis (1878) states that the motor cortex is five-layered. He calls them first, second and third layers, ganglionic layer, and spindle layer. This does not agree with Brodmann's interpretation of Lewis's work.
*28 Campbell (1905) in fact used the following list for cortex in general: Plexiform layer, layer of small pyramidal cells, layer of medium-sized pyramidal cells, external layer of large pyramidal cells, layer of stellate cells, internal layer of large pyramidal cells, layer of spindle-shaped cells. For calcarine cortex he used: Plexiform layer, layer of small pyramidal cells, layer of medium-sized pyramidal cells, layer of large stellate cells, layer of small stellate cells, layer of small pyramidal cells with an ascending axis-cylinder, layer of giant pyramidal cells (solitary cells of Meynert), layer of medium-sized pyramidal cells, layer of fusiform and triangular cells.
*29 The terms used here approximate to those of Bolton (1900), which I reproduce, but not to those of Mott, which are best represented in Table 1.
*30 "Studies of the human cerebral cortex"
*31 Cajal includes the rabbit with the rodents, as does Brodmann himself later (see note *98). Although this is not accepted now, it was at the time - cf Hertwig, quoted several times by Brodmann.

*32 Brodmann thus indicates that he has omitted some words. In his version of Cajal's text he has not copied these "separations" quite accurately, so I have added extra dots where necessary. Cajal actually says that structural simplification involves *not only* the number of nuclei, layers etc., but *also*, and *especially*, the morphology of the neurons. In the second quote, Brodmann misplaces his quotation marks, so I have corrected this.
*33 "The phyletic development of the cerebral cortex"
*34 Throughout, I give the modern English name for the animals Brodmann refers to; the species names he used, and the modern species names, are given in the Glossary on p.281.
*35 The "pallium" is essentially the cortex, the outer "coat" or "mantle" of the cerebral hemisphere.
*36 The opening quotation mark is missing in Brodmann's text.
*37 ie: Haller (1908)
*38 The common bats.
*39 "inneren Feldgebiet"
*40 Brodmann puts "Rhinencephalon" in inverted commas.
*41 "The development of the human brain during the first months"
*42 Brodmann does not give it a name in the text, but in the legend to Figure 4 calls it "Rindenplatte", the modern cortical plate (see "Boulder Committee, 1970, in the "Translator's References". This also applies to notes *43 and *45.
*43 "Randschleier", the modern marginal zone.
*44 "Matrix"
*45 "Innenplatte", the modern ventricular zone.
*46 Brodmann uses this format for expressing magnification and section thickness, although he uses the then common form "μ" for the modern "μm". Oddly, he does not tell us the meaning of the figures until the legend for Fig. 5! His standard section thickness is 10μm, except in Figures 5 (5μm) and 28 (20μm), which may be mistakes.
*47 In the rest of the text Brodmann sometimes uses the Latin, sometimes the German form for the layers; I have standardised on the English form throughout. See also the legend to Fig. 11. It should be noted that Brodmann's terms are more based on those of Meynert (1868) than on the other authors he quotes, especially in that layer II is usually considered by the others to be a "pyramidal" layer. See Table 1.
*48 "Notes on fibrillogenesis and its relationship to myelogenesis". Brodmann's work on neurofibrillar staining helped him understand axonal distribution in the cortex, and was in parallel with the work of the Vogts on myeloarchitectonics, which dealt with myelinated axons in particular.
*49 "Innenplatte"
*50 "Zwischenschicht"
*51 "Rindenplatte"
*52 "The development of the human brain"
*53 As far as possible I have utilised the terminology of the Boulder Committee (1970) for foetal cortical layers.
*54 The first mention of his maps, to be described in detail in Chapter IV.
*55 Not in fact all orders!

*56 I give the common English names here; for the species names used by Brodmann and for modern species names, see the Glossary of Species Names on p.281.
*57 "Cynocephalus mormon": see Glossary of Species Names.
*58 Not, in fact, an order, but a suborder of the primates.
*59 A suborder of the carnivores.
*60 Again, not an order.
*61 In addition to this list, Brodmann mentions studies on other animals in the text: golden mole, ground squirrel, mole rat, tucu tucu, two other species of kangaroo, and one other species of flying fox, making a total of 62 species
*62 Brodmann uses various spellings for "*pennicillata*" - see Fig. 29, and elsewhere.
*63 Brodmann seems to have reversed the definitions of homo- and heterogenetic cortex here!
*64 This view of the claustrum, originating with Meynert in 1868, would not be accepted by many, if any, modern neuroanatomists: Brodmann comes back to it frequently. Krieg (1963), p.243, describing the macaque brain, states: "The insula is easily defined and limited. It consists of that part of the cerebral cortex which is separated by a thin fiber lamina from the putamen, and which is underlaid by the claustrum". But there is no suggestion of the claustrum being *part* of the insular cortex.
*65 This legend is, curiously, placed after the four figures in question.
*66 This section is labelled "1" by Brodmann, apparently by mistake.
*67 "Lamina triangularis"
*68 "Lamina fusiformis"
*69 Although Brodmann refers here to a "Bergkänguruh" ("mountain kangaroo"), the species mentioned in Figures 15 and 82 is "*Onychogale frenata*", a wallaby, but not the rock wallaby (*Petrogale penicillata*), which might be Brodmann's "*Macropus pennicillatus*"! To simplify matters, I have translated "*Macropus rufus*" as "kangaroo", and all Brodmann's other "kangaroos" as simple "wallabies". The full species names are recorded in the Glossary of Species Names.
*70 "Seelenleben"
*71 "nervösen Grau"
*72 "Concerning cortical measurements" - Brodmann (1908e)
*73 "other things being equal"
*74 Brodmann uses the term "Affen" ("monkeys"). They do not form an order, and nor do the prosimians.
*75 Misspelled "Risenzellen".
*76 Once again, there is some doubt about the species of "kangaroo" (here "Bergkänguruh"). See note *69.
*77 The kinkajou is not part of the bear family.
*78 Here, as elsewhere, Brodmann refers to him as "B. Lewis". There is an obvious spelling mistake - "probable" - in the quotation; the phrase does not appear in either Lewis (1878) or (1880), nor in Lewis and Clarke (1878).
*79 Spelled "Jakson", another of several such misspellings.
*80 Once again, Brodmann leaves us guessing as to which publication he refers, but see Jackson (1880).
*81 Lewis (1878)

*82 As noted already, Brodmann sometimes gives no references for his statements. Pierret does not appear in his Literature list, nor can I find the relevant reference. However, for the others in Footnote 6, see the Translator's References.
*83 It is not clear to what the term "*Nucleus angularis*" refers.
*84 Brodmann does not seem to distinguish between neurons and glia in his consideration of cell number and density.
*85 See Berger (1900).
*86 Compare the species names in Figures 50 and 59.
*87 "Histological aspects of the neuron theory"
*88 They are not in the medulla!
*89 "bläschenförmig"
*90 In the text, p.91, this is referred to as a guenon ("Meerkatze").
*91 "*Edwardsi*"
*92 Figure 69 is, in fact, labelled "Macacus rhesus".
*93 In fact, it should be 71.
*94 Is this *Cercopithecus mona*, referred to in Chapter I?
*95 Capuchin monkeys do not form a family in themselves, but are part of the *Cebidae*.
*96 Not considered an order now, but was by Hertwig, to whom Brodmann often refers.
*97 "*Pteropus* edwardsi und edulis": see Glossary of Species Names, p.282.
*98 The rabbit is not a rodent, but was classified as such by Hertwig. See note *31.
*99 Brodmann uses latin forms (eg "*Regio postcentralis*") for the regions, and later for the individual areas, although he intermingles a number of German expressions also. For the sake of consistency I have anglicised all terms, including names of gyri and sulci, basing my choice on the literature, both contemporary and modern.
*100 The olfactory region is also not shown.
*101 "Angulus"
*102 Given here as "Scheidelläppchen", presumably a mistake for "Scheitelläppchen".
*103 Brodmann refers to this sulcus variously as "intraparietalis" and "interparietalis".
*104 In fact Elliot Smith called it "visuo-sensory band".
*105 "area postcentralis oralis"
*106 In fact, his Plate I.
*107 In fact "Z".
*108 Brodmann usually uses the term "postcentralis", but here writes "retrocentralis".
*109 A curiously truncated quote by Brodmann! In fact Elliot Smith continues: "depicted in fig. 1".
*110 Brodmann adds a bracketed note within Elliot Smith's quotation to clarify that the area referred to is indeed the "frontal".
*111 Although Brodmann here terms area 11 "prefrontal area", he later (1913) used "prefrontal region" for the whole of his present "frontal region". This has become of some significance in view of the importance of the prefrontal cortex to concepts of human intelligence and consciousness (see Elston and Garey, 2004).

*112 "gyrus rectus"
*113 This should presumably be "lateral".
*114 "The human brain"
*115 This is presumably a mistake for Figures 85 and 86.
*116 In the figure legend, Brodmann writes "parinsularis", but uses "parainsularis" in the text.
*117 "Areae profundae"
*118 The quotation is from page 252, and really speaks of "thin cortex"!
*119 Brodmann uses the term "Taenia tecta".
*120 "*Caput gyri hippocampi*" - Brodmann seems to be referring to the uncus here.
*121 Respectively, "The human brain" and "The external morphology of the olfactory brain of mammals and man".
*122 The guenons.
*123 The marmosets.
*124 These figures are not unchanged from the third communication: although the areas depicted are the same, the format is different. They do appear, however, in the sixth and seventh communications (Brodmann, 1908a,b).
*125 From the map in Figure 90 the "inferior precentral sulcus" seems to correspond to what has been called the "subcentral dimple", a very small sulcus at the inferolateral end of the precentral gyrus.
*126 There appears to be some confusion here. From the map in Figure 90 it seems likely that Brodmann means to say that the postcentral gyrus extends rostrally beyond the central sulcus onto the *pre*central gyrus.
*127 cf p. 260, where he states that the arcuate sulcus is a "strict boundary". (see note *269)
*128 Probably the "precentral dimple".
*129 Area 17 may indeed be relatively larger in the cercopithicids, and other monkeys, but it can hardly be considered as absolutely larger. In 1913 Brodmann measured many primate species and found that in man area 17 represented some 3% of the total cortical area, while in macaque it was 12%, but that the human area 17 had an average absolute area of 3000 mm2 or more, whereas the macaque had less than 2000 mm2. However, chimpanzees and mandrills actually reached human absolute values. Considering the body size difference, these figures emphasise the relatively large development of area 17 in non-human primates, which is the point Brodmann is trying to make. See Elston and Garey, 2004.
*130 Figures 98 and 99 are almost unchanged, with just an indication to area 51 added to them.
*131 The letters that suddenly appear in the next sentence refer to Figures 100 and 101.
*132 Referred to here as "M. und K."!
*133 Cercopithecids *are* anthropoids! He presumably means great apes.
*134 The illustration of the slow loris does not, in fact, appear until Figures 134 and 135.
*135 By "Area limbica posterior" Brodmann probably means "Area retrolimbica", areas 29 and 30, to be described below.
*136 "homogenen": he presumably means "homologous".

*137 In fact, only areas 31a and 31b belong to the cingulate region, extending along the horizontal branch of the splenial sulcus. Areas 30a and 30b belong to the retrosplenial region, described in the next paragraph, as can be seen in Figure 103. Area 30b runs along the horizontal branch of the splenial sulcus, posterior to area 31a, while area 30a follows the vertical branch of the sulcus.

*138 Although we are not told in the text, the prepiriform cortex is labelled "51" in Figure 103, and the amygdaloid nucleus is covered with six "A"s.

*139 This should presumably be "central sulcus of primates". The so-called cruciate sulcus in primates is, indeed, within area 4, on its medial aspect, as described by Campbell, 1905.

*140 Brodmann presumably means "area 52" here.

*141 Like Mauss, mentioned several times in support of Brodmann's observations, Zunino was another of his colleagues at the Berlin Neurobiological Laboratory (see "Introduction", p. 4)

*142 Misspelled "cystologisch".

*143 Brodmann here draws attention to the importance of the *connections* of the areas he is studying.

*144 Area 9 is described here as having a *complete* belt-like form, but in the next sentence as only a *partial* segment.

*145 See Translator's Introduction, and Henneberg (1910).

*146 See Grünbaum and Sherrington (1901).

*147 Oddly, Brodmann seems to have reversed his definition of "essential" and "non-essential" since introducing the terms in the previous paragraph. See also p.195, note *152.

*148 In Figure 143, what Brodmann refers to as the "striate area" for the cat in fact includes much more than area 17 medially. See Otsuka and Hassler (1962).

*149 Marmosets and lemurs are from the same order - primates.

*150 In fact this is five times more!

*151 "The cerebral cortex of the dolphin"

*152 See note *147.

*153 They are not all orders; eg ungulates, pinnipeds and prosimians.

*154 "General morphology of organisms. Vol. II. General development (or embryology) of organisms"

*155 "The olfactory brain. A comparative anatomical study"; "The external morphology of the olfactory brain of mammals and man"

*156 "*taenia tecta*" and "*stria lanzisi*" - more correctly "lancisi"

*157 "Area praeterminalis". This is Brodmann's area 25.

*158 "Morphological studies of organisational principles of the body in nature, with particular reference to organs"

*159 "Gegenstücke"

*160 "Folgestücke"

*161 "Cortex primitivus, Cortex rudimentarius, Cortex (heterogeneticus) striatus". This "Cortex striatus" is not to be confused with the striate cortex of area 17, the "Area striata".

*162 Brodmann writes "Induseum griseum"

*163 "The structure of the cerebral cortex"

*164 "The phylogenesis of the rhinencephalon, the corpus striatum and the forebrain commissures"

*165 It is difficult to see how these two figures can be "drawn in their natural size relations", when they are not at the same scale.
*166 "Comparative anatomy of the vertebrates with reference to invertebrates"
*167 "Collected popular lectures in the field of development. Part 1. Functional specialisation (literally "Division of labour") in nature and human life"
*168 "Textbook of zoology"
*169 "The battle of basic problems in biology"
*170 "The position of comparative embryology in relation to comparative anatomy"
*171 "Emergence and concept of natural historical method"
*172 "indifferenten"
*173 "Vorratsgebilde"
*174 "Lectures on man's place in creation and in the history of the earth"
*175 Brodmann puts Huxley's thesis as a quotation, but I have been unable to find it given verbatim. However, a very similar statement (his "Pithocometra-thesis") is given by Haeckel (1898), p.12.
*176 "Growth of human body weight at different ages and measurement of volume in organisms"
*177 "Measurement of the surface of the cerebrum"
*178 Henneberg (1910)
*179 "Brain measurement using the compensation polar planimeter"
*180 "Morphological and systematic research in birds. General considerations"
*181 "The fibre architecture of the cerebral cortex in lower monkeys"
*182 Mauss (1911)
*183 I have been unable to find any papers on this topic in this volume. There was much debate at the time as to the homology of the "Affenspalte" - literally "ape sulcus" - with the lunate or simian sulcus of monkeys; see also Elliot Smith, 1904b.
*184 Brodmann was later (1913) to take up the challenge of cortical localisation related to anthropology which, even in those days, raised lively polemic between those who postulated that possible racial differences in brain structure might relate to racial differences in intelligence or other cerebral functions. In fact, the enormous harvest of quantitative data on cerebral cortex of man and other mammals that Brodmann derived from this study far outweighs the racial aspects of his work, although he does not emerge as a supporter of brain structure being a basis for racial characteristics (see Elston and Garey, 2004).
*185 Respectively "The present state of pathological anatomy of the central nervous system" and "Hypothesis of neuronal functional specificity"
*186 "Histological and histopathological research on the cerebral cortex"
*187 "Introduction to the histology and histopathology of the nervous system"
*188 "A case of amyotrophic lateral sclerosis with degeneration of the pyramidal tract"
*189 "An anatomical pathological contribution to the study of cortical motor localisation through three cases of amyotrophic lateral sclerosis with degeneration of the pyramidal tract traced with the Marchi technique from the spinal cord to the cortex"
*190 "Progressive diseases of the motor pathways"
*191 Campbell (1905), p.93. Brodmann's attempt at Campbell's title is "Histological Studies on the Lokalisation of the Cerebral Funktion"

*192 Literally "Thalamus opticus", as was the normal term at the time.
*193 Slightly misquoted from Campbell (1905), p.85.
*194 "On the anatomical basis of idiocy"
*195 Respectively "On the anatomy, nature and development of microcephalic malformations" and "Structure of the cerebral cortex"
*196 "Clinical and anatomical investigations of a special form of familial amaurotic idiocy"
*197 "Familial amaurotic idiocy and related syndromes"
*198 "Pathology and anatomical pathology of different forms of idiocy"
*199 "Anatomical and clinical aspects of Tay-Sachs familial amaurotic idiocy with a consideration of related forms"
*200 "Pathological anatomy of Huntington's chorea
*201 Probably in the sense of a "primary defect"
*202 "Idiocy and epilepsy symptomatic of tuberous or hypertrophic sclerosis"
*203 Probably best understood as "space occupying lesions".
*204 "The question of functional localisation in the cerebral cortex"
*205 Respectively "A developmental deficit caused by the extirpation of circumscribed cortical areas" and "Experimental and pathological anatomical investigations of the relationship of the so-called visual cortex to the infracortical visual centres and to the optic nerve"
*206 "Further communications concerning the influence of unilateral eye destruction on the development of the visual cortex"
*207 "Contribution to knowledge of fibre crossing in the optic chiasma"
*208 "Changes in the optic nerve and tract in diseases of the occipital lobe"
*209 Respectively "Contributions on the histology of the cerebral cortex" and "Experimental anatomical studies on developmental deficits in the occipital lobe of the dog and cat caused by lack of visual stimulation". See note *211.
*210 Respectively "The behaviour of neuroblasts in the occipital lobe in anophthalmia and atrophy of the eyeball and its relation to vision" and "Contributions to knowledge of secondary changes in the primary visual centres and pathways in cases of congenital anophthalmia and atrophy of the eyeball in neonatal infants"
*211 An interesting prediction of the future! See Wiesel and Hubel (1963).
*212 "Contribution to the anatomy of the brain of deaf-mutes"
*213 "The frontal lobe"
*214 "The brain of deaf-mutes"
*215 "Anatomical investigation of the human auditory cortex"
*216 "Deaf-mutism and the auditory pathway". Academic dissertation.
*217 "Zell-Äquivalentbild"
*218 "tektonische Äquivalentbild"
*219 "Textbook of zoology"
*220 See Introduction, pp. 5-7
*221 "Ichbildung"
*222 "Outline of a physiological explanation of psychic phenomena"
*223 "The basis of physiological psychology"
*224 For a modern view of percepts and concepts, see Changeux (1985).
*225 "The functions of the cerebral cortex. Collected communications"
*226 "Comparative anatomy of the forebrain of vertebrates"

*227 Brodmann refers to Edinger and Wallenberg as "Edinger-Wallenberg".
*228 "Brain and soul"
*229 Spelled "Evens" here.
*230 A rather curious reference! L. Asher and K. Spiro were the editors of "Ergebnisse der Physiologie", a journal started in 1902 to report "results" ("Ergebnisse") in a wide variety of physiological disciplines. The reference should be to three papers by von Monakow on this subject - see Monakow (1902, 1904, 1907). The other reference is to Monakow's "Brain pathology".
*231 "The present state of the concept of cerebral localisation"
*232 "strengen Lokalisten"; "Halblokalisten"
*233 "Experimental research on the properties and functions of the nervous system"
*234 Another example of Brodmann's frequent inadequate citations. I have not been able to locate relevant works by Carville, Duret or Soltmann.
*235 "Gegnern" misspelled.
*236 "Sammelpunkte" - literally "rallying-points".
*237 "Bernhard von Gudden's collected and posthumous works"
*238 "The question of localisation of function in the cerebral cortex"
*239 See Grashey (1889).
*240 "Sphären" - literally "spheres".
*241 "Körperfühlsphäre"
*242 "Physiological and clinical brain research. Collected essays"
*243 "Lectures on cerebral localisation". See Ferrier 1890.
*244 "Functional localisation in the cerebral cortex"
*245 "Cortical psychosensorial centres"
*246 "Visual disturbances after damage to the cerebral cortex". The page number is wrong.
*247 "Contributions to cerebral physiology"
*248 "Physiology of the brain"
*249 "Research on functional localisation in the human cerebral cortex"
*250 "New concepts in the question of cerebral localisation"
*251 "Collected essays"
*252 "Memory as a supporting principle in the exchange of organic events"
*253 Literally "medulläre": probably here not actually referring to the "medulla" alone.
*254 "Übungsfaktor" - literally a phenonemon due to exercise or practice, probably best expressed by the modern concept of plasticity.
*255 "Die elektromotorische Region". I have not expressed the idea of excitability by electric stimulation in this sub-heading, as it would have been rather clumsy, and is amply developed in the text.
*256 "Morphological brain centres, with particular reference to morphological areas of the cerebral cortex"
*257 ie the ventro-anterior nucleus
*258 the ventroposterior nucleus
*259 "Information on electrically excitable cortical areas in mammals"
*260 Respectively "Clinical and anatomical contributions to the pathology of the brain" and "The projection of the retina on the calcarine cortex"
*261 "The cortical visual centres"
*262 "Loss of orientation, cortical blindness and mind blindness"

*263 "Riechhirn"
*264 "Remarks on the auditory cortex of the human brain"
*265 literally "Struktur"
*266 Respectively "The present position of aphasia" and "Pathology of the left temporal lobe"
*267 "Zentrenlehre"
*268 "Medianfläche": Brodmann frequently seems to confuse "median" with "medial".
*269 Although on page 127/131 Brodmann states: "The rostral boundary of the (precentral) region coincides approximately, but not exactly, with the arcuate sulcus". (see note *127)
*270 "Extremitätenzone"
*271 "Fokus der Sehsphäre"
*272 This problem was to be resolved later with our understanding of primary and secondary visual cortices, especially in the cat - see Hubel and Wiesel (1962).
*273 "A study of myelinisation in the cerebral hemispheres"
*274 I have not added any references to Brodmann's original bibliography, nor have I corrected or standardised them. Many are only partial, or contain errors. Their format and alphabetical order are inconsistent. It should be noted that Brodmann gave a number of other references as footnotes, and I have retained them in their relevant places in the main text. He also frequently referred in his text to other authors without giving any bibliographic reference; as far as possible I have identified relevant references for these, and incorporated them in the Translator's References pp. 267-280, where corrected versions of Brodmann's bibliographic references will be found.

Index

In this Index certain common terms (such as area, gyrus, layer, lobe, region, sulcus) occur so frequently that I have only given general references to the text or to a specific chapter. The index does not include bibliographic references, figure legends, or footnotes. References to animal species is limited to the most fully described ones (see the Glossary of Species Names, p297, for more details).